Towards a Theory of Development

Towards a Theory of Development

EDITED BY

Alessandro Minelli

Department of Biology, University of Padova, Padova, Italy

Thomas Pradeu

Department of Philosophy, Paris-Sorbonne University, Paris, France

OXFORD
UNIVERSITY PRESS

Great Clarendon Street, Oxford, OX2 6DP,
United Kingdom

Oxford University Press is a department of the University of Oxford.
It furthers the University's objective of excellence in research, scholarship,
and education by publishing worldwide. Oxford is a registered trade mark of
Oxford University Press in the UK and in certain other countries

First Edition published in 2014

Impression: 1

Published in the United States of America by Oxford University Press
198 Madison Avenue, New York, NY 10016, United States of America

British Library Cataloguing in Publication Data
Data available

Library of Congress Control Number: 2013953856

ISBN 978–0–19–967142–7 (hbk.)

ISBN 978–0–19–967143–4 (pbk.)

Printed and bound by
CPI Group (UK) Ltd, Croydon, CR0 4YY

Contents

List of contributors

Wallace Arthur School of Natural Sciences, National University of Ireland, Galway, Ireland

Jonathan Bard Department of Physiology, Anatomy, and Genetics, University of Oxford, UK

Roberto Carrer Department of Mathematics, V. Volterra Institute, San Donà di Piave, Italy

Giuseppe Fusco Department of Biology, University of Padova, Italy

Scott F. Gilbert Department of Biology, Swarthmore College, Swarthmore, PA, USA and Biotechnology Institute, University of Helsinki, Helsinki, Finland

James Griesemer Department of Philosophy, University of California-Davis, Davis, CA, USA

Isaac Hernández Department of Philosophy, University of Toulouse 2, France

Johannes Jaeger Centre de Regulació Genòmica, Barcelona, Spain

Jean-Jacques Kupiec Centre Cavaillès, Ecole Normale Supérieure, Paris, France

Lucie Laplane Department of Philosophy, University of Paris X, Nanterre, Paris, France

Alan C. Love Department of Philosophy, Minnesota Center for Philosophy of Science, University of Minnesota, Minneapolis, MN, USA

Margaret J. McFall-Ngai Department of Medical Microbiology and Immunology, University of Wisconsin-Madison, Madison, WI, USA

Alessandro Minelli Department of Biology, University of Padova, Padova, Italy

Armin P. Moczek Department of Biology, Indiana University, Bloomington, Indiana, USA and National Center for Evolutionary Synthesis, Durham, North Carolina, USA

Michel Morange Centre Cavaillès, Ecole Normale Supérieure, Paris, France

Stuart A. Newman Department of Cell Biology and Anatomy, New York Medical College, Valhalla, NY, USA

Spencer V. Nyholm Department of Molecular and Cell Biology, University of Connecticut, Storrs, CT, USA

Thomas Pradeu Department of Philosophy, Paris-Sorbonne University, Paris, France

Emanuele Serrelli Riccardo Massa Department of Human Sciences, University of Milano Bicocca, Milano, Italy

James Sharpe Centre de Regulació Genòmica, and Institució Catalana de Recerca i Estudis Avançats, Barcelona, Spain

Davide Vecchi Department of Philosophy, Universidad de Santiago de Chile, Santiago, Chile and Instituto de Filosofía y Ciencias de la Complejidad, Santiago, Chile

Michel Vervoort Institut Jacques Monod, Centre National de la Recherche Scientifique, Université Paris Diderot, Sorbonne Paris Cité, Paris, France and Institut Universitaire de France, Paris, France

Foreword: a biologist's view

Laws to understand and explain embryonic development have been in existence for two centuries or more. These laws, which are based on comparative embryology, resulted in conclusions of sufficient generalization that they could be termed a law.

von Baer proposed the law that general features of animals appear in development before more specific features and concluded that general features were those of phyla and specific features, those of species. Such a 'law' does not propose a mechanistic explanation for development and, in von Baer's case, not even an evolutionary explanation. Phylogenetic analyses support von Baer's generalizations but do not provide an explanation for them. Recent identification of shared Hox genes, phylotypic stages, and conserved gene pathways identify shared mechanisms but do not provide a theory of development.

Haeckel's Biogenetic Law that ontogeny recapitulates phylogeny proposes that embryos provide the way to understand evolution through the mechanism of recapitulation. But the Biogenetic Law is not a theory of development, even though it too is based on comparative studies leading to a generalization. If anything, it is a law of evolution. Similarly many of the explanatory terms used in development—regulative development, preformation, programmes, fields, epigenetic landscape, genotype-to-phenotype map, increasing complexity—are metaphors and neither laws nor theories. A theory is a statement of a principle(s) derived from and able to explain facts or events and proposes a mechanism(s) that explains the facts or events. Laws and metaphors describe but do not explain. The 17 chapters in the present ambitious book set out to evaluate past theories of development with the intention of moving towards 'a theory of development'.

The 'theory of development' proposed by Ludwig von Bertalanffy in the late 1920s is most well known from the English translation of his 1928 book *Kritische Theorie der Formbildung*, published in 1933 under the English title *Modern Theories of Development: An Introduction to Theoretical Biology*. The link to theoretical biology shows that Bertalanffy posited that a theory of development would point the way towards a theory of life. As J. Z. Young commented in his back cover blurb: 'This book contains a discussion of the nature of living organisms and the methods by which they should be studied.' Conrad Hal Waddington's four-volume *Towards a Theoretical Biology* (1968–1972) makes the same claim as do a number of the authors in the present book, although theoretical biology has a much wider (and potentially more diffuse) ambit now than it had 80 or even 40 years ago.

von Bertalanffy used embryology to illustrate the approaches required for a theory of biology. He sought to counter the strongly mechanistic approach to biology that dominated research and interpretation by stressing organization, organicism, and a systems approach; von Bertalanffy is acknowledged as the founder of general systems theory. It is no coincidence that his German text was translated and adapted by the biologist turned logician-philosopher and anti-mechanist, Joseph Woodger, whom von Bertalanffy had met in Vienna in 1926 when both came under the influence of the Vienna Circle. For both men, biology would never be physics, and so no general theory would unite the sciences. Woodger was a founder and major participant in the Theoretical Biology Club, which first met in 1932 in the grounds of Woodger's home in Epsom Downs.[1]

Any theory is embedded in and arises from approaches in vogue at the time. von Bertalanffy's book was reissued in 1962 with a new Foreword (von Bertalanffy died in 1972) but with no changes to the text of the 1933 edition. Anti-mechanism, anti-reductionism, organicism, and the development of theoretical approaches to biological organization, such as cybernetics, information theory, and general systems theory, remained sufficient for von Bertalanffy's theory of development.

Although approaches from theoretical biology remain and are discussed in the present book, times have changed. As well represented in this volume, today's theories have different bases, reflecting increased knowledge of (and perhaps even understanding of) the processes of development. For the past half-century, approaches to a theory of development have been gene based. The three editions of the seminal book *Gene Activity in Early Development* by Eric Davidson published in 1968, 1976, and 1986 have been replaced by *Genomic Regulatory Systems: Development and Evolution* (2001), reflecting the transition from genes, to gene cascades/networks, to gene regulatory networks. Gene-based theories need not necessarily conflict with systems-based theories but often are cast as alternative explanations for development, reflecting alternate worldviews. Explanation and potential explication of these tensions are an important aspect of the present volume.

I can hear the murmurs through the ether: but surely any theory of development must be gene based? Not according to theories discussed herein, which include emphases on morphogenetic fields, cell differentiation, natural selection, disparity, plasticity, stem cells, dynamical systems theory, and more. These theories are presented either as explaining one aspect of development or as signposts to a more general theory.

The discussions in various chapters that explicitly or implicitly acknowledge that we have theories of evolution and of heredity but not a theory of development should make us pause and reflect. The diversity of approaches in *Towards a Theory of Development* leads one to ask whether 'a' theory of development is attainable and/or whether such a theory should be 'A theory of ontogeny' (as discussed in Chapters 1 and 2) or incorporated into 'A theory of life?' Given the multitude of processes and mechanisms operating during development/ontogeny and the multiple of scales of the biological hierarchy that emerge during development—genes; cells; germ layers; tissues; organs; epigenetic, functional, environmental, and symbiotic interactions—it may be a good thing that we do not have a theory of development. That said, *Towards a Theory of Development* is a great place to find answers to such questions as whether a theory of development will:

- account for what we now know to be the distributed nature of causation including from the genome and environment, and between cells?
- incorporate, or indeed be based on, epigenetic control of development?
- enhance the role of natural selection on development and speed synthesis of development and evolution?
- have to take stocasticity of development into account?
- account for convergent evolution?
- enhance our understanding of homology and variation?
- provide a framework for constraint and deep homology on the one hand and plasticity, adaptability and evolvability on the other? or
- be a theory based on concepts not yet dreamt of?

Brian K. Hall
November, 2013
Halifax

[1] Present at that meeting were Joseph Woodger, Conrad Hal Waddington, Joseph Needham, Dorothy Needham, J. D. Bernal, and Dorothy Wrinch. Except during World War II, the club met annually until 1952.

Foreword: a philosopher's view

The authors and editors of *Towards a Theory of Development* ask whether, in the near-to-middle term, a theory of (biological) development can plausibly be expected that can serve as a basis for further research on development and explain the main phenomena and processes covered traditionally in that discipline. In pursuit of this ambitious and engaging agenda, they draw on recent research and on past attempts (all inadequate to the task) to provide developmental biology with a core theory that would fulfil these desiderata. The issues involved are discussed from many angles. One of the most important investigates how the many distinct developmental mechanisms and processes revealed by recent biological work, deployed in distinct ways in different sorts of organisms, can be fit together with the findings and theories of evolutionary biology, genetics, and several other disciplines (e.g. ecology, immunology, and systematics). New genomic tools are taken into account, for example, to ask whether developmental systems are, in fact, composite in light of the prevalence of symbiosis and commensalism; when microbes are taken into account, these tools have shown that the developmental processes of most, perhaps all, plants and animals depend on genetic contributions of several distinct organisms whose life histories are interconnected with that of the plant or animal.

Developmental biology has been changing very rapidly over the last few decades. Many of the changes have depended not only on the new technologies of genomics and so-called genetic networks (which include many nodes and signals that are not genes or gene products) but also on biochemical and biophysical tools that have elucidated biophysical and biochemical mechanisms of various sorts that are not, as such, gene based (e.g. gastrulation, to name just one). The work with these tools must be made to fit with the reliable findings of anatomy, developmental biology, evolutionary biology (plus, as some of the authors argue, other disciplines such as ecology and immunology) to work out what must be explained by a theory of development. Many of the authors take strong positions in ongoing controversies regarding the role and value of various kinds of theories in developmental biology and regarding the sort of unification of our understanding of developmental process that is feasible in light of current knowledge. But they do so in a cooperative spirit that helps the reader achieve a synoptic view of the state of the argument on the primary issue to which the book is dedicated. Since experts who work with a variety of developmental systems are included among the authors, there is much that is of interest here for developmental biologists as well as biologists coming from other disciplines. However, since I am a philosopher of biology, not a biologist, I shall speak primarily to a couple of the more philosophical issues prominent in the volume in the remainder of this foreword.

The questions to which the book is addressed sit, quite naturally, at the interface between biology and philosophy of biology. Several 'background' problems are also at issue throughout the book, for example, issues regarding delimitation of the precise domain a theory of development must cover (must it apply to every sort of organism? might it apply to some non-organismal developmental systems?), what sort of theory would be appropriate for achieving this ambitious goal (certainly not one composed solely of universal laws!), what expectations such a theory must meet, and whether the ingredients are now in hand for constructing a theory that could do the job. The positions on these questions represented within the book vary widely. Some authors hold that developmental biology is not the sort of discipline that can be covered by a core theory, others hold that the materials are now in hand to construct such a theory in one or another of several distinct ways, and still others think that a revision of the current boundaries of developmental biology is required before the question can be answered. This foreword does not represent this issue well; it turns on recognizing the need for a comprehensive theory of development to get beyond plants and animals to cover at least 'unusual' developmental systems (note the plant and animal bias in such a description) and unicells (which *do* undergo developmental processes and have a variety of forms of multicellularity).

Setting this last point aside, consider how the regulation of developmental processes in multicellular organisms (and for that matter, in unicellular organisms as well) is coordinated. It is not at all obvious whether or how gene networks and biochemical mechanisms can coordinate developmental processes by strictly genetic and genomic controls. After all, development is altered by environmental regularities and contingencies and by a great variety of signals that are not encoded or contained in the relevant genetic system(s) that affect development. For development of multicellular organisms to be coherent, that is, to respond coherently to the state of various parts of the same developing organism and various environmental conditions, some sort of biophysical and biochemical mechanisms must be utilized that are poised to respond differentially to distant signals, whether from within the organism or from the external environment. Some of these are best understood (at least in part) as consequences of historical constraints (their poised state depends on their history). Some relevant historical constraints are consequences of phylogeny, others are maternally imposed, and still others, consequences of commitments made in earlier steps in development. Such constraints limit the developmental processes available to particular organisms at particular stages of their life cycles and regulated differentially by intracellular, intercellular, hormonal, and 'external' environmental signals at different stages of the life cycle. (Note, too, that a particular signal, occurring at different times and affecting the same mechanism, can have radically different consequences because of the intervening developmental or contingent changes.) On such grounds, it has often been argued that developmental processes cannot be (fully or properly) explained on a strictly genetic basis—but, also, that there is no single theory characteristic of developmental biology that can provide a theoretical basis for understanding development in general. This conundrum is of central importance in this book.

Since most developmental biologists have not been strongly interested in constructing an overarching theory of development and since developmental biology has not been a focal topic for philosophers of biology, the tasks undertaken in this book are quite important for understanding the current status of the theories that are employed in dealing with development. I recommend that the reader bear these three questions in mind:

- What kinds of theories are there in developmental biology?
- What should be expected of a theory of development?
- Is it reasonable on the basis of the currently available evidence to hope to construct a satisfactory theory of development that can guide research for the entire field of developmental biology?

There is no general consensus regarding the answers to these questions at present. The authors present a wide variety of perspectives on these and related issues. Although the authors differ quite widely on the answers, collectively they have made the issues sharper and brought leading lines of evidence that they consider relevant to bear in justifying the directions in which they believe things will move in the next decade or two. Thus, even though the 'big' issues around the leading questions are not resolved, the authors have made considerable progress in articulating and advancing the issues at this very interesting interface between developmental biology, theoretical biology, and the philosophy of biology. The chapters make good contact with each other and have been constructed so that readers with a wide range of backgrounds and interests will be able to follow the arguments. This book deserves to be read by all those interested in the directions in which biology is moving at present.

Richard M. Burian
Professor Emeritus of Philosophy and Science Studies
Virginia Tech
2014
Blacksburg, VA, USA

CHAPTER 1

Theories of development in biology—problems and perspectives

Alessandro Minelli and Thomas Pradeu

What is the role of theories in developmental biology? At first glance, developmental biologists seem rather suspicious of theories, as they do not often use the term 'theory' in their publications, and most of them tend to see theories as too broad and speculative to be really useful in everyday scientific practice. The present volume addresses the question of what theories are, or could be, in developmental biology, and what could be the functions, if any, of such theories. It is crucial to understand that at stake in this debate is the scientific status of developmental biology, and in particular, the respective roles of descriptions, explanations, and predictions in this domain. Moving towards a more theoretical developmental biology may imply that development can be explained and predicted. But is this idea realistic, and, to begin with, is it even desirable? After all, building a theory of development could be impossible in the face of the amazing diversity of developmental processes; or perhaps such a theory, if anything, would be an obstacle rather than an incentive to do actual research, as it would blinker people on other possible ways of seeing developmental processes.

The positions that can be taken on these issues are tremendously diverse, as testified by the chapters of this volume, written by leading developmental biologists, scholars of evo-devo, and philosophers of biology. Instead of trying to deny the divergences between the different chapters, we decided to present them as clearly as possible, so that the readers can make their minds about the role of theories in developmental biology. In this introduction, our aim is not to take sides in the debate over the existence of theories in developmental biology, but rather to briefly summarize the historical roots of this debate, and to show how it is related to some of the most fundamental issues in the field—including the delineation of developmental biology, the very definition of the word 'development', and the characterization of the spatial and temporal boundaries of developmental processes. After the presentation of these issues, we will draw some conceptual and practical consequences as for the construction of theories in developmental biology.

From embryology to developmental biology

The oldest root of the modern scientific study of development is embryology, but the developmental biology of our time is a broader discipline. Classical embryology focussed on a limited segment of ontogeny, i.e. embryonic development, defined as the interval between the fertilization of the egg and 'birth' (egg hatching in the case of the chicken and animals with eggs encased in a chorion). Until the late nineteenth century, embryology was strictly descriptive, only becoming experimental (with the development of mechanical embryology first, then chemical embryology) around the turn of the century. Description, however, was accompanied by theoretical speculations, polarized around the contrasting theories of preformationism and epigenesis (for a history of embryology and a discussion of the preformism vs epigenesis contrast, see e.g. Cobb, 2006;

Towards a Theory of Development. Edited by Alessandro Minelli and Thomas Pradeu

Gilbert, 1991; Horder et al., 1986; Maienschein, 2012; Needham, 1934; Pinto-Correia, 1997; Shirley, 1981).

Aristotle's (384–322 BC) position with respect to animal development can be characterized by and large as epigenetic: in the embryo, many parts of the future animal are not present and are formed progressively, thus accomplishing the organism's *final cause*, eventually developing into the adult. The epigenetic position was generally accepted until the seventeenth century, when this position found its first opponents, who interpreted development as the becoming visible of structures already present in the germ. The rising contrast between the epigenetic and preformist theories of development was accompanied by contrasting interpretations of the generative role of both egg and sperm. Within the epigenetic camp, Reinier de Graaf (1641–1673) contended that only the egg provides for the material origin of new individual, while the sperm would simply contribute a 'fermenting principle' (*aura seminalis*) stimulating the uterine membranes to produce the foetus.

Modern biology has repeatedly demonstrated how seriously our knowledge is biased by the idiosyncratic properties of the most fashionable model species on which observations and experiments are performed. The same bias is already evident in seventeenth century embryology. In his studies on mammals (the rabbit in particular) de Graaf found reasons to defend epigenesis, while Marcello Malpighi (1628–1694) was convinced that his observations on the development of the chick embryo would require an interpretation in terms of preformation. For about one century, preformation became more and more popular. By studying insect development, Jan Swammerdam (1637–1680) defended an even more extended view, within which the pupa was interpreted as a kind of second, 'major' egg, within which the adult is preformed.

In the following century, preformists were knowledgeable scientists like Charles Bonnet (1720–1793), Albrecht von Haller (1708–1777), and Lazzaro Spallanzani (1729–1799), but the epigenetic view of development was about to rise again. According to Georges Louis Leclerc de Buffon (1707–1788) the new organism was not preformed in the egg but was shaped instead by an internal mould (*moule intérieur*), a developmental paradigm in which a few modern commentators would like to see an anticipation of the idea of the genetic programme. An epigenetic perspective was also defended by the excellent microscopist Felice Fontana (1730–1809), whose painstaking observations of both eggs and sperm cells failed to reveal any anticipation of the future structure of the animal, and by Caspar Friedrich Wolff (1733–1794), whose comparative studies of animal and plant development provided the foundation of a *Theoria generationis* (Wolff, 1759). Despite a facile but incorrect translation of the title, this was indeed less a theory of reproduction than a theory of development. According to Wolff, developmental dynamics are due to an internal formative principle (*vis essentialis*) that in the subsequent elaboration by Johann Friedrich Blumenbach (1752–1840), right at the time German biology was taking the distinctive 'Romantic' character of Naturphilosophie (e.g. Nyhart, 1995), appeared as a 'formative drive' (Bildungstrieb, *nisus formativus*), the investigation of which did not seem to offer any hope of obtaining anything beyond a mere description of its putative effects.

The nineteenth century opened as a time of great progress in descriptive embryology, accompanied by sober generalizations about the visible patterns of transformation, as in the chief monograph (1828) of Karl Ernst von Baer (1792–1876). The last decades of the century saw however the first manipulations of embryos and the success of the 'mechanical' approach to development pioneered by Wilhelm Roux (1850–1924). His *Programm und Forschungsmethoden der Entwicklungsmechanik der Organismen* (*Programme and Research Methods in Developmental Mechanics*) was published in 1897 (Roux, 1897). Further progress in experimental embryology led to the discovery of the regulative properties of some embryos, especially those of sea urchins. Ablation of one or two cells from an embryo immediately after the first few cleavage divisions leads to the formation of a larva that is a little smaller, but still regularly organized. The embryo, in a sense, seems to know what it is destined to form, and to generate it despite manipulation, because it follows its intrinsic final principle, its *entelecheia*. This is, at least, what Hans Driesch (1867–1941) was proposing. In the first half of the twentieth century, experimental embryology expanded by adopting an increasingly diverse array of techniques, mechanical and chemical, including transplantation of cell nuclei and the establishment of chimaeric embryos made of cells from two different species.

Eventually, the advent of developmental genetics trained the transformation of embryology into developmental biology (see Burian & Thieffry, 2000). Two main arguments were expressed to favour this transition:

(i) the need to 'molecularize' embryology, with a particular insistence on genes. This was in part related to the opinion, expressed by some actors in the field, that embryology would be about old-fashioned collections of embryos in bottles, whereas molecular developmental biology would be much more up-to-date with regard to the tools it used. Incidentally, this move convinced some biologists that molecular biology would replace theoretical biology, while others, most famously Waddington (1968), rejected this idea and sought an articulation between molecular and theoretical perspectives;
(ii) the need to integrate the work on plant development into the mainstream field of the branch of life science dealing with developmental processes.

To a large extent, the research agenda of the new developmental biology was articulated around the notion of a genetic programme (Goodfield, 1969; Mayr, 1969). This was initially formulated in terms of a multiplicity of 'developmental genes' controlling, in a deterministic way, the processes through which an egg gives eventually rise to an adult animal. The rapidly expanding body of experimental evidence soon directed attention beyond the spatial and temporal patterns of expression for individual genes to the 'cross-talk' of macromolecules through which the expression of 'developmental genes' is controlled (Wolpert & Lewis, 1975). Finally, largely through the works of Eric Davidson and his school, developmental genetics moved a further step forward, engaging in the exacting task of describing whole gene networks deemed to be responsible for specific aspects of the organization of complex organisms (e.g. Davidson, 2003, 2006). However, dissenting views with respect to a strict gene-centric view of development have been increasingly voiced.

The last two decades of the twentieth century have been an extraordinarily fecund period for studies on development. According to Gilbert (1998), the critical transformations of developmental biology during this period led to seven key conceptual breakthroughs. First, the elucidation of the mechanisms of anterior–posterior axis formation in *Drosophila* (for an overview, see Nüsslein-Volhard, 1994) led to the conviction that developmental processes should not anymore be seen as 'mysterious', but as processes that can be explained scientifically. Second, the use of techniques coming from molecular biology showed that the core of development consists of paracrine factors, transcription factors, and signals between them. Third, the discovery of Hox genes paved the way for the conviction that homologous developmental genes and pathways exist between distantly related phyla (e.g. between insects and vertebrates), leading to the growing realization that comparisons between different organisms (and model organisms) were necessary to make further progress in developmental biology (on this question see, in particular, Bolker, 2012 and Minelli, 2009). Fourth, several lines of research led to the idea that development occurs through discrete and interacting *modules* (Raff, 1997; Wagner, 1996). Fifth, the idea that changes in development are responsible for major evolutionary changes spread among biologists. Sixth, this period witnessed a resynthesis of medical genetics and medical embryology. Seventh, the end of the twentieth century marked the beginning of the integration of developmental biology and ecology, with a growing interest in the way environmental factors can impinge on development (on this question, see Gilbert, 2001).

One of the most actively discussed trends in recent studies on development has been, naturally, the integration of development and evolution, or 'evo-devo'. This integrative aim has deep historical roots (Amundson, 2005; Laubichler & Maienschein, 2007; Minelli, 2009), but it is beyond doubt that it has been invigorated by molecular approaches to development. The field of 'evo-devo' raises fundamental issues (Arthur, 2002; Garson et al., 2003; Gilbert et al., 1996; Hall, 1998, 2000; Minelli, 2009; Raff, 1997; Stadler et al., 2001; Wagner, 2000), in particular with regard to how to articulate an explicitly theoretical framework to a domain clearly more reluctant to speak in terms of general theories (e.g. Love, 2013). Many of these key issues are examined in this volume, in particular in the contributions of Arthur, Griesemer, Kupiec, Minelli, Moczek, Morange, and Vervoort.

In the meantime, new classes of facts have been added to the scope of developmental biology, sometimes smoothly (post-embryonic development, including growth and the control of sexual maturation) and sometimes controversially (regeneration, ageing, and other areas discussed in this book). Largely independent until recently has been progress in plant developmental biology. Regretfully, this field is only marginally covered in this book, but we hope that our efforts will stimulate the active involvement of plant scientists in shaping possible general theories of development.

What is development?

Development, literally, means 'unfolding': 'Development [. . .] is literally an unfolding or unrolling of something that is already present and in some sense preformed' (Lewontin, 2000: 5). The term is thus tightly linked to the preformist tradition of the old embryology. Yet, despite the epochal advances of life sciences since the times of Haller and Spallanzani, the metaphor is still largely borne by the widespread reductionist and deterministic view of development as programmed in genes. Indeed, developmental biology has been to a large extent framed in terms of genes, genetic 'blueprint', and genetic programme, with the implicit idea that the final form of the organism is 'already there' in the instructions contained in its genome as early as the egg stage. In this view, environmental influences are no more than 'background conditions'. As Richard Lewontin puts it:

> It is usually said that the epigenetic view decisively defeated preformationism. [. . .] Yet it is really preformationism that has triumphed, for there is no essential difference, but only one of mechanical details, between the view that the organism is already formed in the fertilized egg and the view that the complete blueprint of the organism and all the information necessary to specify it is contained here, a view that dominates modern studies of development. (Lewontin, 2000: 5).

However, dissatisfaction with this neo-preformist concept of development has been growing over the years and non-gene-centric perspectives, if not really popular, have at least achieved the status of valid alternatives. This debate is of dramatic importance for the problem examined in this volume, since it raises the question of how one can account for the regularity of development without presupposing that this order is already contained in a pre-existent structure (Oyama, 2000).

But if development is not an 'unfolding' process, then what is development, and which entities can be said to 'develop'? A well-articulated answer to this seemingly innocuous question is very far from easy. In fact, the most likely reaction to it—one framed in terms of the sequence of structural and functional changes a biological system undergoes from egg to adult—is inapplicable to a diversity of living beings. Adding, in parallel, a sequence of structural and functional changes from seed to mature flowering plant would certainly help to fix the scope of a tentative exploration of the space of developmental trajectories, but not enough. In addition to addressing those developmental sequences that have an egg or seed stage, a general concept of development must also cover animal or plant life histories in which reproduction proceeds not via an egg or a seed, but by means of buds or other 'vegetative' parts. In those systems, there is no egg (or seed) and no embryo; but there is still a developmental story.

Moreover, in those frequent cases where the naïve notion of individuality does not obviously apply, such as, for example, with colonial organisms like corals, with polyembryonic animals, and with clonally reproducing plants, our common understanding of development meets with some difficulties. Another moot point is whether the notion of development only applies to multicellular organisms, or may instead be extended to aspects of the life cycles of a selection at least of unicellular organisms (or, perhaps, applies to one particular level of organization such as, for example, biofilms). By opening our perspective as wide as possible, one may even consider whether our concept of development could, or should, extend even to systems whose building units are other than cells. This would at least be required if we want to include not only infracellular systems such as viruses but also those 'hybrid' systems in which 'environmental' components act as necessary scaffolds for development to be deployed, particularly as, in the course of evolutionary history, some scaffolds become 'internalized' as part of the developing system itself (Griesemer, this volume).

Therefore, determining what development is, at what level(s) it occurs, and which entities in the living world can be said to 'develop', is a tremendously difficult task. Several of the contributions to this volume tackle these thorny issues and try to articulate some tentative answers. Overall, one might be tempted to adopt a definition of development that is straightforwardly broader than the traditional one, though hopefully still operative for everyday research. For example, Moczek (this volume) suggests to define development 'as the sum of all processes and interacting components that are required to allow organismal form and function, on all levels of biological organization, to come into being'. Griesemer (this volume) sees development as the recursive acquisition of the capacity to reproduce, where reproduction involves material propagation of developmental capacities from parents to offspring. Pradeu (this volume) suggests that development is the set of processes that lead to the construction of a novel organismal form. Future investigations will have to say if definitions of these kinds are indeed satisfactory and fruitful to conduct research.

The problematic boundaries of development

Whichever definition of the word 'development' one eventually adopts, a number of questions about the temporal, spatial, and functional boundaries of development must also be addressed (Pradeu et al., 2011).

First of all, to what extent is developmental biology really an autonomous field of research? Is the divide between development and other kinds of biological change, usually described as 'simply' metabolic, always a clear one? Some events are indeed difficult to classify (Minelli, this volume), blurring the boundary between what belongs to 'development' and what belongs to other dynamic biological processes.

Second, when does a developmental trajectory start, and when does it end? Many developmental biologists, including some contributors to this volume, describe development as the process by which a multicellular organism is formed from a unicellular zygote. However, one might remark that it would make little sense to exclude from developmental biology the process through which an animal is formed starting from a bud, as usually happens for the hydra, not to mention development from an unfertilized egg, or the fact that in the plant kingdom corresponding phenomena are quite more common than in animals. Moreover, if on the one hand it may be argued (Minelli, this volume; Song et al., 2006) that the egg cell, as such, undergoes a peculiar kind of development, it is also reasonable to remark (see Newman, this volume): 'Acknowledging both the dispensability of the egg stage for many forms of development, as well as the dramatic specializations that often characterize this stage, we [suggest] that events at the egg stage are indeed of lesser consequence for body plan organization compared to the major morphogenetic and patterning events that are initiated subsequently, in the clusters of cells derived from the egg.' Moreover, in so far as we make reference to a life cycle as the unitary set of events to be described by developmental biology (for a different view see Laplane, 2011; Minelli, 2011a, this volume), we must perhaps consider that along any cycle, life cycles included, any point can be taken as representing the 'start' (Griesemer, this volume).

According to some views (e.g. Gilbert, 2003; Minelli, 2003; Oyama, 2000) development 'never stops'. But even if we do not embrace such an extreme position, how sensible is it to restrict development to the events that take place between fertilization and adulthood? This perspective clearly stresses the importance of the reproductive stage, to the point that all previous stages are only regarded as preparatory. This 'adultocentric' perspective has been criticized from several points of view (e.g. Bonner, 1965; Minelli, 2003, 2009).

A sensible temporal delimitation of developmental sequences is far from being the only kind of problem encountered when fixing the boundaries of development. Indeed, we must additionally address the equally difficult problem of determining the spatial boundaries of developmental entities with respect to what we might regard as their environment. This problem is obviously difficult when the developing system is composite, as in the case of a lichen: shall we separately focus on the development of its fungal and algal components, treating in turn the other partner as a close but nonetheless 'external' component of its environment? Shall we, instead, regard the symbiotic pair as the 'real' unit of development? From an empirical point of view, we must at least acknowledge the multigenomic nature of many, and probably most, developmental systems: evidence is rapidly growing about the close interactions of animals as diverse as mammals and squids (see Nyholm & McFall-Ngai, this volume) and their closely associated microbiomes during more or less extended parts of their ontogeny (Pradeu, 2011). The case of developmental symbioses is clearly most intriguing because the proximate 'other' of each symbiotic partner is another living organism with which it lives in steady intimate contact; but not less difficult to fix are the boundaries of more 'ordinary' systems, where development is conditioned, often to a very specific degree, by interactions at distance with other organisms, as in the case of predator-induced polyphenism (e.g. Pfennig, 1992; Pigliucci, 1998). Some developmental biologists would even include, as a component of the developmental system, abiotic components of what we would otherwise describe as the environment (Gilbert, 2001; Gilbert & Epel, 2009). In this vein, the most radical position is represented by the developmental system theory (DST), according to which any clear-cut distinction between the developing organism and its environment is meaningless, and the true units of development and evolution are 'developmental systems', understood as unitary organism–environment systems (Oyama 2000; Oyama et al., 2001; see also Pradeu, 2010).

Finally, the recursive nature of life cycles, when put in the context of the continuity of life through the generations, in the form of the Darwinian 'common descent with modification', may suggest a closer investigation of the relationships between ontogeny and phylogeny: not, to be sure, as a perhaps unconfessed desire to resuscitate Haeckel's (1866, 1874) idea of ontogeny as a recapitulation of phylogeny but in an effort to integrate these two aspects of change into a unifying framework (Griesemer, this volume) and possibly into a theory of 'ontophylogenesis', as suggested by Kupiec (2009).

Possible steps towards a theory of development

In Moczek's words (this volume), a theory of development would be 'any conceptual framework that is applicable to a wide range of organismal diversity and across levels of biological organization, and which would allow us to identify, understand, analyse, and derive predictions about the nature of development'.

But, do we need a theory of development? It is often said that the only theory in biology is the theory of evolution, and the latter in turn is often equated to a population genetics-based, neo-Darwinian view of the history of life. Theoretical aspects of biology are arguably much wider than that, in particular because theoretical attitudes are found in many other biological fields (see Pradeu, this volume), and because key concepts (e.g. robustness, organization, and individuality) cross over into different biological fields and can shed light on a diversity of approaches to living things. And yet, when trying to construct an explicit theoretical framework in the biological sciences, one must face two major pitfalls. The first one is that sometimes purely mathematical models per se are of no real help in advancing our knowledge of actual biological systems (on this question Griesemer's (2013) distinction between theory in the 'exact' and the 'inexact' sciences is particularly useful). The second is that the quest for generality in building a theory may well fail when it is confronted with the wonderful but perhaps discouraging diversity of living things: the living world is full of oddities, exceptions, and, indeed, unique features. Even in the case of concepts often seen as crucial, such as 'individual', 'gene', 'reproduction', 'sex', etc., huge difficulties arise when attempting to apply them to 'non-classical' living systems (clonal animals, plants, unicellular organisms, etc.). This suggests that the task of building a theoretical framework for development is challenging, though not necessarily such that we should abandon it altogether (Minelli, 2011b).

On the one hand, as argued by Love (2008, this volume) the lack of reference to theories in a discipline should be taken at face value. If we can show that developmental biologists have progressed till now without the need to formulate a theory of their subject matter, why bother now? Some contributors to this volume, Arthur especially, do not object in principle to the possible interest of a theory for developmental biology, but are pessimistic with respect to the actual implementation of a research programme in that direction. Gilbert and Bard (this volume) point the finger at the low degree of predictability in development: 'Development can be understood with hindsight; there are still no formal predictors or rules as to what might happen. Development still lacks the sorts of underlying principles that make physics tractable and laws that gives quantitative predictions; one reason is that there are no elementary particles and another is that any laws have proved elusive.' Others have suggested that development is process, and that 'we entirely lack a theory of organization of process' (Kauffman et al., 2008: 28).

On the other hand, taking for granted that aiming at a theory of development is both feasible and interesting, several contributors to this book are offering food for thought on some key aspects of a possible theory. A first point is, how much space should be reserved in this theory for the control of development by genes? This means to take position in front of a question famously formulated by Wolpert (1994): 'Is the embryo computable?' As is well known, Wolpert defended the view that the answer to this question should in principle be 'yes' (see also Rosenberg, 1997). Reactions against the strongly reductionist view of development were voiced soon by philosophers and biologists alike. To quote Lewontin (2000: 17), 'We could not compute the organism, because the organism does not compute itself from genes.' A strongly motivated criticism of the gene-centric view of development had been expressed even before Wolpert's (1994) provocative paper, e.g. by Nijhout (1990), Oyama (2000), and Waddington (1957, 1975), and this position has been finding increased currency (including Gilbert, 2001; Gilbert & Epel, 2009; Gilbert & Sarkar, 2000; Minelli, 2009; Oyama et al., 2001; West-Eberhard, 2003).

In this book, no author explicitly defends a strictly genetic determination of development, and the opposite view is clearly and repeatedly advocated, e.g. by Moczek, Griesemer, Gilbert and Bard, and Vecchi and Hernández. Let's quote, for example, from Gilbert and Bard (this volume): 'In the wider context, a theory of development cannot be a subset of a theory of genetics because much of development is not run by the genome. Genomic activity is neither cell- nor tissue-autonomous but acts as a resource to be activated by signals from other tissues. Any theory of the development of a tissue involves the prior history of that tissue, knowledge of the tissue's environment, and a description of the geometry of that tissue's environment.' The question, however, is not simply an empirical one, as shown by Laubichler and Wagner (2001; see also Vecchi and Hernández, this volume): 'There are many good philosophical reasons to reject the epistemological legitimacy of purely molecular generalizations in developmental biology. The most general is that to assume that the molecular is the most fundamental ontological level in biology, and therefore the most fundamental level of biological analysis, are two controversial theses probably without sound metaphysical grounding. The second reason stems from empirical considerations: the same morphogenetic function can be molecularly realized in different ways, suggesting that morphogenetic function is developmentally more stable than its molecular basis.'

Once accepted that genes are not the only determinants of development, the next step is to identify the other causal agents involved in the process. Biologists should not have serious problems accepting the common multigenomic nature of developing systems, at least in the form of regular interactions between animals and the associate microbiome (see Nyholm & McFall-Ngai, this volume). Other perspectives are more demanding, but not less sensible. Newman, as better specified below in a summary of his chapter, focusses on the developmental role of the physical properties of living matters. Griesemer (this volume) insists instead on the widespread hybrid nature of developing systems, where the developing system regularly interacts with a 'scaffold', often represented by a conspecific (e.g. the maternal body in respect to the developing embryo of a viviparous animal) but can otherwise be an organism of different nature, or an inanimate body and even an artefact.

A different question is whether we can imagine a theory of development completely decoupled from a theory of evolution. Arthur (this volume) argues that evolutionary biology, as a branch of the life sciences with a long and consolidated tradition in theoretical conceptualization, should be taken into account at least for its exemplary value; but Moczek (this volume) does not hesitate to state that 'a theory of development should be nested within a theory of developmental evolution'. A more radical suggestion has been recently advanced by Kupiec (2009), who defends a holistic view of development and evolution as aspects of one and the same chain of events ('ontophylogenesis').

One may thus wonder why evolutionary biology and developmental biology have taken so long to open dialogue following decades of mutual isolation (see e.g. Amundson, 2005 for a historical and conceptual perspective). Besides long-entrenched, but principally minor causes such as schools' and scientists' research agendas, and the objective difficulties caused by the very limited experimental approaches to developmental genetics available till the late 1970s, developmental biology and evolutionary biology are indeed separated by an objective divide, as admirably shown by Lewontin (2000: 8–10). Lewontin contrasts the *variational* conception of change in evolutionary biology (a change in the proportional representation of the different variants) with the *transformational* conception of change in developmental biology (a collective change due to the fact that each individual undergoes during its lifetime the same law-like history). In turn, this dichotomy explains why evolutionists have been interested in the *differences* between individual organisms and the differences between closely related species, while developmental biologists have long focussed on what is *invariant* from organism to organism, and even from species to species. The tension between the two disciplines is still well present in their opposite attitudes towards the use of model organisms: while for developmental biologists an ideal model organism to be used in the lab should not exhibit individual variation, it is precisely the variation within populations that evolutionary biologists have found to be the prime target of their studies (Love, 2010).

Overall, key questions raised by this volume include:

(i) *Is developmental biology descriptive or explanatory?* Is it possible to explain development, and, if so, how? What epistemological category should be used to account for development: 'mechanisms', 'principles', 'laws', 'models', 'hypotheses', or 'theories'? In particular, are theories indispensable, or at least useful, for building explanations and predictions? Moreover, at what level should developmental explanations be situated: at the level of genes, cells, intercellular communications, fields, etc.?

(ii) *What is the domain of extension of developmental biology?* The possible construction of any theory of development immediately raises the question, to which entities does this theory apply? This is thus one of the most direct ways to raise the fundamental issue of which entities in the living world can be said to 'develop' (for example: can one speak of development in the case of unicellular organisms?)

(iii) *When and where does development start and stop?* Delineating the temporal and spatial boundaries of development is both crucial and immensely awkward.

(iv) *To what extent is development 'constant'?* Many have insisted that what needs to be explained in developmental biology is the extraordinary stability and robustness of development, so much so that some have tended to describe development as a necessary process, as famously illustrated by at least some uses of the 'programme' metaphor. Yet, for others, development is much more *contingent* (Oyama, 2000), and/or *stochastic* (Kupiec, this volume), than has been traditionally considered. And if indeed development is much more contingent and unpredictable than previously thought, how does this impact our capacity to offer a scientifically promising account of development?

(v) *On what basis should developmental biology be unified?* Today's developmental biology is a strongly divided field: specialists of one model organism do not often talk to specialists of other model organisms, plant embryology is hardly considered by mainstream developmental biologists, proponents of the genetic programme of development tend to look haughtily at more classical embryological approaches, etc. It seems important to collectively determine what these different approaches have in common. Shedding light on the unity of developmental biology might be important for a better understanding of what development itself is, and perhaps adopting a 'theoretical stance' in developmental biology might help to fulfil this aim.

(vi) *How should developmental biology be articulated with other fields of the life sciences?* Though the need to articulate development and evolution has now long been recognized, the concrete realization of this articulation remains to be accomplished. However, other articulations are equally important, in particular those with ecology, immunology, medicine, or even current genetics and proteomics.

Overview of the chapters of this volume

In 'Regenerating theories in developmental biology' (Chapter 2), Thomas Pradeu defends the view that developmental biology does offer theories. This claim is sustained by three main arguments: first, several other biological fields routinely formulate theories, and these fields do not seem to possess any characteristic that developmental biology would lack; second, embryology and developmental biology have offered theories in the past; and third, several developmental biologists (admittedly not numerous, but often among the most influential ones) have insisted on the role of theories in their field and have themselves formulated theories. Proposing to define a 'theory' as a structured set of testable explanatory and predictive hypotheses, Pradeu shows that such a definition is widely applicable to current biological fields, and in particular that it makes it possible to speak of *developmental theories*. He analyses some examples of such theories in recent developmental biology and suggests some avenues for the further construction of developmental theories and their articulation into a growingly unifying picture of developmental processes.

In his chapter 'The erotetic organization of developmental biology' (Chapter 3), Alan C. Love moves from the observation that many presentations of developmental biology make no reference to a theory or theories of development. Instead of seeing this absence as an indicator of immaturity or an invitation to reconstruct a theory in different guise, this, Love argues, reveals the 'erotetic' (or question-based) organization of developmental biology. This organization consists of problem agendas each of which corresponds to a class of phenomena such as pattern formation, differentiation, growth, and morphogenesis. Hence, developmental biology does not offer theories and needs not to.

In 'The concept of mechanism in development' (Chapter 4), Johannes Jaeger and James Sharpe argue that it should be possible to enumerate and classify different developmental mechanisms that can achieve the same biological function (e.g. create a stripe of gene expression in a static or growing tissue) due to the finite number of dynamic behaviours that can be implemented by regulatory systems. Comparing these different mechanisms would help discover the design principles of the regulatory networks and thus constitute the basis for a theory of development that characterizes and explains the regularities and recurring motifs observed in organismal morphology. This chapter introduces and defines a concept of developmental mechanism based on the conceptual framework of dynamical systems theory. Equivalent mechanisms are defined as sharing the same topology of their phase portraits: they have the same number and geometrical arrangement of attracting states, saddle points, and basins of attraction, and undergo structurally stable bifurcations as systems parameters change over time. This constitutes a first, tentative step towards a more general geometrical theory of developmental mechanisms and the complex map from genotype to phenotype.

Issues of causality in developmental biology are addressed by Davide Vecchi and Isaac Hernández in Chapter 5 ('The epistemological resilience of the concept of the morphogenetic field'). Specifically, they revisit the ontological status of the old concept of the morphogenetic field, whose main causal features were its holistic character and its non-genetic informational nature. The advent of developmental genetics made the concept redundant. Vecchi and Hernández analyse Wolpert's concept of positional information and its transformation into a genetic concept within the preformationist programme of 'computational embryology', as well as the conception of causality involved in Crick's source-sink model of morphogen diffusion. The renaissance of the concept of the morphogenetic field is subsequently illustrated, based either on Turing's reaction–diffusion model, or on the processes of biochemical oscillation and cellular synchrony. The authors of this chapter suggest that these recent interpretations of the morphogenetic process vindicate the epistemological relevance of the concept of the morphogenetic field, even though a lack of clarity concerning its ontological nature persists.

Moving from a perspective that focusses on the physical aspects of morphogenesis, as complementary to the genetic ones and even historically and materially prior to them, Stuart A. Newman describes in Chapter 6 ('Physico-genetics of morphogenesis: the hybrid nature of developmental mechanisms') the hybridity that can be found within developmental processes. Like all parcels of matter, embryos bear the morphological signatures of their material nature. In the case of multicellular animals, these correspond to 'generic' motifs such as cavities, immiscible layers, segments, tubes, and appendages. Early stage animal embryos and organ primordia are simultaneously soft matter and excitable media, and are thus inherently hybrid. In addition, developing tissues behave in a multi-scale fashion, continually and simultaneously changing their composition and organization on several temporal and spatial scales, which also contributes to their hybridity. Newman also provides examples of well-characterized generic determinants of the organization of tissue masses, as well as of regulative processes that have resulted from evolutionary complexification. He concludes that a rigorous physico-genetic theory of development is possible, but only if the hybridity and historicity of these systems are taken into account and the ideal of a unitary set of laws like those applicable to monoscale physical systems is abandoned.

In Chapter 7 ('The landscape metaphor in development'), Giuseppe Fusco, Roberto Carrer, and Emanuele Serrelli review and discuss the use of the landscape metaphor in development by analysing its relationship with experimental work and theoretical modelling. Defining a landscape as a mathematical function of multiple variables, the authors show how this can be interpreted as a dynamical system. From this perspective they analyse Waddington's epigenetic landscape metaphor and other landscape representations occurring in current developmental biology literature. The authors then go on to discuss the somewhat parallel stories of Wright's fitness landscapes and Waddington's developmental landscapes. After briefly discussing developmental landscapes in the context of visualization in science, with a focus on theoretical work in developmental biology, they conclude that landscapes seem to be too limited a form of abstraction to stand as a central metaphor in the search for a comprehensive theory of development.

In Chapter 8 ('Formalizing theories of development: a fugue on the orderliness of change'), Scott Gilbert and Jonathan Bard look at developmental biology as performance. Each animal inherits a score (the DNA), mechanisms for interpreting the score, and mechanisms for improvisation should the score be deficient. Developmental causation is found to be both upwards from the genome, downward from the environment, and laterally between cells. Developmental plasticity, organicism, phenotypic heterogeneity, symbiotic co-development, and cytoplasmic localization are examples of causation from the environment downward. Stereocomplementary relationships are the key components of most developmental interactions. These interactions can be placed into a formal language of graph theory, thus emphasizing the distributed nature of causality in development. This graph approach portrays dynamic processes as the drivers of developmental momentum, which can be viewed as modules whose underlying networks are genomic subroutines.

In Chapter 9 ('General theories of evolution and inheritance, but not development?'), Wallace Arthur moves from a critical dissection of the title of this book, noting that it implicitly assumes that there is not at present a single, overarching theory of development but that such a theory may yet be possible. Arthur remarks that old theories in embryology (preformation, epigenesis) were not mechanistic, whereas a 'modern' theory needs to be. A possible route to such a modern theory is then taken through an examination of patterns of interconnection between causal links, including hierarchical, combinatorial, feedback, and modular ones. According to Arthur, the development of animals probably includes all these sorts of interconnections, with the overall pattern forming a complex network. The question of whether there might be alternative types of networks is then considered, noting that a pair of alternatives is often useful in revealing the nature of both (e.g. particulate vs blending inheritance). The tentative conclusion is that no obvious pair of alternative types of causal networks underlying development can be currently discerned.

In 'Cell differentiation is a stochastic process subjected to natural selection' (Chapter 10), Jean-Jacques Kupiec criticizes the classical determinist theories of development in the light of evidence showing that gene expression is a stochastic phenomenon. Kupiec reviews these data as well as the mechanisms causing stochastic gene expression (SGE), shows that it creates the diversity of gene expression patterns accompanying cell differentiation, and outlines a model of cell differentiation based on SGE. In this model, cell differentiation is seen as a two-sided mechanism operating at each stage of development. On the one hand, SGE produces a diversity of gene expression patterns in differentiating cell populations. On the other hand, gene expression is stabilized in differentiated cells by means of cellular interactions. SGE is caused by random molecular interactions between chromatin proteins and DNA, making chromatin dynamic stochastic, whereas epigenetic chromatin modifications triggered by signal transduction between cells are responsible for the stabilization of SGE. This model underlies a new theory in which individual development results from natural selection operating on embryonic cell populations.

According to Michel Morange in Chapter 11 ('From genes to gene regulatory networks: the progressive historical construction of a genetic theory of development and evolution'), different models coexist within the field of evo-devo that give genes more or less important roles in the control of development and evolution. The gene regulatory network (GRN) model proposed by Eric Davidson is one of the most ambitious, as a precise knowledge of these GRNs would enable a reconstruction of the evolutionary path that led to present living forms and to describe the space of possibles presently accessible to evolution. Morange describes the different steps that led to the progressive emergence of the GRN model and examines the reasons why Eric Davidson considers his model incompatible with Modern Synthesis In particular, Morange discusses whether the description of GRNs will be sufficient to account for the evolution of organisms, or whether this description will have to be complemented by a rejuvenated version of modern synthesis, or by an entirely new theory of innovation.

In Chapter 12 ('Reproduction and scaffolded developmental processes: an integrated evolutionary perspective'), James Griesemer expands conceptual boundaries beyond the traditional view of biological development as a cellular process by considering molecular development in the life cycle of the retrovirus HIV-1 and in other complex life cycles such as those of malaria parasites. Implied in his view is a new concept of hybrid developmental entities that arise during the process of reproduction from sources providing for different developmental capacities: according to Griesemer, biological reproduction mostly involves the temporary formation of a hybrid consisting of a developmental system and scaffolds that facilitate development. Development is characterized as a process of acquiring and refining the capacity to reproduce by means of the material overlap of parts between parents and offspring. The chapter also formulates a view of scientific theories as comprising a set of core principles, a family of models, and a theoretical perspective that coordinates model-building and empirical investigation.

In 'Comparison of animal and plant development: a right track to establish a theory of development?' (Chapter 13), Michel Vervoort suggests that there could be common principles underlying development in organisms as different as animals and plants, indeed shared by all organisms displaying development. Together these principles would constitute a unified, or unifying, theory of development. Vervoort suggests that such a theory of development should be based on a comparison between developmental processes convergent among remotely related branches of the tree of life (plants and animals in particular). Vervoort discusses the possible obstacles and pitfalls of this approach, and examines a case study of such comparisons.

In Chapter 14 (Towards a theory of development through a theory of developmental evolution'), Armin P. Moczek explores the relationship between a theory of development and a theory of developmental evolution. The first part of the chapter reviews points of tension between different perspectives on the importance of understanding development in order to understand organismal form, function, and evolution. In the second part Moczek argues that to formulate a theory of development that could serve as a conceptual mediator to revise existing disconnects, this theory should be nested within a theory of developmental evolution. Such a theory of development could be built in two steps: first, by accumulating empirical evidence, thus identifying and linking together developmental products and processes; and second, by organizing this information focussing on the development of homologues, a nested hierarchy of homologues, and a description of patterns and causes of variation within homologues. The strategy would allow a conceptualization of development that is biologically realistic, can be refined alongside a growing base of empirical knowledge, is flexible enough to incorporate both homology (descent) and variation (modification), and is capable of bridging to relevant conceptual frameworks in adjacent biological fields.

In Chapter 15 ('Developmental disparity'), Minelli argues that a theory of development should apply to the widest diversity of existing life cycles and forms of change. The fact that life cycles are often multigenerational conflicts with the widespread approach of identifying a life cycle with the developmental sequence of an individual animal or plant. Focus is next placed on the multigenomic nature of many developmental systems, and on unicells, which often undergo processes that deserve to be described as developmental. Minelli criticizes adultocentrism (i.e. development seen as change targeted to obtaining the adult condition) and the opinion that only functional changes deserve to be regarded as developmental. Changes pertaining to senescence, regeneration and also—to some extent—to pathogenesis are argued instead to pertain to development as well. The chapter also discusses the divide between development and metabolism and the existence, or the relevance, of aspects of development other than morphological change, to conclude by contrasting the properties of development that emerge from a survey of its disparity across different forms of life with the 'current view' of development, especially in its prevailing adultocentric variety.

In Chapter 16 ('Identifying some theories in developmental biology: the case of the cancer stem cell theory') Lucie Laplane addresses whether it is possible to distinguish theories from non-theories in developmental biology and then goes on to describe how it might be done. First, Laplane argues that the classical conceptions of scientific theories cannot help in identifying theories in a particular field of science. She then shows how a particular interpretation of the *vera causa* principle offers simple and operational criteria for the identification of theories. This interpretation then leads to the minimal definition of a scientific theory as a scientific hypothesis, derived from empirical premises, that has been proven able to explain independent classes of facts. The author then discusses an example of an existing theory in developmental biology (the cancer stem cell theory), highlighting the fact that the conception of theories that is presented in this chapter is both simple and operational.

In the last chapter of this volume ('Animal development in a microbial world'), Spencer V. Nyholm and Margaret J. McFall-Ngai discuss the developmental role of microbial partners that live in association with animals. Recent, growing evidence indicates that symbiosis is the rule rather than the exception among animals. The chapter first considers the selective forces that have shaped the development of form and function in the immune and digestive systems over evolutionary history and discusses how these forces are likely to have been strongly influenced by the radiation of animal phyla in the microbe-rich marine environment. In the second part of the chapter, Nyholm and McFall-Ngai show how microbes can influence each stage of development, from the egg through maturation to the adult animal. Particular emphasis is placed on recent findings about the impact of symbiosis on gut development, immune reactions, brain development, and normal behaviour of the animal host.

Acknowledgments

Our first acknowledgments go to the contributors to this volume, who have been extraordinarily responsive and reliable during the whole ontogeny—and the few local metamorphoses—of this volume. In addition, we would like to thank Ian Sherman, our editor at Oxford University Press, who provided incredibly useful suggestions and feedback, as well as his assistant, Lucy Nash, who has done a remarkable job in keeping us on track concerning both the schedule and the overall organization of this volume, Elizabeth Farrel, for her careful and sensible copyediting, and Saranya Manohar, for her standing and friendly assistance during the book's production.

References

Amundson, R. (2005). *The Changing Role of the Embryo in Evolutionary Thought*. Cambridge University Press, Cambridge.

Arthur, W. (2002). The emerging conceptual framework of evolutionary developmental biology. *Nature*, **415**, 757–63.

Bolker, J. (2012). Model organisms: there's more to life than rats and flies. *Nature*, **491**, 31–3.

Bonner, J.T. (1965). *Size and Cycle: An Essay on the Structure of Biology*. Princeton University Press, Princeton.

Burian, R.M., and Thieffry, D. (2000). Introduction to the special issue 'From embryology to developmental biology'. *History and Philosophy of the Life Sciences*, **22**, 313–23.

Cobb, M. (2006). *The Egg and Sperm Race. The Seventeenth-Century Scientists Who Unravelled the Secrets of Sex, Life and Growth*. Free Press, London.

Davidson, E.H. (2003). *Genomic Regulatory Systems: Development and Evolution*. Academic Press, San Diego.

Davidson, E.H. (2006). *The Regulatory Genome*. Academic Press-Elsevier, Amsterdam.

Garson, J., Wang, L., and Sarkar, S. (2003). How development may direct evolution. *Biology and Philosophy*, **18**, 353–70.

Gilbert, S.F., ed. (1991). *A Conceptual History of Modern Embryology*. The Johns Hopkins University Press, Baltimore.

Gilbert, S.F. (1998). Conceptual breakthroughs in developmental biology. *Journal of Biosciences*, **23**, 169–76.

Gilbert, S.F. (2001). Ecological developmental biology: developmental biology meets the real world. *Developmental Biology*, **233**, 1–12.

Gilbert, S.F. (2003). *Developmental Biology*, 7th ed. Sinauer Associates, Sunderland.

Gilbert, S.F., and Epel, D. (2009). *Ecological Developmental Biology*. Sinauer Associates, Sunderland.

Gilbert, S.F., and Sarkar, S. (2000). Embracing complexity: organicism for the 21st century. *Developmental Dynamics*, **219**, 1–9.

Gilbert, S.F., Opitz, J.M., and Raff, R.A. (1996). Resynthesizing evolutionary and developmental biology. *Developmental Biology*, **173**, 357–72.

Goodfield, J. (1969). Theories and hypotheses in biology. In R.S. Cohen and M.W. Wartofsky, eds., *Boston Studies in the Philosophy of Science*, Vol. 5. Springer, Dordrecht, pp. 421–49.

Griesemer, J. (2013). Formalization and the meaning of 'theory' in the inexact biological sciences. *Biological Theory*, **7**, 298–310.

Haeckel, E. (1866). *Generelle Morphologie der Organismen: Allgemeine Grundzüge der organischen Formen-Wissenschaft, mechanisch begründet durch die von Charles Darwin reformirte Descendenz-Theorie*. Reimer, Berlin.

Haeckel, E. (1874). *Anthropogenie: Keimes und Stammes-Geschichte des Menschen*. Engelmann, Leipzig.

Hall, B.K. (1998). *Evolutionary Developmental Biology*, 2nd ed. Chapman and Hall, London.

Hall, B.K. (2000). Evo-devo or devo-evo—does it matter? *Evolution and Development*, **2**, 177–8.

Horder, T.J., Witkowski, J.A., and Wylie, C.C., eds. (1986). *A History of Embryology*. Cambridge University Press, Cambridge.

Kauffman, S., Logan, R.K., Este, R., Goebel, R., Hobill, D., and Shmulevich, I. (2008). Propagating organization: an enquiry. *Biology and Philosophy*, **23**, 27–45.

Kupiec, J.-J. (2009). *The Origins of Individuals*. World Scientific, Singapore.

Laplane, L. (2011). Stem cells and the temporal boundaries of development: toward a species-dependent view. *Biological Theory*, **6**, 48–58.

Laubichler, M.D., and Maienschein, J., eds. (2007). *From Embryology to Evo-devo: A History of Developmental Evolution*. MIT Press, Cambridge.

Laubichler, M.D., and Wagner G.P. (2001). How molecular is molecular developmental biology? A reply to Alex Rosenberg's reductionism redux: computing the embryo. *Biology and Philosophy*, **16**, 53–68.

Lewontin, R. (2000). *The Triple Helix: Gene, Organism and Environment*. Harvard University Press, Cambridge.

Love, A.C. (2008). Explaining the ontogeny of form: philosophical issues. In S. Sarkar and A. Plutynski, eds., *A Companion to the Philosophy of Biology*. Blackwell, Malden, pp. 223–47.

Love, A.C. (2010). Idealization in evolutionary developmental investigation: a tension between phenotypic plasticity and normal stages. *Philosophical Transactions of the Royal Society of London B*, **365**, 679–90.

Love, A.C. (2013). Theory is as theory does: scientific practice and theory structure in biology. *Biological Theory*, **7**, 325–37.

Maienschein, J. (2012). Epigenesis and preformationism. In *The Stanford Encyclopedia of Philosophy*. <http://plato.stanford.edu/archives/spr2012/entries/epigenesis/> (last accessed 17 September 2013).

Mayr, E. (1969). Comments on 'Theories and hypotheses in biology'. In R.S. Cohen and M.W. Wartofsky, eds., *Boston Studies in the Philosophy of Science*, Vol 5. Springer, Dordrecht, pp. 452–6. Reprinted as: Theory formation in developmental biology, in E. Mayr (1976) *Evolution and the Diversity of Life. Selected Essays*. Harvard University Press, Cambridge, pp. 377–82.

Minelli, A. (2003). *The Development of Animal Form*. Cambridge University Press, Cambridge.

Minelli, A. (2009). *Forms of Becoming: The Evolutionary Biology of Development*. Princeton University Press, Princeton.

Minelli, A. (2011a). Development, an open-ended segment of life. *Biological Theory*, **6**, 4–15.

Minelli, A. (2011b). A principle of developmental inertia. In B. Hallgrímsson and B.K. Hall, eds., *Epigenetics: Linking Genotype and Phenotype in Development and Evolution*. University of California Press, San Francisco, pp 116–33.

Needham, J. (1934). *A History of Embryology*. Cambridge University Press, Cambridge.

Nijhout, H.F. (1990). Metaphors and the role of genes in development. *BioEssays*, **12**, 441–46.

Nüsslein-Volhard, C. (1994). Of flies and fishes. *Science*, **266**, 572–74.

Nyhart, L.K. (1995). *Biology Takes Form: Animal Morphology and the German Universities, 1800–1900*. University of Chicago Press, Chicago.

Oyama, S. (2000). *The Ontogeny of Information: Developmental Systems and Evolution*, 2nd ed. Duke University Press, Durham.

Oyama, S., Griffiths, P., and Gray, R., eds. (2001). *Cycles of Contingency: Developmental Systems and Evolution*. MIT Press, Cambridge.

Pfennig, D.W. (1992). Proximate and functional causes of polyphenism in an anuran tadpole. *Functional Ecology*, **6**, 167–74.

Pigliucci, M. (1998). Developmental phenotypic plasticity: where internal programming meets the external environment. *Current Opinion in Plant Biology*, **1**, 87–91.

Pinto-Correia, C. (1997). *The Ovary of Eve: Egg and Sperm and Preformation*. University of Chicago Press, Chicago.

Pradeu, T. (2010). The organism in developmental systems theory. *Biological Theory*, **5**, 216–22.

Pradeu, T. (2011). A mixed self: the role of symbiosis in development. *Biological Theory*, **6**, 80–8.

Pradeu, T., Laplane, L., Morange, M., Nicoglou, A., and Vervoort, M. (2011). The boundaries of development. *Biological Theory*, **6**, 1–3.

Raff, R. (1997). *The Shape of Life*. University of Chicago Press, Chicago.

Rosenberg, A. (1997). Reductionism redux: computing the embryo. *Biology and Philosophy*, **12**, 445–70.

Roux, W. (1897). *Programm und Forschungsmethoden der Entwicklungsmechanik der Organismen*. Engelmann, Leipzig.

Shirley, R. (1981). *Matter, Life, and Generation: Eighteenth-Century Embryology and the Haller-Wolff Debate*. Cambridge University Press, Cambridge.

Song, J.L., Wong, J.L., and Wessel, G.M. (2006). Oogenesis: single cell development and differentiation. *Developmental Biology*, **300**, 385–405.

Stadler, B.M.R., Stadler, P.F., Wagner, G.P., and Fontana, W. (2001). The topology of the possible: formal spaces underlying patterns of evolutionary change. *Journal of Theoretical Biology*, **213**, 241–74.

Waddington, C.H. (1957). *The Strategy of the Genes*. Allen and Unwin, London.

Waddington, C.H. (1968). Theoretical biology and molecular biology, in C.H. Waddington, ed. *Towards a Theoretical Biology*, Vol. 1, *Prolegomena*. Edinburgh University Press, Edinburgh, pp. 103–8.

Waddington, C.H. (1975). *The Evolution of an Evolutionist*. Cornell University Press, Ithaca.

Wagner, G.P. (1996). Homologues, natural kinds, and the evolution of modularity. *American Zoologist*, **36**, 36–43.

Wagner, G.P. (2000). What is the promise of developmental evolution? Part I: Why is developmental biology necessary to explain evolutionary innovations? *Journal of Experimental Zoology (Molecular Development and Evolution)*, **288**, 95–8.

West-Eberhard, M.J. (2003). *Developmental Plasticity and Evolution*. Oxford University Press, Oxford.

Wolff, C.F. (1759). *Theoria Generationis*. Hendelianis, Halle.

Wolpert, L. (1994). Do we understand development? *Science*, **266**, 571–72.

Wolpert, L., and Lewis, L.J. (1975). Towards a theory of development. *Federation Proceedings*, **34**, 14–20.

CHAPTER 2

Regenerating theories in developmental biology

Thomas Pradeu

Introduction

Developmental biology offers descriptions and explanations of the processes involved in the development of living entities. But does it formulate *theories*? If it does not, does this raise difficulties? If it does, in which ways is the formulation of theories useful? In general terms, is it important, for a given scientific field, to express theories? The dominant view among current biologists and philosophers of biology is that developmental biology does not offer theories, or at the very least is not structured by theories (Love, 2008, this volume; Arthur, this volume; Gilbert and Bard, this volume; see also Goodfield, 1969), and needs not to. Yet I would like to argue in this chapter that developmental biology does offer theories. The initial focus will not be on whether or not developmental biology offers a single, general, theory, though the last section will offer some thoughts about this issue.

First of all, what is 'development?' This apparently naïve question has in fact no straightforward answer. Delineating development from a conceptual point of view is a daunting task (Pradeu et al., 2011). Most of the time, in their publications and in everyday practice, developmental biologists do not define what they mean by 'development'. When they do, they usually insist on the fact that development refers to the *construction* of an organism (e.g. Davidson, 1991; Gilbert & Raunio, 1997; Gilbert, 2003, 2013 ('the formation of an orderly body from relatively homogenous material'); Wolpert, 1991 ('the moulding of form'); see also Minelli and Pradeu, this volume). Classically, developmental biologists have seen this construction as lasting from fertilization to the acquisition of the capacity to reproduce, while others see it as ending only with death (e.g. Gilbert, 1994, 2013; see also Oyama, 2000) and/or criticize dominant adultocentric approaches to development (Bonner, 1965; Minelli, 2003, 2009, 2011a, this volume; see also the interpretation of the capacity to reproduce defended by Griesemer, this volume). A majority of developmental biologists would concur on the explananda of their field: cell differentiation, morphogenesis (including organogenesis), and growth as the main aspects (e.g. Bonner, 1974:130–2 (though Bonner also emphasizes the possible biases of these notions); Gilbert 2013; Robert 2004; Shostak 1991; Wolpert 2011), to which many specialists of the field add the acquisition of the capacity to reproduce as well as regeneration (e.g. Gilbert, 2013; Slack, 2013; Wolpert et al., 2011), and also, for some, evolution and environmental integration (Gilbert (2013) lists these six questions as the key ones addressed by current developmental biology). Moreover, there is a consensus on the key steps of animal development (fertilization, cleavage, gastrulation, organogenesis, metamorphosis (in some species), maturity, and gametogenesis (on these steps and their place in the agenda of developmental biology, see Love, 2008, and this volume; these steps, however, are not all universal: see Minelli, this volume)) and on the need for the field of developmental biology to account for these steps.

Here I understand development as the set of processes that leads to the construction of a novel organismal form. This definition identifies

Towards a Theory of Development. Edited by Alessandro Minelli and Thomas Pradeu

morphogenesis and organogenesis as crucial components of any developmental process. This view is broad enough to incorporate crucial mechanisms of late organogenesis, such as metamorphosis, or local late development of new tissues in mammals, for example, tertiary lymphoid organs (Eberl, 2005; Pradeu, 2011). Nevertheless, it is more precise than conceptions that equate 'development' with all forms of metabolic change from fertilization to death. Of course, other definitions of 'development' could be offered, but this one seems operational enough and will be sufficient to serve our purpose here. With this definition in mind, let us now return to the central question raised above: does developmental biology offer theories?

This chapter argues that developmental biology does indeed offer theories. This view rests on the conviction that both biologists and philosophers of biology, when they reject the idea that developmental biology could offer theories, have in mind a conception of a theory that is excessively demanding. Therefore, the path chosen here is not an easy one: my aim is to suggest a concept of a scientific theory that is demanding (in the sense that it is not trivial and may lead to fruitful results in a given scientific field), but not *too* demanding (to avoid the often expressed idea that, for instance, quantum mechanics and evolutionary biology could offer theories, but developmental biology would not be ready yet to do so).

The structure of this chapter is as follows. The first section describes the widespread suspicion of theories in current developmental biology and philosophy of developmental biology. The second section offers a critique of this suspicion, in particular through a detailed analysis of biological fields in which people routinely speak of 'theories', and a historical account of the different uses in the past of the term 'theory' in embryology and developmental biology. The third section offers a plausible conception of what a theory may be in biology. The fourth section argues that developmental biology does offer theories of development, and it suggests some avenues for the construction of developmental theories and their articulation into a growingly unifying picture.

The suspicion towards theories in developmental biology

The idea that developmental biology might suggest theories, let alone a general theory of development, is usually seen with the most extreme suspicion. This suspicion is expressed both by biologists and philosophers (Arthur, this volume; Gilbert and Bard, this volume; Love, 2008, this volume). Developmental biologists, in their current practice, seldom use the term 'theory'. Table 2.1 shows how frequently the term 'theory' appears in titles and abstracts of several of the main journals in developmental biology. This table confirms that developmental biologists do not use this term on a regular basis (Table 2.2 offers a comparison with other biological fields).

What could explain the scarcity of mentions of theories in today's developmental biology? It is often claimed that theories are too abstract (i.e. remote from experimental practice), too ambitious ('grand' theories, which pretend to explain everything, do not explain much), or too general and therefore unable to cope with the extreme diversity and complexity of developmental phenomena across species (e.g. Bard and Gilbert, this volume; see also Freeman, 2002; Goodfield, 1969). In addition, many biologists have the feeling that if our best physical theories are seen as the archetype of what a scientific theory should be, then developmental biology cannot suggest 'theories' in this strong sense, since its statements are less universal, less abstract, and less likely to be expressed in a mathematical language.

Philosophers of biology, for their part, have tended to be suspicious of the notion of theory because they see it as forged in the context of a philosophy of science dominated by logical positivism and philosophy of physics (Callebaut, 2013). It has often been suggested that there are no theories and laws in biology, mainly because of the historical character of this domain (e.g., Smart, 1963—an idea discussed in Hull, 1974; see also Beatty, 1995). In addition, many philosophers of biology consider that classic views about scientific laws and theories should be discarded as illegitimate applications to biology of problems framed in the terms of the philosophy of physics, ignorant of the specificities

Table 2.1 Number of occurrences of the word 'theory' in the titles and abstracts of papers published in several of the main journals in developmental biology. The first column indicates the name of the journal, as well as the period examined.

	'Theory' in article title	'Theory' in article abstract
Annual Review of Cell and Developmental Biology (1985–2012)	0	2
Current Opinion in Genetics and Development (1995–July 2013)	3 (including two about the neutral theory of evolution)	27
Current Topics in Developmental Biology (1966–July 2013)	4	35
Development (1987–2012)	6	92
Development Genes and Evolution[1] (1894–July 2013)	6 (mainly in 19th century papers)	NA
Developmental Biology (1959–July 2013)	2	32
Developmental Cell (2001–July 2013)	1	7
Genes and Development (1987–July 2013)	2	11
Seminars in Cell and Developmental Biology (1996–July 2013)	3	10

[1]This journal has been published since 1894. From 1894 to 1923, it was published as *Archiv für Entwicklungsmechanik der Organismen*. Then, its name was changed five times. Since 1996, the journal has been published under the title *Development Genes and Evolution*.

Table 2.2 Number of occurrences of the word 'theory' in the titles and/or abstracts of papers published in the *Proceedings of the National Academy of Sciences of the United States of America* (*PNAS*) in each field of the biological sciences between 1915 (first issue of the journal) and July 2013. The second column indicates the total number of papers published in each field, and the third column shows the percentage of papers having the word 'theory' in theirs titles and/or abstracts (please note that papers in *PNAS* can sometimes be attributed to several categories).

Domain	Papers with 'theory' in title or abstract	Total number of papers in the domain	Percentage
Applied biological sciences	6	791	0.76
Biochemistry	49	8513	0.58
Biophysics and computational biology	234	4046	5.78
Cell biology	21	4172	0.50
Developmental biology	5	1648	0.30
Ecology	106	1116	9.50
Evolution	180	2391	7.53
Genetics	32	4175	0.77
Immunology	10	3404	0.30
Microbiology	4	3039	0.13
Neuroscience	107	5839	1.83
Physiology	23	1509	1.52
Psychological and cognitive sciences	75	954	7.86
Systems biology	9	146	6.16

of biology as a science (Hull, 1969; Sterelny & Griffiths, 1999; this view was shared by the evolutionary biologist Ernst Mayr (1969a; 1969b)). Hull (1974) considers that biology can offer laws, but those are almost always to be understood as accidental generalizations; moreover, biology can have theories, but, Hull argues, only evolutionary biology expresses well-articulated theories (Hull, 1992), leaving all the rest of biology, including embryology, with no 'true' theories. In the general philosophy of science, a major debate occurred in the 1960s and continued until the 1980s between the 'syntactic' and the 'semantic' approaches to theories: the 'syntactic' (often dubbed 'classic') view sees a theory as an abstract calculus and correspondence rules that give an empirical content to this calculus (Nagel, 1961), while the 'semantic' view sees theories as classes of models (Suppe, 1977; Suppes, 1967; van Fraassen, 1972). This debate, as many others in the general philosophy of science, appears extremely remote from the actual questioning and practices of the life sciences. Several philosophers of biology did attempt to apply the classic debate about the syntactic and/or the semantic conceptions to 'theories' and reduction in molecular biology (Schaffner, 1968, 1993) and to the theory of evolution by natural selection (Lloyd, 1988; Ruse, 1973; Thompson, 2007; see also the work of evolutionary biologist Mary Williams (1970)), but these attempts have generated no consensus and have seemingly not been of much use to biologists. In addition, many historians and philosophers have shown that, contrary to what is often believed, the dominant view of evolution, which stems from the Modern Synthesis, does not constitute a single and unified theory, but rather a set of heuristic, and to some extent heterogeneous, statements (Gayon, 1998; Love, 2013). Recently, lively discussions have occurred about models, modelization, and idealization in biology (e.g. Downes, 1992; Godfrey-Smith, 2006; Weisberg, 2006; Wimsatt, 2007; see also Morrison, 2007), often in the footsteps of Richard Levins (1966); however, most of them have tended to set 'theories' aside. Overall, many philosophers of biology consider that evolution by natural selection constitutes a genuine 'theory' and that, in contrast, non-evolutionary fields would not offer bona fide theories, but rather, at best, general descriptions situated at different levels (molecular, cellular, tissue, etc.) A more recent version of this claim has been articulated by several proponents of a mechanistic framework, who suggest that molecular biology, cell biology, neurology, etc. offer 'mechanisms' and 'schemata' (i.e. aggregations of mechanisms) rather than 'theories' in the sense usually retained by general philosophy of science (Machamer et al., 2000; Darden & Tabery, 2009; see also Craver, 2002, although this text expresses a more balanced view on theories and suggests the notion of 'mechanistic theories' for molecular biology; Griesemer (2011a, 2011b), for his part, proposes a useful distinction between 'mechanistic' and 'quantitative' theories). Among philosophers with a special interest in developmental biology, Alan Love (2008, this volume) has forcefully defended the idea that this field does not, and needs not, offer 'theories', but rather is driven by the formulation of 'research agendas' (please note, however, that Love's line of thought is independent of the 'mechanistic' trend in current philosophy of biology). As will become clear in the rest of this chapter, I agree to a very large extent with Love's attention to scientific practices, though our conclusions diverge radically about the role theories have, or may have, in developmental biology.

Overall, the suspicion about scientific theories expressed in biology and in the philosophy of biology rests on a double assumption. The first assumption is that traditional debates in the general philosophy of science (in particular the 'syntactic' vs 'semantic' debate) would have given us an adequate conception of what a theory is. The second assumption is that physics (or, rather, physics as seen by biologists) would offer the archetype of what a scientific theory should be. But it is very likely that these assumptions are inadequate. The next section offers three reasons to cast doubts on the idea that developmental biology would not offer theories.

Critique of this suspicion

The first reason for being sceptical on the dominant view presented in the previous section is that several biological fields routinely offer theories. Indeed, it is perfectly usual to talk about theories in evolutionary biology (e.g. Gould, 2002; Wilson, 1997), ecology (Scheiner & Willig, 2008), genetics (e.g. Morgan, 1917), biochemistry (Weber (2002) gives important examples), molecular biology (Culp & Kitcher (1989) give important examples), immunology (e.g. Bretscher & Cohn, 1970; Burnet, 1959; Jerne, 1974), neurology (e.g. Dayan & Abbott, 2001), etc. (Several examples of biological theories are analysed in Krakauer et al., 2011).

It is often claimed that theories are found in evolutionary biology but not in the 'mechanistic' fields of the life sciences, that is, molecular biology, cellular biology, physiology, developmental biology, etc. (together, these domains are often described as 'experimental biology', but this label is problematic, as evolutionary biology can also perfectly be experimental). Table 2.2 shows, for each biological domain, the number and percentage of papers published in the *Proceedings of the National Academy of Sciences of the United States of America* (*PNAS*) that have the word 'theory' in their titles and/or abstracts (*PNAS* was chosen because it is representative of a high-profile scientific journal covering all scientific fields). It appears that evolution and ecology mention theories more often than other domains, but several key areas of mechanistic biology, in particular biophysics, computational biology, systems biology, and neurosciences do mention theories relatively often. It should be emphasized, for the sake of comparison, that physics, probably the most theoretically oriented of all fields in experimental sciences, has 206 occurrences of 'theory' in the titles and abstracts of a total of 1765 papers in the same journal, that is, 11.67%—much less, probably, than many people, and especially philosophers of science, would have expected. In contrast, developmental biology has only five such occurrences, none of which are in the article title or are specific to developmental biology (for instance, it is in fact the theory of evolution that is mentioned).

So, it seems that some biological domains (evolution, ecology, systems biology, etc.) routinely mention theories, while others (immunology, neurosciences, etc.) do not really talk routinely of theories, but have nonetheless well-identified and commonly discussed theories. Developmental biology, in contrast, seems to mention very few theories, and none of them seems properly specific to development. What could explain the fact that developmental biology does not formulate theories? Several criteria could be envisioned, including the level of generality (a theory is often supposed to have an extensive domain, that is, to apply widely across diverse entities), the level of abstraction, the complexity of the subject, the use of models or idealizations, or the role played by mathematics. But none of these criteria resists a close examination. Let us take the example of the clonal selection theory in immunology (Burnet, 1959): it has a limited domain (it applies only to jawed vertebrates), it is not formulated in mathematical terms, it is not highly abstract, it is not characterized by a particular use of models and idealizations, and its subject (the generation of lymphocytes) is certainly complex, but not necessarily more so than the explanation of limb development. And yet it is recognized by virtually everyone in the field as a 'theory'. Many theories in neurosciences, physiology, or biochemistry are in a similar situation. Therefore, it is likely that, if it is true that current developmental biologists fear to speak about theories in their field, this is just a contingent fact, related to the impression they have that a theory is such a demanding notion that their field cannot fit its requirements.

The second reason for putting into question the view that developmental biology could not offer theories is a straightforward historical one: in the past, developmental biology and its 'ancestor', embryology (much can be said about the differences between embryology and developmental biology: see for instance Burian and Thieffry, 2000 and Gilbert, 1998; however, the idea that embryology has been transformed into developmental biology is undisputed) have undoubtedly formulated 'theories'. Without going back to the long controversy between preformation and epigenesis (see e.g. van Robert, 2004; Speybroeck et al., 2002), this is particularly true of 19th century embryology. Some of these embryological theories are comparative, and later on evolutionary, in nature. An important example is offered by Karl von Baer, who formulated the 'germ-layer theory' and a 'law', articulated in four parts, about the development of metazoans (Brauckmann, 2008; on this theory, its influence, and its opposition to the 'recapitulation theory', see Ospovat, 1976). The germ-layer theory states that: (i) during the development of animals, the cells that come from the egg cell constitute embryonic layers (the ectoderm, mesoderm, and endoderm), and (ii) in every animal species, the same organs or structures develop from the same layers (von Baer, 1828). Other embryological theories are related to the idea of 'recapitulation', that is, to the idea that development of advanced species passes through stages represented by adult organisms of more primitive species. This view is often attributed to Haeckel (1867), but was in fact diversely expressed by several authors. This view was later criticized, but it probably played a significant role in the promotion of comparative embryology and in the conviction that it was crucial, in establishing phylogenies, to pay attention to embryonic stages (for a historical analysis, see Hoßfeld & Olsson, 2003; Rinard, 1981). Other theories in 19th century embryology are mechanistic in nature. For the *Entwicklungsmechanik* trend, in particular, the task of embryology was to uncover the laws followed by all ontogenetic processes across species (Roux, 1895). In other words, it sought to discover *general mechanisms* of development, that is, mechanisms that could be observed in all metazoans. Two key general concepts, which have been discussed ever since, are the gradient and the organizer (Gilbert et al., 1996). The highly general mechanistic frameworks suggested by Roux and his followers constitute theories, which must be understood within a context characterized by a particular attention to theories, vividly described by Klaus Sander (Sander, 1991b; for an overall characterization of Roux's programme for a mechanistic developmental biology, see Sander, 1991a). In these different examples, we find two important components of theories that will reappear later in this chapter: evolutionary generalizations and mechanistic generalizations: what is at stake is to make comparisons across species and, if possible, to find results that hold for several species, as well as finding out some fundamental mechanisms of how development occurs.

The building of theories did not stop with the 19th century. During the second half of the 20th century, several influential theories about developmental processes were formulated. Five of them will briefly be mentioned here. The genetic program of development constitutes a first example. Mayr (1969a: 379), in particular, sees the genetic program, based on the Jacob–Monod model of gene regulation, as 'the "general theory of development" which embryologists are looking for so assiduously'. Though the idea of a developmental program has been criticized in different ways, it is still present in several contemporary texts (e.g. Levine & Davidson, 2005; Wolpert, 1991, 1994; for a critique see, in particular, Minelli, 2003; Oyama, 2000; Robert, 2004). A second example is offered by a paper by Wolpert and Lewis (1975), beautifully entitled *Towards a theory of development*. It seemed absolutely clear to these two prominent developmental biologists that it was possible to articulate a general theory of development, based on 'general principles' rooted in positional information and a proper understanding of the genetic program. The third example is the tradition started with Turing's formal model of morphogenesis through reaction–diffusion (Turing, 1952). This tradition had many followers in the 1980s and 1990s. The reaction–diffusion model received recently striking experimental confirmations (Bansagi et al., 2011; Maini et al., 2006; Sick et al., 2006). The 'source-sink model' constitutes a fourth example: formulated by Crick (1970) and partially related to Wolpert's positional information, it has given rise to experimental tests and was recently corroborated by Yu et al. (2009) (see Newman, this volume; Vecchi and Hernandez, this volume). Finally, one last example has been the use of thermodynamics to suggest a 'thermodynamic theory of growth' (Zotin, 1972, following in the footsteps of Prigogine (1967) and Prigogine and Wiame (1946); see also Zotin and Zotin 1997). These five theories were influential when they were suggested; and, as we will see, some of them are still very influential in today's developmental biology.

Again, what has changed in the field of developmental biology and might explain this shift from a 'theory-friendly' field (up to the end of the 1970s, approximately) to one in which it seems inappropriate to evoke theories? One might think that, starting with the 1970s, developmental biology entered a descriptive phase, during which developmental biologists have primarily tried to offer genetic and molecular descriptions of local developmental mechanisms in a given organism, often taken from a limited number of model species (Gilbert, 2009; Gilbert & Tuan, 2001). Many have expressed the necessity to keep in mind the 'big picture' (theories, high-level concepts, comparisons with other biological fields, etc.), but also emphasized that it was difficult to find time to do it. As Bonner (2013: ix) said: 'Many biologists, and I am one of them, live two lives at the same time. In one they work day to day in the laboratory, or in the field. This is what keeps them in touch with their subjects—the real world that they find so fascinating. The other life is a concern for the big picture: how it all fits together' (cited by Callebaut, 2013: 416). My suggestion is that, even though experimental work and the elucidation of molecular details are key to the progress of developmental biology, the worry for the 'big picture' is as important, and as fruitful, as ever (see also Fraser & Harland, 2000; Laubichler & Wagner, 2001).

The third reason for putting into question the view that developmental biology does not offer theories is that, presently, *some* very influential developmental biologists do talk about theories in their research. As a first example, Stuart Newman proposes a physico-genetic theory of morphogenesis, according to which some key aspects of morphogenesis have emerged through 'generic' physical processes, to which subsequent genetic processes were added (Newman, this volume; Newman & Comper, 1990). This framework helps identify developmental mechanisms (both generic and genetic) shared by all animals, or even by all multicellular organisms. It is partly convergent with theories of self-rearrangement in embryogenesis (Brodland, 2002) and self-organization theories (e.g. Kauffman, 1971, 1993). Secondly, since Wolpert (1968, 1969) has suggested the notion of 'positional information', several theories of positional information have been suggested (Jaeger et al., 2008; Jaeger & Martinez-Arias, 2009; see also the mechanistic theory based on dynamical systems theory suggested by Jaeger and Sharpe, this volume). The concept of positional information constitutes the basis for a spatial explanation of embryonic regulation. According to Wolpert's initial views, positional values correspond to a morphogen's gradient that is measurable by responding cells. Wolpert (1968) famously proposed the 'French flag model' to illustrate his concept of positional information. Recently, Jaeger et al. (2008) have offered what they call a 'relativistic' theory of positional information, to which we shall return in 'Towards developmental theories'. Thirdly, Eric Davidson has developed for several decades an explicit theory of the regulation of gene activity in development (Britten & Davidson, 1969; Davidson, 2006; Davidson et al., 2002. See also Garcia-Deister, 2011; Morange, this volume). This theory seeks to identify general molecular mechanisms of gene expression, and it leads directly to comparative approaches (Davidson, 1991). Fourthly, in a decisive step in the elucidation of homeotic genes, Lewis (1978) proposed what he describes as a 'model' of the determination of body segments by the combinatorial action of homeotic genes. Strictly speaking, his suggestion is a set of structured hypotheses, characterized in particular by six 'rules' describing how bithorax complex genes are regulated (Lewis, 1978: 569–570). In this sense, Lewis's proposal fits what will be called a 'theory' in the next section. Finally, the need for a theory of development has also been expressed by Alessandro Minelli: 'What is at stake is the prospect of moving at last toward a *scientific theory of development*' (2003: 2; emphasis in the original). Recently, Minelli (2011b) has suggested a 'null model' for development, according to which developmental processes are deviations from a local self-perpetuation of cell-level dynamics which can be called 'developmental inertia.'

Many other examples could be mentioned: theories of neurulation (Clausi & Brodland, 1993; Gordon, 1985); theories of neurogenesis (Simpson et al., 2009); morphogen gradients theories (Freeman, 2002; Green, 2002; Lander et al., 2002); the chondral modelling theory (which says that joint congruence is maintained in mammalian limbs throughout postnatal ontogeny because cartilage growth in articular regions is regulated in part by mechanical load; Congdon et al., 2012; Frost, 1979; Hamrick, 1999); cell lineage theory (Bjerknes, 1993; Croxdale et al., 1992); cell fate theory (Garcia-Ojalvo & Martinez Arias, 2012); cell migration theories (McLennan et al., 2012); theories of animal muscle development (Fukushige et al., 2006); segmental theories of the formation of the vertebrate head (Kuratani et al., 1999); theories of pigmentation pattern (Kondo & Shirota, 2009); theories of flower development (Alvarez-Buylla et al., 2010); the germ-layer theory and the consequences of the discovery of the neural crest on embryonic theories (Hall, 2008); etc.

To all these examples, one could be tempted to add the 'developmental systems theory' (Oyama, 2000; Oyama et al., 2001). This view has offered a fruitful conceptual and experimental framework with which to conduct research in several fields, from development to evolution and psychological studies. Yet, as acknowledged by its main architects, the developmental systems theory should not actually be understood as a scientific theory, strictly speaking, but rather as a 'perspective', a way of seeing developmental processes (Oyama et al., 2001: 1–2; see also Barberousse et al., 2011).

Even some textbooks occasionally mention developmental theories. For example, Shostak (1991) discusses the germ-layer theory, the gastrea theory, and sees Spemann's proposal on induction and the concept of organizer as a theory (Shostak, 1991: 577 ff). Several textbooks mention 'hypotheses' that are in fact highly complex and well-structured sets of hypotheses, for example, induction and positional information (Wolpert et al., 2011). Finally, even when they adopt quite different perspectives, several textbooks repeatedly use the idea that there are 'principles of development' (e.g. Gilbert, 2000; Wolpert et al., 2011).

Many of the theories mentioned above focus on the most important questions raised by developmental biology (differentiation, morphogenesis, genetic regulation of development, etc.), are testable, and open avenues for original and fruitful predictions. In addition, several of them have been extremely influential in the field. For instance, in a landmark book, Jonathan Slack (1991) devotes a whole chapter, significantly titled 'Theoretical embryology', to dynamical system theories, gradient models, diffusion of morphogen theories, reaction–diffusion models, clock models, etc. Therefore, though it is true that the majority of developmental biologists do not routinely mention theories in their daily research (Love, 2008, this volume), developmental biology does possess several general theories, rooted in broad-ranging concepts (e.g. positional information), which give rise to continuous discussions and experimental assessments by key actors of this domain.

To sum up this section, theories are found not only in biological fields other than developmental biology but also in the past and present of developmental biology. For these reasons, the idea that developmental biology would be an atheoretical domain seems very difficult to defend. Naturally, one possibility here would be to say that these biological theories are not 'true' theories because, for example, they do not fit the syntactic or semantic conceptions of theories, or because they are not law-like theories. But why should scientific theories be asked to conform to the utterly unrealistic demands of these conceptions? I suggest that the most fruitful path is to try to construct a notion of a scientific theory that will both be demanding (not everything in science should count as a theory) and fit the present and past uses of the word 'theory' by biologists.

Characterization of theories in biology

The previous section suggests that biology does offer theories but in a sense quite remote from what mainstream philosophy of science, or more precisely the syntactic and semantic conceptions, have claimed. This should not in fact surprise us, as those conceptions are to be understood as methodological *reconstructions* of scientific theories, and not as descriptions or definitions of what scientific theories are (e.g. Nagel, 1961: 90; Suppes, 1967: 63–64). Our focus here is not the logical reconstruction of theories, but the identification of a meaning of 'theory' that will enable us to state if biological sciences formulate theories. What is needed in our case is a conception of a 'theory' that will both be consistent with the usage of this notion by the scientific community and shed light on the roles that theories play in science.

As a reasonably demanding definition of a scientific theory, it is possible to suggest that a theory is a structured set of testable explanatory and predictive hypotheses (Pradeu, 2009). An isolated hypothesis is one specific statement, not a set of statements; in addition a hypothesis can be weakly corroborated. A theory, in contrast, is a hierarchical organization of several hypotheses (they can be dependent one on the other; one can be the consequence of another; etc.) which are corroborated to a high degree. In other words, a theory is the articulation of corroborated explanatory and predictive statements. Most of the time, a theory contains abstract concepts, related to entities that cannot be directly observed as such in the world (e.g. 'force' in physics, 'fitness' in evolutionary biology, 'positional information' in developmental biology).

I agree with Love (2013, this volume)—and with other people who emphasize the importance of scientific practices, in particular Griesemer (2000, 2013, this volume)—that a crucial question is to determine whether the conceptual tools constructed by philosophy of science are useful or not for the scientific community and correspond to something this community deems to be important. Of course, philosophers of science can develop a reconstruction activity, meaning that they endeavour to use logical instruments to reconstruct scientific theories in order to clarify their structure or perhaps characterize the ideal that every scientific theory should aim at (this is, at least to some extent, what some proponents of the 'syntactic' conception tried to do). But, particularly in the philosophy of biology, a domain that has been from its inception very close to biology itself, it seems extremely important to take into account biologists' discourse and the effect of our conceptual analyses on biologists themselves. But I disagree with Love on the role played by theories in biology, and in developmental biology in particular. In my view, theories are crucial in science, because they do several key things.

Scientific problems need to be well-defined, well-structured, limited in scope, and, once they are organized into problem agendas (Love, 2008, this volume), they constitute a decisive step to do science and to make some progress in a given domain. But, in order to make significant progress, science needs not only problems; it needs *answers* as well. I suggest that what theories do in a given domain is bring answers; by being novel, daring, testable, and hence often wrong, theories give rise to decisive refinements in science.

It is important to realize that a theory is the best open gate to possible worlds. Possibility is key in science. It is certainly important for scientists to describe what they observe and to organize this knowledge into concepts, categories, and explanations. But it is equally important to project themselves into new answers as to how phenomena *could possibly occur*, and correlatively, how their field could *possibly* be deeply reorganized in the near future in the face of newly described phenomena. This kind of projection into possible worlds occurs especially in the case of major breakthroughs in a given scientific domain. Part of what theories do is to open up new possibilities.

So, what exactly are the roles that are played by scientific theories and cannot be played by something else? I would like to emphasize here two critical roles of theories. First, theories make possible *explanations* and *predictions* (e.g. Hempel, 1965, Nagel, 1961). This is in contrast with descriptions, including molecular descriptions, which are often seen by experimental scientists as what their domain should focus on. Nobody doubts that descriptions are indispensable (in particular, nobody doubts that the 'molecularization' of several biological fields, notably developmental biology, has led to major progress). Yet the accumulation of descriptions often leaves us without a clear idea of what exactly is being investigated, what should be found if this or that experiment was done, and how all these isolated, local descriptions might fit together into a unified picture. Faced by the recent accumulation of flows of data, some biologists have insisted that only well-structured and unified theories can impose an order on this disconcerting diversity: 'Biology urgently needs a theoretical basis to unify [what modern technology has to offer us] and it is only theory that will allow us to convert data to knowledge' (Brenner, 2010: 207). Of course, some order can be brought about by problem agendas; but theories, because they take the form of answers to these problems, offer a stricter order. As a daring set of testable explanatory and predictive statements, a theory, if it is to be vindicated, articulates both a possible order and the consequences of this order. In rough terms, a scientific theory says: if I am accurate, then this should happen, while this cannot happen. Naturally, by making such claims, that is, by suggesting such an order, a theory runs the risk of being proved wrong. But this is a perfectly normal and fruitful way for science to change. By making predictions, a theory tells us where to look and what to test (Waddinton, 1968). Therefore, a theory offers explanations and makes predictions in a much more audacious, and hence productive way, than other types of answers. Sometimes, the content of the predictions is unexpected, or even counterintuitive; one of the things theories can do is to compel us to modify our intuitive ontology and/or what we have up until now taken for granted scientifically. As evolutionary biologist George C. Williams said:

> [The] most obviously fruitful role [of theory] is in providing explicit direction for research. From theory we can deduce conclusions not previously reached and that are occasionally counterintuitive. . . . (Williams, 1988: 297)

Interestingly, examples of counterintuitive conclusions coming from theoretical insights can be found in the developmental biology literature (e.g. Freeman, 2002 about gradient diffusion). A related role of theories is that, by offering highly general and daring answers to identified problems, they immediately stimulate *challenges*. As soon as a theory is formulated, scientists will be tempted to assess it, to find exceptions or inadequacies, to ask for more precise definitions of the key terms of the theory, etc. Testability is a prominent feature of science, and theories help favouring testability (to say this is not to deny the difficulties associated with scientific testability; see, e.g. Duhem, 1906; Kuhn, 1970; Quine 1980). That one of the main roles of scientific theories is to formulate specific and testable explanations and predictions is an idea shared by philosophers and historians of science as different as James Griesemer (2013), David Hull (1988: 466), Thomas Kuhn (1970), and Ernest Nagel (1961: 93).

A second role of scientific theories is *unification*. A theory unifies into a simple and coherent picture a diversity of heterogeneous phenomena (Nagel, 1961: 89). By suggesting a common structure and common principles, a theory makes some connections between phenomena that, before the formulation of the theory, had seemed distinct, with no obvious link one to the other. The canonical example of such a theoretical unification is, naturally, Newton's theory of mechanics, which succeeded in accounting for phenomena that had been hitherto considered as very distinct (planetary motion, freely falling bodies, tidal action, etc.). But the same thing occurs in all scientific domains, including developmental biology (as, for example, a theory of gene regulation constructed by Eric Davidson and colleagues that aims at exhibiting the common principles of this regulation across phyla (e.g. Britten & Davidson, 1969; Davidson, 2006)) and other 'mechanistic' domains of the life sciences (for a recent attempt in immunology, see Pradeu et al., 2013). It is important to emphasize the difference between unification and universality. Contrary to what is often believed, a scientific theory is not necessarily universal, in the sense that it would hold in the entire universe without spatiotemporal restrictions. In fact, most scientific theories are *not* universal, and one could even argue that *no* theory is universal. Even highly general theories of physics have exceptions and hold only within certain conditions (this has been one of the main reasons for criticizing the notion of scientific 'laws' understood as universal statements: Cartwright, 1980; Giere, 1999). A theory holds for a given *domain*, and what is important is to delineate clearly this domain (the clonal selection theory in immunology has been shown above to be an example of a scientific theory with a quite narrow domain). Most theories do not seek universality, but rather, they seek unification; that is, they aim at gathering seemingly heterogeneous phenomena under a unique framework of common principles and explanations. A related role of scientific theories is comparison. Because of their condensed formulations and the need to always better define their domain, theories call for comparisons, articulations, and confrontations. These comparisons can occur within a given field (for instance within developmental biology), but also among different biological fields (a key example is, of course, the articulation between evolution and development, or 'evo-devo'), or even among different scientific fields (e.g. Newman, this volume, shows how physics of soft matter can shed light on the past and present of the process of development). These comparisons can be tremendously helpful, since they can suggest new ideas or analogies; but in addition, they often play a key role in the unification process just described, as when, for instance, two competing theories are reconciled under a common framework (for example corpuscular and wave theories of light) or a theory of a given domain inspires and deeply modifies another domain (for example when microbiology resorts to ecological theory to better understand the inner ecological system constituted by commensal gut bacteria: Costello et al., 2012).

These two key roles—the formulation of explanations and predictions, and the unification process—demonstrate the indispensability of theories in science, and in particular in biology, provided that a simple, not excessively demanding, definition of 'theory' is adopted. Thus, the convergence between the perspective defended here and that presented by Love (this volume) is clear: I agree with Love that theories as they have traditionally been defined in philosophy of science are not necessary to produce explanations, predictions, unifications, etc. What I suggest, though, is that Love's view is still too dependent on the traditional definition of a theory and that, if one is ready to adopt a more 'relaxed' conception, then theories are found in many biological sciences.

According to the view defended here, therefore, constructing explanatory, predictive, unifying, and comparative theories should be seen as a legitimate goal in the biological sciences, and one that is in fact much more often realized than sceptics would think. As a recent report of the National Research Council said:

> Theory is already an inextricable thread running throughout the practice of biology, as it is in all science. Biologists choose where to observe, what tool to use, which experiment to do, and how to interpret their results on the basis of a rich theoretical and conceptual framework. Biologists strive to discern patterns, processes, and relationships in order to make sense of the seemingly endless diversity of form and function. Explanatory theories are critical to making sense of what is observed—to order biological phenomena, to explain what is seen and to make predictions, and to guide observation and suggest experimental strategies. (National Research Council, 2008: 13–14)

To sum up, theories are here viewed as structured sets of testable explanatory and predictive hypotheses. This definition is demanding (not just anything can be called a scientific theory), but not too demanding (it does not make the formulation of theories an unreachable horizon in scientific research). In particular, this definition makes possible the identification of key theories in biology, including developmental biology. According to the view defended here, the formulation of theories is immensely useful for the advancement of a given scientific field, as theories are well-articulated answers to specific problems that make possible explanations, (testable) predictions, and unifications to a degree that cannot be reached without theories. The next section examines developmental theories and the way in which different theories may be articulated in order to build a more unified picture of development.

Towards developmental theories

The word 'mystery' is used again and again in the scientific literature to describe development (e.g. Barinaga, 1994; Travis, 2013; Wolpert, 1991). This sense of mystery unfortunately suggests that today's developmental biology would not be very different from embryology as described by Spiegelman in 1958:

> I have found it difficult to avoid the conclusion that many of the investigators concerned with morphogenesis are secretly convinced that the problem is insoluble. I get the feeling that many of the intricate phenomena described are greeted with a sort of glee as if to say, 'My God, this is wonderful, it is so complicated we will never understand it.' (Spiegelman, 1958: 491)

Though the feeling that development is mysterious is understandable, as development has obviously fascinated humans at least since Aristotle, we should resist it. In Spiegelman's footsteps, we could say that 'the phenomena of morphogenesis can hardly be as complicated as implied by the welter of apparently unrelated observations constituting the literature of embryology' (Spiegelman, 1958: 491).

A first way to dissolve the apparent mysteries of development is to transform them into scientific *problems*. But I suggest that the second, equally important, step is to suggest *answers* to these problems. Here is how, in 1998, Scott Gilbert described the embryology of the beginning of the 1980s: 'Fifteen years ago, embryology was what could be characterized as the only field of science that celebrated its questions more than its answers. We had the greatest problems one could imagine . . . But we had very few answers' (Gilbert, 1998: 169). The construction of present-day developmental biology needs to focus on answers, and not on problems or problem agendas only. As suggested above, theories are the best articulated answers to scientific problems, and as such they can greatly improve our current understanding of development. In this section, we examine how theories of development can be constructed, how they can play a decisive role in the field of developmental biology, and how they can be progressively articulated together in order to build a unifying picture of development. Hence, the attitude on which this section rests is in fact quite similar to that of Bonner forty years ago:

> By this rocking back and forth between the reality of experimental facts and the dream world of hypotheses, we can slowly move toward a satisfactory solution of the major problems of developmental biology. So our watchword in this last chapter, which considers the grand themes of development, is never to admit mystery, defeat, or chaotic complexity, but with Calvinistic zeal put such easy, backwards thoughts to one side and bravely make a hypothesis at each breach. (Bonner, 1974: 219).

The previous sections have shown that biological fields in general can, and do, offer theories. They have also emphasized that developmental biology has offered theories in the past, and still offers theories today. We can now show, through the analysis of some examples, that these developmental theories fit the general definition of theories as structured sets of testable explanatory and predictive hypotheses. But a preliminary remark is that we should not be too surprised by the capacity of developmental biology to offer theories. In fact, developmental biology is in a particularly favourable situation to offer theories, for several reasons: it has a long experimental and mechanistic history (in particular *Entwicklungsmechanik*); it has gathered a huge quantity of data, at many different levels (molecules, cells, tissues, organs, as well as the whole organism); and development is characterized by at least some highly general mechanisms (Nüsslein-Volhard, 1994; Wolpert, 1994), sometimes subject to mathematical analysis and modelization.

A remarkable example of a theory in developmental biology has recently been offered by Jaeger et al. (2008). They call their theory the 'relativistic theory of positional information.' As stated explicitly by the authors, this theory must be understood in contrast with the 'classic' theory of positional information that originated in Wolpert's (1968, 1969) conception of positional information. According to the classic theory, the establishment and interpretation of positional values are independent of each other: the cells are 'passive' in that they simply measure the morphogen gradient but do not themselves influence the developmental field. The classic theory, however, fails to account satisfactorily for important phenomena, including size regulation and developmental robustness. Using diverse experimental data, established in particular in *Drosophila* and in neural tube patterning in vertebrates, Jaeger et al. (2008) show that positional specification actually depends on *regulative feedback* from responding cells. Most fundamentally, the relativistic theory of positional information suggests the existence of a dynamic metric that allows cells to measure their relative position within a developing field that itself changes in response to the activity of those cells. The authors then describe in more detail this metric using the tools of dynamical systems theory. The take home message is that it is possible to explain and predict how cells during early development, far from being simple passive 'receivers' of positional information coming from a morphogen gradient, participate actively in the process through feedback mechanisms that affect the developmental field.

What can be deduced from this specific example and other examples of developmental theories analysed in the course of this chapter? It seems now clear that establishing structured sets of testable explanatory and predictive hypotheses is a major goal of today's developmental biology. Indeed, we find in current developmental biology all the characteristic features of scientific theories identified above. First, developmental biologists try to *explain* and *predict*. Explanations usually take the form of the elucidation of *mechanisms*, often of a high level of generality. For instance, Jaeger et al. (2008: 3176) assert that their aim is to find a 'precise, mechanistic understanding of regulative phenomena and developmental robustness' (see also Jaeger and Sharpe, this volume). They also draw consequences and predictions from their theory, in particular about how taking into account feedback mechanisms will impact our understanding of specific developmental processes. More generally, developmental biologists regularly make bold testable predictions (e.g. Bansagi et al., 2011; Maini et al., 2006; Sick et al., 2006). Second, developmental biologists give an important role to *testability*, meaning that they regularly try to test hypotheses and theories (e.g. Kuratani et al., 1999; Jaeger et al., 2004; Yu et al., 2009). In a recent and fascinating work, Collart et al. (2013)

measured the abundance of replication initiation factors in *Xenopus laevis* embryos, with the explicit aim to test a hypothesis put forward in 1982 (Newport and Kirschner 1982). In their original paper, Newport and Kirschner (1982) sought to better understand the precise timing of biochemical events in early morphogenesis. Using *Xenopus* as their system, they showed through precise experimental arguments that the timing of the midblastula transition (MBT) depended not on cell division, on time since fertilization, or on a counting mechanism involving the sequential modification of DNA, but instead, on reaching a critical nuclear to cytoplasmic ratio. They concluded: 'We can speculate that the nucleus may act only passively to titrate some component of the cytoplasm, the removal of which in turn could initiate several cytoplasmic events that collectively make up the MBT' (Newport & Kirschner 1982: 684). Thirty years later, Collart et al. (2013) tried to assess this hypothesis; they showed that four DNA replication factors (Cut5, RecQ4, Treslin, and Drf1) function as key regulators of cell cycle duration during development and thus are key for the MBT and more generally for the normal development of *Xenopus*, confirming the hypothesis of Newport and Kirschner (1982). Third, unification is often an explicit goal of theoretical work in developmental biology. For instance, Wolpert (1994), reflecting on what the future of developmental biology would be, called for integration and for the identification of an 'underlying logic' of developmental processes (see also Jaeger et al. (2008) and Jaeger and Sharpe, this volume):

> Almost certainly there will be new ways of integrating particular aspects of development, and so we will learn, for example, the logic underlying the apparently varied mechanisms for generating periodic structures and the reasons for the variety of mechanisms for setting up the axes in early development. (Wolpert,1994: 572)

Along the same lines, McMahon (1974) emphasizes the importance of expressing many things about various developmental processes using a very limited number of structuring statements, an idea expressed by many others in the field (e.g. Garcia-Ojalvo & Martinez Arias, 2012).

We can conclude that current developmental biology offers theories in the sense suggested in this chapter and that these theories express all the characteristic features of a scientific theory detailed above. But can developmental biology offer a unique, general, theory for development as a whole? I suggest that what is presently at stake in the field of developmental biology is not to find a single general theory of development (interestingly, in this volume, while Bard and Gilbert reject this idea, Jaeger and Sharpe make a stimulating attempt to construct a general theory of development, one rooted in dynamical systems theory). According to the perspective defended here, the aim is, rather, to construct several theories, in the demanding, but not too demanding, sense proposed above: theories should be structured sets of testable explanatory and predictive hypotheses. The identification or construction of several such theories will subsequently be the basis for intertheoretic comparisons. Let's get back to our example of the relativistic theory of position information to better understand how this process works. First of all, this theory is intrinsically in opposition to another theory, namely the 'classic' theory of positional information, thus expressing a kind of theoretical tension, the importance of which is emphasized by Wallace Arthur (this volume). Second, the authors of this theory try to situate it with regard to other theoretical frameworks, including Turing's morphogen gradient theory (Gurdon & Bourillot, 2001; Turing, 1952) (this is 'intrafield' theoretical comparison, that is, comparison of theories within developmental biology). Third, the authors relate their theory with other biological theories, in particular instantiations of developmental systems theory in other biological fields (this is 'intradomain' theoretical comparison, that is, comparison of theories within biology). Fourth, they draw an interesting comparison between the move from classic theory of positional information to relativistic theory of positional information on the one hand, and the move from classical mechanics to Einstein's theory of relativity on the other hand (this is 'interdomain' theoretical comparison, that is, comparison of theories between distinct scientific domains, here biology and physics). Two other important examples of intertheoretic comparisons are 'evo-devo' (a fruitful way to make progress in this domain will be through the confrontation of *specific* developmental theories with *specific* evolutionary theories) and the elucidation of generic morphogenetic mechanisms understood thanks to physical theories (Newman, this volume). Ultimately, it is probably these kinds of articulations and confrontations that will lead to the emergence of general theoretical frameworks in developmental biology and beyond, and to an elucidation of what the evasive notion of 'development' means.

Conclusion

Scientific theories should not be seen as the logical apparatuses depicted by many 20th-century philosophers of science, in particular through the syntactic and semantic conceptions. This chapter has suggested a more modest and practice-oriented conception of scientific theories, defined as structured sets of testable explanatory and predictive hypotheses. With such a definition in mind, it becomes clear that biology does offer theories and can benefit from the important advantages associated with the formulation of theories (in particular, explanations, predictions, testability, and unification).

As a matter of fact, even if current literature rarely mentions theories, developmental biology has always offered, and still offers, theories. In my view, one can say of developmental biology, word for word, what Scheiner and Willig (2008: 21) say about ecology: 'pessimism about the theoretical foundations of ecology is ill founded: ecology has had a robust theoretical framework for many years. We ecologists simply have not recognized that fact, in part because we have misunderstood the nature and form of a comprehensive theory.' The taste for constructing theories in developmental biology should be regenerated, and in turn those theories will undoubtedly regenerate our understanding of development.

Acknowledgments

I would like to thank Alessandro Minelli, who is simply the best co-editor anyone can dream of, as well as all the contributors to this volume. In addition, many thanks to the members of the 'What is development?' research group in Paris at the Institut d'histoire et de philosophie des sciences et des techniques, and especially to Lucie Laplane for her work on how development is understood in the main textbooks and monographs in the field. Special thanks to Alan Love for a dialogue that proved to be very helpful to me. For comments on earlier versions of this chapter, I would like to thank warmly James Griesemer, Alexandre Guay, Lucie Laplane, Alan Love, Alessandro Minelli, Virginie Orgogozo, Karine Prévot, and Marion Vorms.

References

Alvarez-Buylla, E.R., Azpeitia, E., Barrio, R., Benítez, M., and Padilla-Longoria, P. (2010). From ABC genes to regulatory networks, epigenetic landscapes and flower morphogenesis: making biological sense of theoretical approaches. *Seminars in Cell & Developmental Biology*, **21**, 108–17.

Bánsági, T., Vanag, V.K., and Epstein, I.R. (2011). Tomography of reaction-diffusion microemulsions reveals three-dimensional Turing patterns. *Science*, **331**, 1309–12.

Barberousse, A., Merlin, F., and Pradeu, T. (2010). Introduction: reassessing developmental systems theory. *Biological Theory*, **5**, 199–201.

Barinaga, M. (1994). Looking to development's future. *Science*, **266**, 561–4.

Beatty, J. (1995). The evolutionary contingency thesis. In G. Wolters and J.G. Lennox, eds., *Concepts, Theories, and Rationality in the Biological Sciences*, Universitätsverlag Konstanz, Konstanz, University of Pittsburgh Press, Pittsburgh, pp. 45–81.

Bjerknes, M. (1993). Theory of cell lineage graphs and their application to the intestinal epithelium. *Seminars in Developmental Biology*, **4**, 263–73.

Bonner, J.T. (1965). *Size and Cycle: An Essay on the Structure of Biology*. Princeton University Press, Princeton.

Bonner, J.T. (1974). *On Development: The Biology of Form*. Harvard University Press, Cambridge, MA.

Bonner, J.T. (2013). *Randomness in Evolution*. Princeton University Press, Princeton.

Brauckmann, S. (2008). The many spaces of Karl Ernst von Baer. *Biological Theory*, **3**, 85–9.

Brenner, S. (2010). Sequences and consequences. *Philosophical Transactions of the Royal Society of London B*, **365**, 207–12.

Bretscher, P., and Cohn, M. (1970). A theory of self-nonself discrimination. *Science*, **169**, 1042–1049.

Britten, R.J., and Davidson, E.H. (1969). Gene regulation for higher cells: a theory. *Science*, **165**, 349–57.

Brodland, G.W. (2002). The differential interfacial tension hypothesis (DITH): a comprehensive theory for the self-rearrangement of embryonic cells and tissues. *Journal of Biomechanical Engineering*, **124**, 188–97.

Burian, R.M., and Thieffry, D. (2000). Introduction to the special issue 'From Embryology to Developmental Biology'. *History and Philosophy of the Life Sciences*, **22**, 313–23.

Burnet, F.M. (1959). *The Clonal Selection Theory of Acquired Immunity*. Cambridge University Press, Cambridge.

Callebaut, W. (2013). Naturalizing theorizing: beyond a theory of biological theories. *Biological Theory*, **7**, 413–29.

Cartwright, N. (1980). Do the laws of physics state the facts? *Pacific Philosophical Quarterly*, **61**, 75–84.

Clausi, D.A., and Brodland, G.W. (1993). Mechanical evaluation of theories of neurulation using computer simulations. *Development*, **118**, 1013–23.

Collart, C., Allen, G.E., Bradshaw, C.R., Smith, J.C., and Zegerman, P. (2013). Titration of four replication factors is essential for the *Xenopus laevis* midblastula transition. *Science*, **341**, 893–6.

Congdon, K.A., Hammond, A.S., and Ravosa, M.J. (2012). Differential limb loading in miniature pigs (*Sus scrofa domesticus*): a test of chondral modeling theory. *Journal of Experimental Biology*, **215**, 1472–83.

Costello, E.K., Stagaman, K., Dethlefsen, L., et al. (2012). The application of ecological theory toward an understanding of the human microbiome. *Science*, **336**, 1255–62.

Craver, C. (2002). Structures of scientific theories. In P. Machamer and M. Silberstein, eds., *The Blackwell Guide to the Philosophy of Science*, Blackwell, Malden, pp. 55–79.

Crick, F. (1970). Diffusion in embryogenesis. *Nature*, **225**, 420–22.

Croxdale, J., Smith, J., Yandell, B., et al. (1992). Stomatal patterning in *Tradescantia*: an evaluation of the cell lineage theory. *Developmental Biology*, **149**, 158–67.

Culp, S., and Kitcher, P. (1989). Theory structure and theory change in contemporary molecular biology. *British Journal for the Philosophy of Science*, **40**, 459–83.

Darden, L., and Tabery, J. (2009). Molecular biology. In E.N. Zalta, ed., *The Stanford Encyclopedia of Philosophy*, http://plato.stanford.edu/entries/molecular-biology/. Accessed 30 August 2013.

Davidson, E.H. (1991). Spatial mechanisms of gene regulation in metazoan embryos. *Development*, **113**, 1–26.

Davidson, E.H. (2006). *The Regulatory Genome: Gene Regulatory Networks in Development and Evolution*. Academic Press, San Diego.

Davidson, E.H., Rast, J.P., Oliveri, P., et al. (2002). A genomic regulatory network for development. *Science*, **295**, 1669–78.

Dayan, P., and Abbott, L.F. (2001). *Theoretical Neuroscience: Computational and Mathematical Modeling of Neural Systems*. MIT Press, Cambridge, MA.

Downes, S.M. (1992). The importance of models in theorizing: a deflationary semantic view. In D. Hull, M. Forbes and K. Okruhlik, eds., *PSA: Proceedings of the Biennial Meeting of the Philosophy of Science Association*, Vol. 1. Philosophy of Science Association, East Lansing, pp. 142–53.

Duhem, P. (1906). *La Théorie Physique: Son Objet, Sa Structure*. Paris: Marcel Rivière; tr. P.P. Wiener as *The Aim and Structure of Physical Theory*. Princeton University Press, Princeton.

Eberl, G. (2005). Inducible lymphoid tissues in the adult gut: recapitulation of a fetal developmental pathway? *Nature Reviews Immunology*, **5**, 413–20.

Fraser, S.E., and Harland, R.M. (2000). The molecular metamorphosis of experimental embryology. *Cell*, **100**, 41–55.

Freeman, M. (2002). Morphogen gradients, in theory. *Developmental Cell*, **2**, 689–90.

Frost, H.M. (1979). A chondral modeling theory. *Calcified Tissue International*, **28**, 181–200.

Fukushige, T., Brodigan, T.M., Schriefer, L.A., et al. (2006). Defining the transcriptional redundancy of early bodywall muscle development in *C. elegans*: evidence for a unified theory of animal muscle development. *Genes & Development*, **20**, 3395–406.

García-Deister, V. (2011). The old man and the sea urchin genome: theory and data in the work of Eric Davidson, 1969–2006. *History and Philosophy of the Life Sciences*, **33**, 147–63.

Garcia-Ojalvo, J., and Martinez Arias, A. (2012). Towards a statistical mechanics of cell fate decisions. *Current Opinion in Genetics & Development*, **22**, 619–26.

Gayon, J. (1998). *Darwinism's Struggle for Survival: Heredity and the Hypothesis of Natural Selection*. Cambridge University Press, Cambridge.

Giere, R.N. (1999). *Science Without Laws*. University of Chicago Press, Chicago.

Gilbert, S.F. (1994). *Developmental Biology*, 4th ed. Sinauer Associates, Sunderland.

Gilbert, S.F. (1998). Bearing crosses: a historiography of genetics and embryology. *American Journal of Medical Genetics*, **76**, 168–82.

Gilbert, S.F. (2000). *Developmental Biology*, 6th ed. Sinauer Associates, Sunderland.

Gilbert, S.F. (2003). *Developmental Biology*, 7th ed. Sinauer Associates, Sunderland.

Gilbert, S.F. (2009). The adequacy of model systems for evo-devo: modeling the formation of organisms/modeling the formation of society. In A. Barberousse, M. Morange, and T. Pradeu, eds., *Mapping the Future of Biology: Evolving Concepts and Theories*. Springer, Dordrecht, pp. 57–68.

Gilbert, S.F. (2013). *Developmental Biology*, 10th ed. Sinauer Associates, Sunderland.

Gilbert, S.F., Opitz, J.M., and Raff, R.A. (1996). Resynthesizing evolutionary and developmental biology. *Developmental Biology*, **173**, 357–72.

Gilbert, S.F., and Raunio, A.M., eds. (1997). *Embryology: Constructing the Organism*. Sinauer Associates, Sunderland.

Gilbert, S.F., and Tuan, R.S. (2001). New vistas for developmental biology. *Journal of Biosciences*, **26**, 293–98.

Godfrey-Smith, P. (2006). The strategy of model-based science. *Biology and Philosophy*, **21**, 725–40.

Goodfield, J. (1969). Theories and hypotheses in biology. In R.S. Cohen and M.W. Wartofsky, eds., *Boston Studies in the Philosophy of Science*, Vol. 5. Springer, Dordrecht, pp. 421–49.

Gordon, R. (1985). A review of the theories of vertebrate neurulation and their relationship to the mechanics of neural tube birth defects. *Journal of Embryology and Experimental Morphology*, **89 (Supplement)**, 229–55.

Gould, S.J. (2002). *The Structure of Evolutionary Theory*. Harvard University Press, Cambridge, MA.

Green, J. (2002). Morphogen gradients, positional information, and *Xenopus*: interplay of theory and experiment. *Developmental Dynamics*, **225**, 392–408.

Griesemer, J. (2000). Development, culture and the units of inheritance. *Philosophy of Science*, **67 (Proceedings)**, S348–S368.

Griesemer, J. (2011a). Heuristic reductionism and the relative significance of epigenetic inheritance in evolution. In B. Hallgrímsson and B.K. Hall, eds., *Epigenetics: Linking Genotype and Phenotype in Development and Evolution*. University of California Press, Berkeley, pp. 14–40.

Griesemer, J. (2011b). The relative significance of epigenetic inheritance in evolution: Some philosophical considerations. In S. Gissis and E. Jablonka, eds., *Transformations of Lamarckism: From Subtle fluids to Molecular Biology*. MIT Press, Cambridge, MA., pp. 331–44.

Griesemer, J. (2013). Formalization and the meaning of 'theory' in the inexact biological sciences. *Biological Theory*, **7**, 298–310.

Gurdon, J.B., and Bourillot, P.Y. (2001). Morphogen gradient interpretation. *Nature*, **413**, 797–803.

Haeckel, E. (1867). *Generelle Morphologie der Organismen*. Georg Reimer, Berlin.

Hall, B.K. (2008). The neural crest and neural crest cells: discovery and significance for theories of embryonic organization. *Journal of Biosciences*, **33**, 781–93.

Hamrick, M.W. (1999). A chondral modeling theory revisited. *Journal of Theoretical Biology*, **201**, 201–8.

Hempel, C.G. (1965). *Aspects of Scientific Explanation*. Free Press, New York.

Hoßfeld, U., and Olsson, L. (2003). The road from Haeckel: the Jena tradition in evolutionary morphology and the origins of 'Evo-Devo'. *Biology and Philosophy*, **18**, 285–307.

Hull, D. (1969). What philosophy of biology is not. *Journal of the History of Biology*, **2**, 241–68.

Hull, D. (1974). *Philosophy of Biological Science*. Prentice-Hall, Englewood Cliffs.

Hull, D. (1988). *Science as a Process: An Evolutionary Account of the Social and Conceptual Development of Science*. University of Chicago Press, Chicago.

Hull, D. (1992). Individual. In E.F. Keller and E. Lloyd, eds., *Keywords in Evolutionary Biology*. Harvard University Press, Cambridge, pp. 180–7.

Jaeger, J., Irons, D., and Monk, N. (2008). Regulative feedback in pattern formation: towards a general relativistic theory of positional information. *Development*, **135**, 3175–83.

Jaeger, J., and Martinez-Arias, A. (2009). Getting the measure of positional information. *PLoS Biology*, **7**, e1000081.

Jaeger, J., Surkova, S., Blagov, M., et al. (2004). Dynamic control of positional information in the early *Drosophila* embryo. *Nature*, **430**, 368–71.

Jerne, N.K. (1974). Towards the network theory of the immune system. *Annales d'Immunologie (Inst. Pasteur)*, **125**, 373–89.

Kauffman, S. (1971). Gene regulation networks: a theory for their global structure and behaviors. *Current Topics in Developmental Biology*, **6**, 145–82.

Kauffman, S. (1993). *The Origins of Order: Self-Organization and Selection in Evolution*. Oxford University Press, Oxford.

Kondo, S., and Shirota, H. (2009). Theoretical analysis of mechanisms that generate the pigmentation pattern of animals. *Seminars in Cell & Developmental Biology*, **20**, 82–9.

Krakauer, D.C., Collins, J.P., Erwin, D., et al. (2011). The challenges and scope of theoretical biology. *Journal of Theoretical Biology*, **276**, 269–76.

Kuhn, T. (1970). *The Structure of Scientific Revolutions*, 2nd ed. with postscript. University of Chicago Press, Chicago.

Kuratani, S., Horigome, N., and Hirano, S. (1999). Developmental morphology of the head mesoderm and reevaluation of segmental theories of the vertebrate head: evidence from embryos of an agnathan vertebrate, *Lampetra japonica*. *Developmental Biology*, **210**, 381–400.

Lander, A.D., Nie, Q., and Wan, F.Y. (2002). Do morphogen gradients arise by diffusion? *Developmental Cell*, **2**, 785–96.

Laubichler, M.D., and Wagner, G.P. (2001). How molecular is molecular developmental biology? A reply to Alex Rosenberg's reductionism redux: computing the embryo. *Biology and Philosophy*, **16**, 53–68.

Levine, M., and Davidson, E.H. (2005). Gene regulatory networks for development. *Proceedings of the National Academy of Sciences USA*, **102**, 4936–42.

Levins, R. (1966). The strategy of model-building in population biology. *American Scientist*, **54**, 421–31.

Lewis, E.B. (1978). A gene complex controlling segmentation in *Drosophila*. *Nature*, **276**, 565–70.

Lloyd, E. (1988). *The Structure and Confirmation of Evolutionary Theory*. Greenwood, Westport. Reprinted (1994). Princeton University Press, Princeton.

Love, A. (2008). Explaining the ontogeny of form: philosophical issues. In S. Sarkar and A. Plutynski, eds., *A Companion to the Philosophy of Biology*. Blackwell, Malden, pp. 223–47.

Love, A. (2013). Theory is as theory does: scientific practice and theory structure in biology. *Biological Theory*, **7**, 325–37.

Machamer, P., Darden, L., and Craver, C.F. (2000). Thinking about mechanisms. *Philosophy of Science*, **67**, 1–25.

Maini, P.K., Baker, R.E., and Chuong, C. (2006). The Turing model comes of molecular age. *Science*, **314**, 1397–98.

Mayr, E (1969b). Footnotes on the philosophy of biology. *Philosophy of Science*, **36**, 197–202.

Mayr, E. (1969a). Comments on 'Theories and hypotheses in biology'. In R.S. Cohen and M.W. Wartofsky, eds., *Boston Studies in the Philosophy of Science*, Vol. 5. Springer, Dordrecht, pp. 452–6. Reprinted as 'Theory formation in developmental biology' in E. Mayr (1976) *Evolution and the Diversity of Life. Selected Essays*. Harvard University Press, Cambridge, pp. 377–82.

McLennan, R., Dyson, L., Prather, K.W., et al. (2012). Multiscale mechanisms of cell migration during development: theory and experiment. *Development*, **139**, 2935–44.

McMahon, D. (1974). Chemical messengers in development: a hypothesis. *Science*, **185**, 1012–21.

Minelli, A. (2003). *The Development of Animal Form*. Cambridge University Press, Cambridge.

Minelli, A. (2009). *Forms of Becoming: The Evolutionary Biology of Development*. Princeton University Press, Princeton.

Minelli, A. (2011a). Development, an open-ended segment of life. *Biological Theory*, **6**, 4–15.

Minelli, A. (2011b). A principle of developmental inertia. In B. Hallgrímsson and B.K. Hall, eds., *Epigenetics: Linking Genotype and Phenotype in Development and Evolution*. University of California Press, San Francisco, pp. 116–33.

Morgan, T.H. (1917). The theory of the gene. *The American Naturalist*, **51**, 513–44.

Morrison, M. (2007). Where have all the theories gone? *Philosophy of Science*, **74**, 195–28.

Nagel, E. (1961). *The Structure of Science: Problems in the Logic of Scientific Explanation*. Routledge and Kegan Paul, London.

National Research Council (2008). The role of theory in advancing 21st-century biology: catalyzing transformative research, http://www.nap.edu/openbook.php?record_id=12026, The National Academies.

Newman, S.A. and Comper, W.D. (1990). 'Generic' physical mechanisms of morphogenesis and pattern formation. *Development*, **110**, 1–18.

Newport, J. and, Krischner, M. (1982). A major developmental transition in early *Xenopus* embryos: I. Characterization and timing of cellular changes at the midblastula stage. *Cell*, **30**, 675–86.

Nüsslein-Volhard, C. (1994). Of flies and fishes. *Science*, **266**, 572–4.

Ospovat, D. (1976). The influence of Karl Ernst von Baer's embryology, 1828–1859: a reappraisal in light of Richard Owen's and William B. Carpenter's 'Palaeontological application of "Von Baer's law"'. *Journal of the History of Biology*, **9**, 1–28.

Oyama, S. (2000). *The Ontogeny of Information: Developmental Systems and Evolution*, 2nd ed. Duke University Press, Durham.

Oyama, S., Griffiths, P. and Gray, R., eds., (2001). *Cycles of Contingency: Developmental Systems and Evolution*. MIT Press, Cambridge.

Pradeu, T. (2009). *Les limites du soi: Immunologie et identité biologique*. Presses Universitaires de Montréal, Montréal.

Pradeu, T. (2011). A mixed self: the role of symbiosis in development. *Biological Theory*, **6**, 80–8.

Pradeu, T., Jaeger, S., and Vivier, E. (2013). The speed of change: towards a discontinuity theory of immunity? *Nature Reviews Immunology*, **13**, 764–9.

Pradeu, T., Laplane, L., Morange, M., et al. (2011). The boundaries of development. *Biological Theory*, **6**, 1–3.

Prigogine, I. (1967). *Introduction to Thermodynamics of Irreversible Processes*. Wiley, New York.

Prigogine, I., and Wiame, J.M. (1946). Biologie et thermodynamique des phénomènes irréversibles. *Experientia*, **2**, 451–3.

Quine, W.V. (1980). Two dogmas of empiricism. In W. V. Quine, *From a Logical Point of View: Nine Logico-philosophical essays*, 2nd ed. Harvard University Press, Cambridge, pp. 20–46.

Rinard, R.G. (1981). The problem of the organic individual: Ernst Haeckel and the development of the biogenetic law. *Journal of the History of Biology*, **14**, 249–75.

Robert, J.S. (2004). *Embryology, Epigenesis and Evolution: Taking Development Seriously*. Cambridge University Press, Cambridge.

Roux, W. (1895). Einleitung. *Archiv für Entwicklungsmechanik der Organismen* **1**, 1–42.

Ruse, M. (1973). *The Philosophy of Biology*. Hutchinson University Press, London.

Sander, K. (1991a). Wilhelm Roux and his programme for developmental biology. *Roux's Archives of Developmental Biology*, **200**, 1–3.

Sander, K. (1991b). Wilhelm Roux and the rest: developmental theories 1885–1895. *Roux's Archives of Developmental Biology*, **200**, 297–9.

Schaffner, K.F. (1968). Theories and explanations in biology. *Journal of the History of Biology*, **2**, 19–33.

Schaffner, K.F. (1993). Theory structure, reduction, and disciplinary integration in biology. *Biology and Philosophy*, **8**, 319–47.

Scheiner, S.M., and Willig, M.R. (2008). A general theory of ecology. *Theoretical Ecology*, **1**, 21–8.

Shostak, S. (1991). *Embryology: An Introduction to Developmental Biology*. Harper Collins Publishers, New York.

Sick, S., Reinker, S., Timmer, J., et al. (2006). WNT and DKK determine hair follicle spacing through a reaction-diffusion mechanism. *Science*, **314**, 1447–50.

Simpson, H.D., Mortimer, D., and Goodhill, G.J. (2009). Theoretical models of neural circuit development. *Current Topics in Developmental Biology*, **87**, 1–51.

Slack, J.M.W. (1991). *From Egg to Embryo: Regional Specification in Early Development*, 2nd ed. Cambridge University Press, Cambridge.

Slack, J.M.W. (2013). *Essential Developmental Biology*, 3rd ed. Wiley-Blackwell, Sussex.

Smart, J.J.C. (1963). *Philosophy and Scientific Realism*. Routledge & Kegan Paul, New York.

Speybroeck, L., de Waele, D., and van de Vijver, G. (2002). Theories in early embryology. *Annals of the New York Academy of Sciences*, **981**, 7–49.

Spiegelman, S. (1958). Discussion. In W.D. McElroy and B. Glass, eds., *A Symposium on the Chemical Basis of Development*. Johns Hopkins Press, Baltimore, p. 491.

Sterelny, K., and Griffiths, P.E. (1999). *Sex and Death: An Introduction to the Philosophy of Biology*. University of Chicago Press, Chicago.

Suppe, F., ed. (1977). *The Structure of Scientific Theories*, 2nd ed., University of Illinois Press, Urbana.

Suppes, P. (1967). What is a scientific theory? In S. Morgenbesser, ed., *Philosophy of Science Today*. Basic Books, New York, pp. 55–67.

Thompson, P. (2007). Formalizations of evolutionary biology. In M. Matthens and C. Stephens, eds., *Handbook of the Philosophy of Science: Philosophy of Biology*. Elsevier, Amsterdam, pp. 485–523.

Travis, J. (2013). Mysteries of development. *Science*, **340**, 1156.

Turing, A.M. (1952). The chemical basis of morphogenesis. *Philosophical Transactions of the Royal Society of London B*, **237**, 37–72.

van Fraassen, B.C. (1972). A formal approach to the philosophy of science. In R. Colodny, ed., *Paradigms and Paradoxes: The Challenge of the Quantum Domain*. University of Pittsburgh Press, Pittsburgh, pp. 303–66.

von Baer K.E. (1828). *Ueber Entwickelungsgeschichte der Thiere, Beobachtung and Reflexion*. Bornträger, Königsberg.

Waddinton, C.H. (1968). Theoretical biology and molecular biology. In C.H. Waddington, ed., *Towards a Theoretical Biology*, Vol. 1. Edinburgh University Press, Edinburgh, pp. 103–8.

Weber, M. (2002). Incommensurability and theory comparison in experimental biology. *Biology and Philosophy*, **17**, 155–69.

Weisberg, M. (2006). Forty years of 'The strategy': Levins on model building and idealization. *Biology & Philosophy*, **21**, 623–45.

Williams, G.C. (1988). Retrospect on sex and kindred topics. In R. Michod. and B.R. Levin, eds., *The Evolution of Sex: An Examination of Current Ideas*. Sinauer Associates, Sunderland, pp. 287–98.

Williams, M.B. (1970). Deducing the consequences of evolution. *Journal of Theoretical Biology*, **29**, 343–55.

Wilson, D.S. (1997). Altruism and organism: disentangling the themes of multilevel selection theory. *The American Naturalist*, 150, S122–34.

Wimsatt, W. (2007). *Re-Engineering Philosophy for Limited Beings: Piecewise Approximations to Reality*. Harvard University Press, Cambridge.

Wolpert, L. (1968). The French flag problem: a contribution to the discussion on pattern development and regulation. In C.H. Waddington, ed., *Towards a Theoretical Biology*, Vol. 1. Edinburgh University Press, Edinburgh, pp. 125–33.

Wolpert, L. (1969). Positional information and the spatial pattern of cellular differentiation. *Journal of Theoretical Biology*, **25**, 1–47.

Wolpert, L. (1994). Do we understand development? *Science*, **266**, 571–2.

Wolpert, L. (1991). *The Triumph of the Embryo*. Oxford University Press, Oxford.

Wolpert, L. (2011). *Developmental Biology: A Very Short Introduction*. Oxford University Press, Oxford.

Wolpert, L., and Lewis, L.J. (1975). Towards a theory of development. *Federation Proceedings*, **34**, 14–20.

Wolpert, L., Tickle, C., Jessell, T., *et al.* (2011). *Principles of Development*, 4th ed. Oxford University Press, Oxford.

Yu, S.R., Burkhardt, M., Nowak, M., et al. (2009). Fgf8 morphogen gradient forms by a source-sink mechanism with freely diffusing molecules. *Nature*, **461**, 533–6.

Zotin, A. I. (1972). *Thermodynamic Aspects of Developmental Biology*. Karger, Basel.

Zotin, A.A., and Zotin, A.I. (1997). Phenomenological theory of ontogenesis. *International Journal of Developmental Biology*, **41**, 917–21.

CHAPTER 3

The erotetic organization of developmental biology

Alan C. Love

No theory of development?

Developmental biology is the science of explaining how a variety of interacting processes generate the heterogeneous shapes, size, and structural features of an organism as it develops from embryo to adult, or more generally throughout its life cycle (Love, 2008b; Minelli, 2011a). Although it is commonplace in philosophy to associate sciences with theories such that the individuation of a science is dependent on a constitutive theory or group of models, it is uncommon to find presentations of developmental biology making reference to a theory or theories of development. For example, in the third edition of *Essential Developmental Biology* (Slack, 2013), three families of approaches are described (developmental genetics, experimental embryology, and molecular and cell biology), and the appendix contains a catalogue of 'key molecular components' (genes, transcription factor families, inducing factor families, cytoskeleton, cell adhesion molecules, and extracellular matrix components); however, no standard theory or group of models provides a theoretical scaffolding to the book nor is any mentioned.

The absence of any reference to a theory of development or some set of core explanatory models is prima facie puzzling. Why is it so difficult to identify or isolate a constitutive set of models or overarching theory for the science of developmental biology? Three interpretations of this situation are possible. The first is that the lack of reference to theories does not indicate the absence of a theory; one can reconstruct a theory (or theories) of developmental biology out of the relevant discourse (e.g. from allied molecular models). The second interpretation is that the lack of reference to theories indicates an immaturity in developmental biology. Since mature sciences always have systematic theories, we should seek to build one for the domain of ontogeny. The third and perhaps least expected interpretation is that the lack of reference to theories should be taken at face value. This is what I will argue for in this chapter.

My argument has two dimensions, one positive and one negative. The negative dimension consists in showing why the first two options are not preferable in this context. On the one hand, claiming that developmental biology is immature is not a viable interpretation. Developmental biology is not a nascent science, groping about for some way to explain its phenomena: 'some of the basic processes and mechanisms of embryonic development are now quite well understood' (Slack, 2013: 7). The philosophical impetus to offer this kind of interpretation arises out of commitments to a conception of science that presumes theories are abstract systems that revolve around a small set of laws or core principles. On the other hand, holding that developmental biology already has a theory or theories costumed in different guise—not referred to as such by developmental biologists—is a *possible* interpretation. It arises out of a cognate conception of science (sciences must have theories) that has been expanded to allow for different understandings of theory structure, such as constellations of models without laws, even though the expectation is that

Towards a Theory of Development. Edited by Alessandro Minelli and Thomas Pradeu

these different conceptions of theory play similar organizing roles in terms of guiding research. This expansion of how to understand the nature of scientific theories is welcome (Griesemer, 2013; Love, 2013), but the assumption of what theories are doing on this interpretation should be challenged and rejected on methodological grounds in the case of developmental biology. An analysis of the reasoning in a science should exhibit epistemic transparency and not postulate 'hidden' reasoning structure (Love, 2012). A philosophical account should make it clear how the scientists themselves access the relevant features of the reasoning in order to engage in scientific practices, such as evaluating the cogency of explanations or inductive inferences.

The positive dimension of my argument consists in showing why the third option is preferable, and it has three facets. First, I offer an overview of textbook presentations of developmental biology to show that the representations of empirical content in the science, as well as its organization, do not rely on a theory or central set of models. Instead, developmental biology is organized primarily by stable, broad domains of problems (*problem agendas*) that correspond to abstract representations of ontogenetic phenomena (e.g. differentiation, growth, and morphogenesis), which operate as multilayered *explanandum* types (Love, 2008a, b). This is consistent with what might appear to be a key objection: how does one interpret the theoretical aspects of developmental biology (e.g. positional information models of pattern formation) and its utilization of theories from other domains (e.g. biochemistry)? Following earlier work (Waters, 2007), I make a distinction between theory-informed science (using theoretical knowledge) and theory-directed science (having a theory that directs inquiry and organizes knowledge), and argue that developmental biology is theory-informed but not theory-directed. This means theories need not be wholly absent from developmental biology but—when present—they play roles very different from standard philosophical expectations. Second, I offer an account of how problems play the organizing and guiding role typically presumed to come from theories, which involves explicating an ontology of the erotetic ('pertaining to questioning') units relevant to interpret developmental biology. Finally, I explore a recent research paper on mammary gland morphogenesis to specifically illustrate my positive argument (Goel et al., 2011). I conclude with a summary of the advantages gained by relinquishing the assumption that sciences must have theories that guide or organize inquiry and embracing the erotetic organization of developmental biology.

Against maturity and for sciences without theories

Against maturity

In his infamous book *Against Method* (Feyerabend, 1975), Paul Feyerabend argued against the presumption that there are universal methodological rules that govern scientific inquiry. In parallel, I argue against 'maturity' in the sense of a single conception of what it means for a science to be mature that is applicable across all areas of scientific inquiry. The association of maturity with a science *having* a central, organizing theory is a longstanding idea. William Whewell expressed the sentiment in the 19th century when he said: 'It is necessary to begin in every science with the Laws of Phenomena; but it is impossible that we should be satisfied to stop short of a Theory of Causes' (Whewell, 1989: 177). Darwin captured this more colloquially when he recalled how reading Malthus in 1838 helped coalesce his ideas on species origins ('Here, then, I had at last got a theory by which to work'; Darwin, 1959: 120) or advanced his ideas on inheritance ('In scientific investigations, it is permitted to invent any hypothesis, and if it explains various large and independent classes of facts it rises to the rank of a well-grounded theory'; Darwin, 1988: 9). In the 20th century, Oppenheim and Putnam (1958) articulated a picture of science with various levels constituted by theories about different compositional domains (elementary particles, atoms, molecules, cells, organisms, and societies), all to be related by reduction. Throughout, maturity and theory are presumed to be co-instantiated: 'mature sciences . . . [are] those sciences in which theoretical considerations contribute significantly' (Boyd, 1983: 221).

The best-known discussion of maturity is found in the work of Thomas Kuhn (1996). Mature science is equated with normal science that operates within the bounds of a particular paradigm. Although 'paradigm' is a slippery notion, Kuhn's 'disciplinary matrix' sense of paradigm is in view when thinking of maturity: 'some accepted examples of actual scientific practice—examples which include law, theory, application, and instrumentation together—provide models from which spring particular coherent traditions of scientific research' (Kuhn, 1996: 10). The difficulty is that sciences we have good reason to consider mature, such as contemporary molecular biology, exhibit features at odds with normal science. Most famously, much of biology doesn't have anything like laws (Beatty, 1995) and yet demonstrates coherent traditions of research. Additionally, Kuhn held that normal science does not bring forth or even notice new phenomena (except sometimes as anomalies): 'No part of the aim of normal science is to call forth new sorts of phenomena; indeed those that will not fit the box are often not seen at all' (Kuhn, 1996: 24). However, molecular biology actively explores new phenomena (e.g. intrinsically disordered proteins or prions).

If we set aside these inconsistencies and focus only on theories, Kuhn clearly assumes a single theoretical framework for each disciplinary matrix because normal science presumes one: 'scientists [doing normal science do not] normally aim to invent new theories, and they are often intolerant of those invented by others' (Kuhn, 1996: 24). In this respect he adheres to the assumption that each science must have an associated theory for the guidance and organization of inquiry. The only difference is whether the relevant unit is a paradigm, which includes the theory *and other items*, or just the theory understood as something like an axiom system with a (small) set of universal laws as core principles: 'A theory is a relatively small body of general laws that work together to explain a large number of empirical generalizations, often by describing an underlying mechanism common to them all' (Rosenberg, 1985: 121). Assumptions like these have been challenged in recent philosophical analysis. It was once thought that each science must have laws in order to offer genuine explanations, but now this is seen as an unnecessary stricture (Giere, 1999; Woodward, 2003). Is the expectation that a science have a theory or group of theories that accomplishes the task of organizing and guiding inquiry of similar vintage?

Notice that relaxing the conception of theories does not obviate the question because the issue is whether a theory or group of models must provide the organizational scaffolding for a science. On alternative conceptions of theory structure (e.g. theories as structured sets of explanatory and predictive hypotheses; see Pradeu, this volume), the nature of theories is much more flexible, but the burden of guiding inquiry remains. From where does this expectation that a science have a systematic theory, paradigm, or structured set of hypotheses that governs inquiry receive its justification? Some of it derives from an intuitive expectation of what counts as a mature science in the first place. Kuhn acknowledged a wide swathe of successful inquiry, including Aristotelian dynamics and Ptolemaic astronomy; positivist conceptions, derived from a tradition begun in the 19th century (e.g. Whewell, 1989), often appealed to mathematized versions of Newtonian mechanics or allied physical sciences as the pinnacle of maturity. The circle of reasoning in this tradition is relatively tight: mature sciences exhibit systematic theoretical frameworks because systematic theoretical frameworks indicate maturity. This outlook has been directed at the history of embryology specifically (Rosenberg, 2006), where broad theoretical frameworks like preformation and epigenesis were present (Maienschein, 2005). (Contemporary developmental biology has rejected this theoretical dialectic or seriously redefined it.) If we find empirically successful and coherent traditions of research without a theory or systematic framework providing this guidance, then the science cannot be mature.

One might shrug off these positivist appeals to maturity by averring to more flexible conceptions of theory and theory structure. But why retain the expectation that these different perspectives on theories should accomplish the same epistemic tasks? Arguably, it is a philosophical prejudice about the structure of knowledge that is not plausible in light of the diversity of contemporary research practices across the sciences. By 'prejudice' I mean a 'preconceived opinion not based on reason or actual experience' or 'unreasoning preference or objection' (OED). The absence of a general theory or some set of theories providing guidance to ongoing inquiry in developmental biology may continue to seem philosophically perplexing, and therefore my strategy is therapeutic—defuse the worry by challenging the prior expectation that a science must be anchored by theories.

Some scientists seem to favor this philosophical response. For example, Minelli has argued that we need a 'comprehensive theoretical account of development' with a principle of 'developmental inertia', analogous to the principle of inertia in Newtonian mechanics (Minelli, 2011a, b). Maturity is not the motivation for this claim, and two other reasons are salient: organization and guidance in the face of a welter of biochemical detail and the need to forge a synthesis between evolution and development. Regarding the former, the 'curse of detail' is one of the costs of developmental biology's meteoric success over the past three decades: 'The principal challenge today is that of exponentially increasing detail' (Slack, 2013: ix). Is developmental biology the embryological instantiation of one damn thing after another? What provides the guiding rails for it to proceed? Is it an unruly frontier of fragmentary hypotheses? Something must provide this kind of organization and guidance to developmental biology, given the overwhelming number and kind of details involved. I agree, but it need not be theories that accomplish the task.

Regarding calls for a synthesis of evolution and development, these often assume that having a developmental theory is a precondition for synthesis (Sommer, 2009). Just as evolutionary theory provides coherent structure to the variety of models and concepts in the study of evolution, so also studies of ontogeny require a framework that structures its methodologically diverse and conceptually disparate investigations. 'Our troubles . . . derive from our standing lack of an explicit theory of development' (Minelli, 2011a: 4). But this line of argument relies on the degree to which evolutionary theory exhibits the supposed structure to which developmental biologists should aspire. The actual practice of using different elements associated with evolutionary theory indicates a much looser framework with chameleon qualities that is responsively adjusted to the diverse investigative aims of evolutionary researchers (Love, 2010, 2013). Although there are frequent references to 'evolutionary theory', its individuation and components are not as explicit and systematic as is sometimes assumed. Thus, it is not clear that evolutionary theory is a valid template, and we return to the question of where the expectation for theories to govern and guide science comes from and whether it is justified. I contend that a productive way forward is to relinquish a prior expectation that sciences must have theories or theories of a certain kind (e.g. where maturity is associated with theory-directed inquiry). Instead, sciences that display empirical success and fecundity should be studied to discover what features are responsible, without assuming that those features will be the same for all sciences. We should be against maturity in this global, normative sense, which emerges from a preconception that all sciences require some kind of theoretical framework to organize their investigative endeavors.

For sciences without theories

Once we have relinquished a prior expectation that sciences must have theories or theories of a certain kind, a natural question is: what then organizes scientific inquiry? Philosophical theorizing about the nature of science has been colored by a set of concepts that held special importance in the modern period: laws of nature, scientific truth, and scientific rationality (Giere, 1999). This picture of science came under scrutiny in the work of Thomas Kuhn and others, who emphasized how the practice of science (both past and present) did not conform to these categories and required philosophers to reconceptualize scientific reasoning. For example, the demand that genuine explanation always involves universal, exceptionless laws of nature has been undermined (Woodward, 2003: Ch. 4), which helps account for Giere's choice of a title—*Science Without Laws*. The moral is simple and yet also radical: 'Science need not be understood in these terms and, indeed, may be better understood in other terms' (Giere, 1999: 4).

'Theory' is another concept that has played a special role in philosophical analyses of science. It is prominent in almost all of the main issues discussed in philosophy of science: How are theories confirmed? Should we believe in theoretical entities? How do we decide between competing theories? How are observations biased by theories? Are theories underdetermined by evidence? What is the structure of scientific theories? Theory is arguably the predominant concept used to conceptualize science, implicitly or explicitly. It is difficult to imagine how a science would proceed without one: 'all philosophers of science agree that science can, does, and should come up with theories' (Bromberger, 1992: 5). And this agreement extends to newer, more flexible conceptions of theories. Must we understand science in terms of theory? Can some sciences be understood in other terms? The relative absence of invocations of theory from most of the discourse in developmental biology and some other areas of molecular and cell biology may be an indicator of their distinctive epistemological architecture.

Here it is important to distinguish two different philosophical questions. The first question can be formulated as: *what is a scientific theory*? Here we tread familiar though still contested ground about whether the syntactic or the semantic conception of theory structure best captures systematic representation of knowledge in the sciences (Halvorson, 2012; Morrison, 2007; Suppe, 1977; Thompson, 2007), or whether some alternative conception of theories is required. Once it is understood what theories are, then this speaks to how sciences should proceed and the role theories play in organizing this activity. For example, the nature of theories affects how they are confirmed or have an impact on observations. A second question can be distinguished from the first: *what is the function of scientific theories*? This question hews closely to scientific practice by analysing the use of theories to achieve specific epistemic goals. If we try to understand what theories are through the lens of what theories do, the structure of scientific theories is cast in a new light (Griesemer, 2013; Love, 2013). Epistemic goals may differ across sciences and thus be attended by differing theory structure or by theories that function in distinct ways. The governing concern is that the operative discourse of different sciences must be consulted to understand why they succeed (or fail) in various ways and to what degree the concept of theory is important.

A concentration on the operative discourse of different sciences means that we actively countenance the possibility that particular concepts might be more appropriate and illuminating in analysing some areas of science rather than others. Many analyses of scientific methodology sank under the weight of the assumption that science is a single kind of activity, distinguishable from non-scientific inquiry, and therefore its diverse manifestations could be characterized by the same set of concepts. (Critics simply highlighted counter examples from the practices of various sciences not consulted in the formulation of a favored philosophical account.) Thus, the fact that a single theory or structured group of theories is not responsible for organizing developmental biology does not imply that theories never organize sciences or are somehow unimportant. But it does imply that we cannot presume that the presence of a central theory or family of explanatory models is necessary for organizing knowledge and inquiry in a science. To do so involves assuming a single measuring rod, and this may indicate more about philosophical prejudice than the actual state of inquiry and knowledge for particular sciences.

The justification for this privileging of the operative discourse of actual scientific practice in order to understand the epistemological architecture of an area of science—in this case, developmental biology—derives from a philosophical criterion of epistemic transparency or accessibility (Love, 2012). This criterion is based on the premise that the successes of scientific inquiry must be available to those engaged in the practice of that science (i.e. scientists). If we postulate hidden structure not present in scientific discourse to account for inductive inference, explanation, or other forms of reasoning, then we risk obscuring how scientists themselves access this structure to evaluate the reasoning (Woodward, 2003: Ch. 4). The result is that the successes of the sciences become mysterious when viewed from the vantage point of its participants. It also can insulate philosophical analyses from features of scientific practice that deviate from the hidden structure sought; they are prevented from facing real counter examples. Epistemic accessibility demands a descriptive correspondence between philosophical accounts of science and scientific practice. Features that are pervasive in scientific practice should be prominent in any adequate philosophical account of the science under scrutiny.

The epistemic accessibility criterion does not mean that every claim made by any scientist should be taken with the same credence when accounting for scientific epistemology. Instead, the ruling concern is *pervasive* features of practice. The problem with assuming laws are required for explanation was their relative absence from a variety of successful sciences routinely offering explanations, not that no scientists ever appeal to laws as explanatory. Therefore, the epistemic accessibility criterion involves an assumption of ubiquity within a domain of science. It is pervasive features of scientific practice that should be prominent in philosophical accounts of sciences, developmental biology or otherwise. Thus, it is not surprising that the desire for a theory can be found among some developmental biologists—'Developing a theory is of utmost importance for any discipline' (Sommer, 2009: 417; cf. Minelli, 2011a, b). But the fact that these calls are rare and isolated means we should not assume theories are actually needed to govern and organize inquiry within the domain.

Therefore, when we consider different presentations of developmental biology, we need to ask how its epistemic units (e.g. models, questions, concepts, etc.) function as organizational scaffolding (or otherwise) in the operative discourse. It is clear from the empirical analysis of word usage across multiple scientific discourses that there is tremendous heterogeneity with respect to theory as an epistemic unit. In a study of 781 articles published in *Science* over an entire year (Overton, 2013), 'theory' does not appear in the keywords of the 27 articles pertaining to developmental biology, whereas it appears in more than 40% of the 34 articles pertaining to evolution (Figure 3.1). A different study of the titles and abstracts in *Proceedings of the National Academy of Sciences USA* over the past 100 years found something similar: 'theory' appeared in less than half a percent of the articles devoted to ontogeny (5 out of 1,648; Pradeu, this volume). This relative dearth of the word 'theory' from articles focused on development is initial evidence that theories play less of a functional role in developmental biology, and the variation observed across different disciplines lends credence to the variability of theory as the primary epistemic unit in different areas of science. Tellingly, a major National Research Council study on the importance of theory in biology did not have a developmental biologist on the committee (NRC, 2008). The report labors to emphasize the importance of theory in biological sciences despite

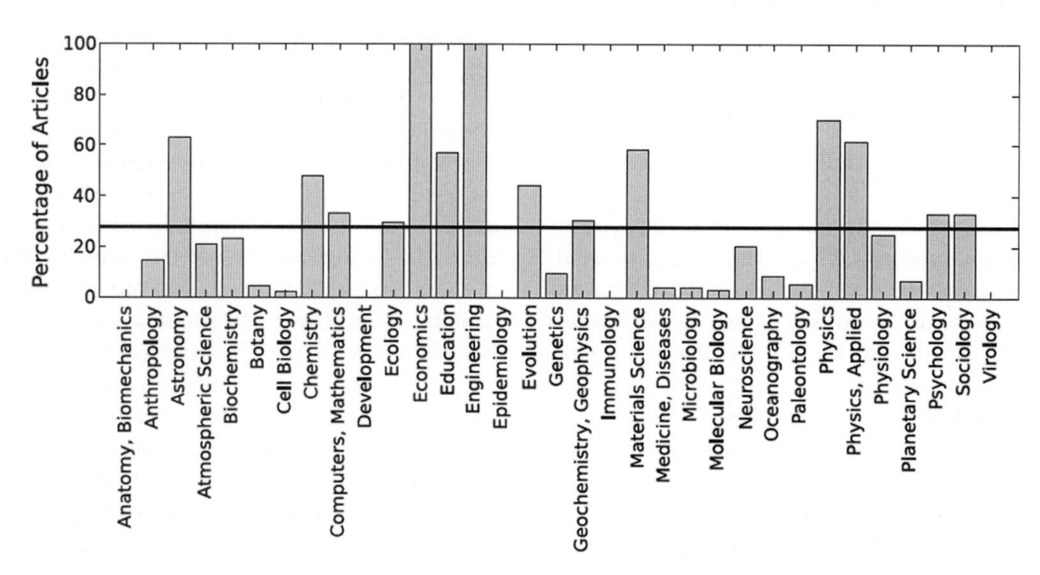

Figure 3.1 The relative frequency of 'theory' (theory, theories, theoretical) in article keywords for different domains of scientific inquiry (Overton, 2013, Figure 6). The solid line indicates overall average occurrence. The data were drawn from 781 articles in regular issues of the journal *Science* between 24 September 2010 and 23 September 2011. Of these, 27 articles were categorized as developmental biology. The material is reproduced with kind permission from Springer Science + Business Media.

it being underappreciated and insufficiently discussed. That the absence might be informative of epistemological practices is not countenanced.

But just because a theory or set of structured hypotheses is not organizing a science does not mean that we never observe theories operating therein. Following Waters (2007), I distinguish theory-informed science (using theoretical knowledge) from theory-directed science (having a theory that directs or guides inquiry and organizes knowledge). Denying the former is untenable. For example, developmental biology uses theoretical knowledge from biochemistry when appealing to morphogen gradients to explain how segments are established or chemical thermodynamics when invoking reaction–diffusion mechanisms to explain pigmentation patterns. It also uses theoretical knowledge derived from *within* developmental biology, such as positional information models. Different kinds of theory inform developmental biology, but these theories or models do not organize research in developmental biology—they are not necessary to structure its knowledge and direct investigative activities. Developmental biologists are not focused on confirming and extending the theory of reaction–diffusion mechanisms, nor are they organizing their research around positional information.

> Regeneration necessarily involves regional specification of pattern and as the relevant interactions are local ones this means that cryptic information about pattern must be present throughout the mature structure. This is sometimes called positional information. The term was once widely used to indicate developmental pattern information but has fallen out of use as understanding of its molecular basis has increased. (Slack, 2013: 385)

This theoretical knowledge is used in building explanations but does not provide rails of guidance for how to proceed in a research program. All sciences may use theoretical knowledge, but this is not the same as all sciences having a theory providing direction and organization.

The major questions tackled by most philosophers have presumed that theories of some kind direct and guide sciences or play the primary organizational role. Theories are central because scientists are testing them, confirming them, constructing them, and making observations in light of them. If a science doesn't have a theory, then supposedly it is difficult to see the rationale for its activities. Therefore, an affirmative answer to the question of whether some sciences can be understood in other terms is about whether a theory directs inquiry and organizes knowledge, not whether theoretical knowledge is present or absent. Sciences do not need theories to structure knowledge, guide inquiry, or otherwise provide organizational scaffolding, and developmental biology is one science where something else accomplishes this task. Importantly, this does not mean that there are absolutely no theories within developmental biology, especially in terms of newer conceptions where theories fulfill scientific goals different from what many have expected (e.g. structuring knowledge and guiding inquiry). The erotetic organization of developmental biology does not exhaust what needs to be said about its epistemology, but since these kinds of theory accomplish different epistemic tasks, an account of their roles in the study of ontogeny will be complementary to the analysis that follows.

Textbook presentations of developmental biology

Erotetically organized

In seeking to understand the organizing structure of developmental biological research in epistemically accessible terms, textbooks are one source to explore because they must capture substantial community consensus in order to be effective and widely adopted. This consensus often includes predominant biases or norms, in addition to an acceptable coverage of concepts and empirical content (Winther, 2006: 478–9). (Textbooks also simplify experimental practices and distort historical development.) The role of textbooks is to codify knowledge in an area of science and provide a distilled format for transferring it to novices. Stasis and change in textbooks with respect to thematic coverage, methodological emphases, or explanatory preferences can indicate deep knowledge commitments because a large number of researchers vet them prior to publication. Although the presentation of theories in textbooks is expected to differ from what is found in everyday investigative practice, it would be odd for theories to play an important role in a science and never be mentioned in corresponding textbooks. Evolutionary biology textbooks talk about the theory of natural selection and economics textbooks talk about microeconomic theory, both of which play major roles in organizing knowledge in these domains. Do we find similar kinds of theories in developmental biology textbooks?

Consider two editions of Jonathan Slack's *Essential Developmental Biology* (Slack, 2006, 2013; Table 3.1). As noted above, three families of approaches are described (developmental genetics, experimental embryology, and molecular and cell biology) and a catalogue of 'key molecular components' is gathered in the appendix (genes, transcription factor families, inducing factor families, cytoskeleton, cell adhesion molecules, and extracellular matrix components), but no standard theory or structured set of hypotheses is mentioned or provides structure to the presentation of the material. Instead, Chapter 2 ('Common Features of Development' (2nd edition); 'How Development Works' (3rd edition)) sets out four main types of processes, also described as clustered groups of problems, which occur during embryonic development: regional specification (pattern formation), cell differentiation, morphogenesis, and growth. (In the third edition, a fifth category of developmental time is described, though without the same solidity as the other four: 'Somehow the component processes of development are coordinated in time. But

Table 3.1 Textbook structure for the 2nd and 3rd editions of *Essential Developmental Biology* (Slack, 2006, 2013). Apart from a few rearrangements (e.g. 'Tissue organization and stem cells'), there is remarkable consistency, and the organization is not accomplished via particular theories or models.

Slack 2006 (2nd edition)		**Slack 2013 (3rd edition)**	
Groundwork	The excitement of developmental biology	Groundwork	The excitement of developmental biology
	Common features of development		How development works
	Developmental genetics		Approaches to development: developmental genetics
	Experimental embryology		Approaches to development: experimental embryology
	Techniques for the study of development		Approaches to development: cell and techniques
Major Model Organisms	Model organisms	Major Model Organisms	Model organisms
	Xenopus		*Xenopus*
	The zebrafish		The zebrafish
	The chick		The chick
	The mouse		The mouse
	Drosophila		*Drosophila*
	Caenorhabditis elegans		*Caenorhabditis elegans*
Organogenesis	Tissue organization and stem cells	Organogenesis	Techniques for studying organogenesis and postnatal development
	Development of the nervous system		Development of the nervous system
	Development of mesodermal organs		Development of mesodermal organs
	Development of endodermal organs		Development of endodermal organs
	Drosophila imaginal discs		*Drosophila* imaginal discs
Growth, Regeneration, and Evolution	Growth, aging, and cancer	Growth, Regeneration, Evolution	Tissue organization and stem cells
	Regeneration of missing parts		Growth, aging, and cancer
	Evolution and development		Regeneration of missing parts
			Applications of pluripotent stem cells
			Evolution and development

developmental time remains the most mysterious aspect of the process'; Slack, 2013: 7.) These broad clusters are then fleshed out along a standard timeline of early development, highlighting gametogenesis, fertilization, cleavage, gastrulation, and axis specification. Axis specification, which involves both regional specification and cell differentiation, is elucidated in terms of developmental control genes (transcription factors) and morphogen gradients. Some of these question clusters, such as morphogenesis, are subdivided into different types: condensation, invagination, and epiboly, among others (Figure 3.2). Growth is explicated in terms of different kinds of cell division mechanisms (cleavage, asymmetrical, exponential growth, and stem cells). With these clusters of problems that attach to the various types of phenomena described, the next several chapters detail the different experimental approaches utilized to dissect how they work (cell and molecular biology, developmental genetics, and experimental embryology) and the model organisms in which they are dissected (nematodes, fruit flies, zebrafish, African clawed frog, chicken, and mouse). Subsequent chapters cover later aspects of development, such as organogenesis, with different systems treated in depth by tissue layer (e.g. mesodermal: somitogenesis, kidney development, vasculogenesis, limb development, cardiogenesis; endodermal: gut, liver, lung, pancreas). Differentiation and growth are detailed in terms of stem cells and proliferative tissues, regeneration, aging, and cancer. Potential applications to evolutionary questions are treated in the last chapter.

Throughout the structured presentation of developmental biology given by Slack (Table 3.1), no specific theory, set of hypotheses, or dominant model is invoked to organize these different domains of investigation. Instead, there are broad clusters of questions that reflect generally delineated phenomena (differentiation, specification, morphogenesis, and growth) whose explanatory characterization sets the agenda of research. These phenomena can be subdivided into different types (e.g. Figure 3.2), partitioned in terms of whether they occur earlier or later in development (e.g. early cleavage versus later organogenesis) or in specific tissue layers (e.g., gut formation vs somitogenesis), and combined to investigate more specific phenomena (differentiation, specification, and morphogenesis in the ontogeny of the nervous system). More analytically, the space of problems is characterized in terms of five variables: *abstraction*, *variety*, *connectivity*, *temporality*, and *spatial composition*. The values given to these variables structure the constellation of research questions within the broad problem agendas corresponding to generally delineated phenomena. For example, research questions oriented around events in zebrafish gastrulation are structured in a way that differs from the research questions oriented around vertebrate neural crest cell migration because they involve different values for the five variables: abstraction (zebrafish vs vertebrates), temporality (earlier vs later), spatial composition (tissue layer interactions vs a distinctive population of cells), variety (epiboly vs epithelium to mesenchyme transition), and connectivity (gut formation and endoderm vs organogenesis and ectoderm/mesoderm). And these configurations can be adjusted readily in response to shifts in the values for different variables. Some of these variables also help to structure problems in other sciences (e.g. abstraction and connectivity), but this specific combination of variables structures developmental biology, especially the juxtaposition of temporality and spatial composition.

The tenth edition of Scott Gilbert's Developmental Biology recently appeared. A consistent feature over multiple editions of the book is that developmental biology is constituted by a set of core questions:

There are two fundamental questions in developmental biology. How does the fertilized egg give rise to the adult body? And how does that adult body produce yet another body? These two huge questions can be subdivided into seven general categories of questions scrutinized by developmental biologists:

The question of differentiation. A single cell, the fertilized egg, gives rise to hundreds of different cell types . . . Since every cell of the body (with very few exceptions) contains the same set of genes, how can this identical set of genetic instructions produce different types of cells? How can a single cell . . . generate so many different cell types?

The question of morphogenesis. How can the cells in our body organize themselves into functional structures? . . .

The question of growth. . . . How do our cells know when to stop dividing? How is cell division so tightly regulated?

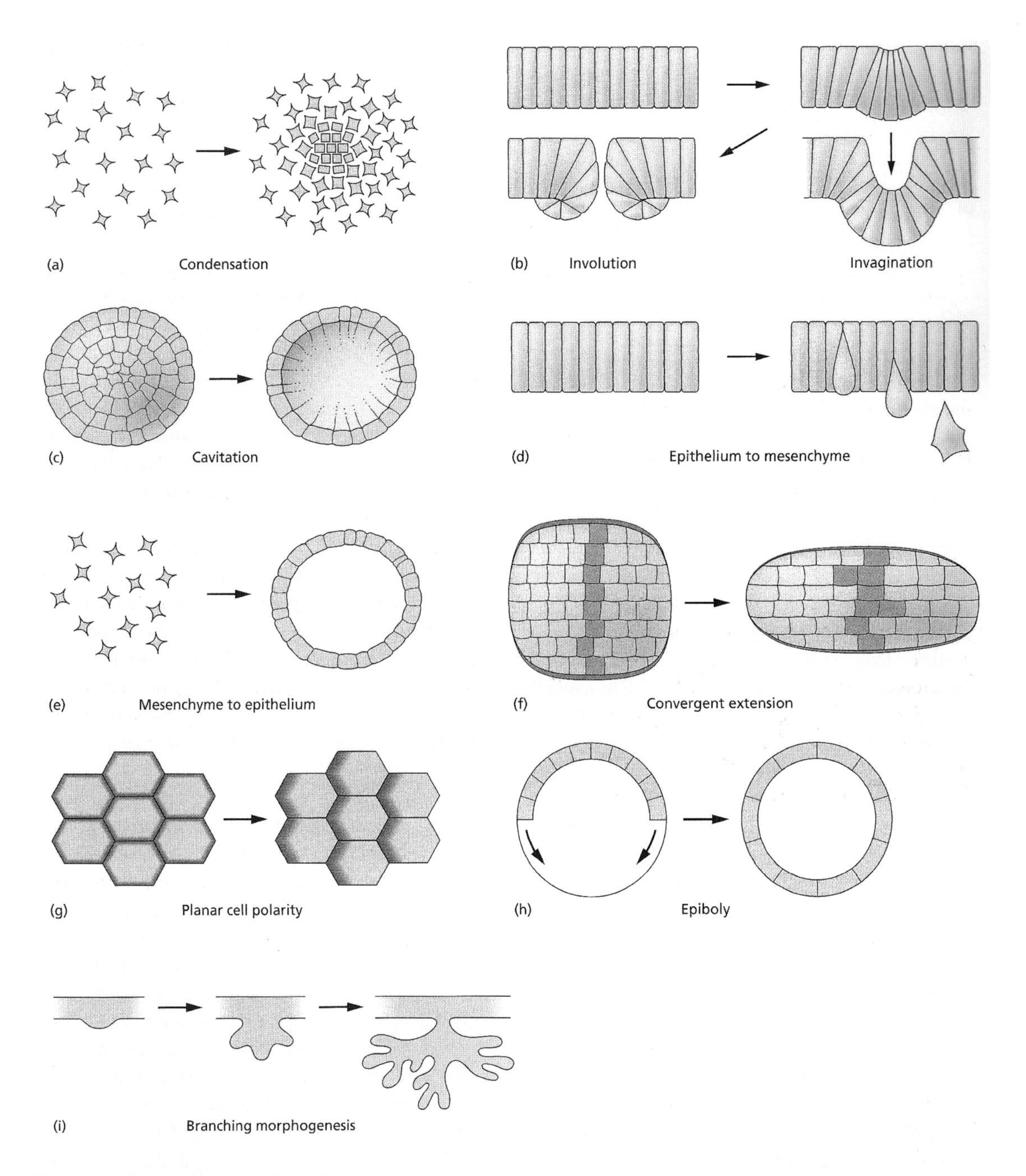

Figure 3.2 Different types of morphogenesis, one of the core problem agendas that organizes developmental biology (Slack, 2013: 22). This illustrates several of the variables that characterize the structure of a problem agenda, including abstraction (these are more concrete exemplars of morphogenesis), temporality (these processes occur at different times of development), connectivity (the connections with other aspects of development is variable, such as branching morphogenesis with mammary gland formation or condensation with bone formation), and variety. This material is reproduced with permission of John Wiley & Sons, Inc.

The question of reproduction. . . . How are germ cells set apart from the cells that are constructing the physical structures of the embryo, and what are the instructions in the nucleus and cytoplasm that allow them to form the next generation?

The question of regeneration. Some organisms can regenerate their entire body. . . . there are some cells in our bodies—*stem cells*—that are able to form new structures even in adults. How do the stem cells retain this capacity and can we harness it to cure debilitating diseases?

The question of evolution. . . . How do changes in development create new body forms? Which heritable changes are possible, given the constraints imposed by the necessity that the organism survive as it develops?

The question of environmental regulation. The development of many (perhaps all) organisms is influenced by cues from the environment that surrounds the embryo or larvae. . . . How is the development of an organism integrated into the larger context of its habitat? (Gilbert, 2010: 2–3)

The structure and ordering of these different questions is remarkably stable from edition to edition. In 1997, only five of the headings under the two main questions are listed: differentiation, morphogenesis, growth, reproduction, and evolution (Gilbert, 1997: 2–3). Environmental regulation appears in subsequent editions (Gilbert, 2006: 4–5), in part because of the author's increasing interest in this area (Gilbert, 2001; Gilbert & Epel, 2009). Regeneration only appears explicitly as a question in the ninth edition, although it has been recognized as a central developmental problem for more than a century (Sunderland, 2010). The appearance of these seven themes is marked across three editions of the textbook in Table 3.2.

The identified domains of questions tightly correspond to those found in Slack's textbook treatment (Slack, 2006, 2013). Gilbert has hierarchical aspects to his clusters of questions, akin to Slack's presentation, as well as similar variables that characterize their relationships: abstraction (e.g. morphogenesis vs condensation), variety (e.g. differentiation vs growth), connectivity (e.g. regional specification and cell migration in organogenesis), temporality (e.g. fertilization vs gut formation), and spatial composition (e.g. mesodermal derivatives vs endodermal derivatives). This anatomy of problems, with explicit epistemological structure from different values for these variables, operates to organize the science of development; investigators from different disciplines can be working on the same problem but asking different questions that require distinct but complementary methodological resources. Knowledge and inquiry in developmental biology are intricately organized, just not by a central theory or group of models, and this erotetic organization is epistemologically accessible to the participating scientists. While theoretical knowledge, especially as drawn from biochemistry and molecular biological mechanisms, as well as mathematical models (e.g. allometric theory or reaction–diffusion models), is ubiquitous (*theory-informed*), the clusters of problems that reappear across the textbooks and correspond to different types of phenomena provide the governing architecture (*not theory-directed*), which can be characterized explicitly according to the variables described.

Could there be some core principle that operates as a unifying thread throughout these problems? *Differential gene expression* plays a prominent role in explaining the paradox of genomic equivalence (Gilbert, 2010): how can heterogeneity emerge from a set of cells that have the same number and kind of chromosomes containing the same genes? If different sets of genes are activated or repressed in different combinations through embryogenesis, then heterogeneous, complex features of form can emerge from homogeneous, simpler structures. Differential gene expression is labeled as a paradigm in an earlier edition (Gilbert, 2006), though it doesn't match either of Kuhn's standard conceptions (a disciplinary matrix representing all of the components of consensus in an area of science, or a set of exemplary solutions to empirical or theoretical problems that display the modus operandi of a science). The paradigm label disappears in the ninth edition even though the content related to differential gene expression remains: differential gene transcription, selective nuclear RNA processing, selective messenger RNA translation, and differential protein modification (Gilbert, 2010: 31).

Another standard developmental biology text, *Principles of Development* (Wolpert et al., 1998, 2010; Table 3.3), invokes the terminology of 'principles', although these are not explanatory generalizations from core models but include statements like: 'Genes control cell behavior by specifying which proteins

Table 3.2 Textbook structure for the 5th, 8th, and 9th editions of *Developmental Biology* (Gilbert, 1997, 2006, 2010). Apart from cosmetic rearrangements, there is tremendous stability to the constituents and their organization, which does not involve an explicit or systematic theoretical architecture.

Gilbert 1997 (5th edition)		**Gilbert 2006 (8th edition)**		**Gilbert 2010 (9th edition)**	
An Introduction to Developmental Biology	An introduction to animal development	Principles of Developmental Biology	Developmental biology: the anatomical tradition	Questions: Introducing Developmental Biology	Developmental anatomy
	Genes and development: introduction and techniques		Life cycles and the evolution of developmental patterns		Developmental genetics
	The cellular basis of morphogenesis: differential cell affinity		Principles of experimental embryology		Cell-cell communication in development
Patterns of Development	Fertilization: beginning a new organism		The genetic core of development	Specification: Introducing Cell Commitment and Early Embryonic Development	Fertilization
	Cleavage: creating multicellularity		The paradigm of differential gene expression		Early development in selected invertebrates
	Gastrulation: reorganizing the embryonic cells		Cell-cell communication in development		The genetics of axis specification in *Drosophila*
	Early vertebrate development: neurulation and the ectoderm	Early Embryonic Development	Fertilization: Beginning a New Organism		Amphibians and fish
	Axonal specificity		Early development in selected invertebrates		Birds and mammals
	Early vertebrate development: mesoderm and endoderm		The genetics of axis specification in *Drosophila*	The Stem Cell Concept: Introducing Organogenesis	The emergence of the ectoderm
Mechanisms of Cellular Differentiation	Transcriptional regulation of gene expression: transcription factors and the activation of specific promoters		Early development and axis formation in amphibians		Neural crest cells and axonal specificity
	Transcriptional regulation of gene expression: the activation of chromatin		The early development of vertebrates: fish, birds, and mammals		Paraxial and intermediate mesoderm
	Control of development by differential RNA processing and translation	Later Embryonic Development	The emergence of the ectoderm: central nervous system and epidermis		Lateral plate mesoderm and the endoderm
Specification of Cell Fate and Embryonic Axes	Autonomous cell specification by cytoplasmic determinants		Neural crest cells and axonal specificity		Development of the tetrapod limb
	The genetics of axis specification in *Drosophila*		Paraxial and intermediate mesoderm		Sex determination
	Specification of cell fate by progressive cell-cell interactions		Lateral plate mesoderm and endoderm		Postembryonic development
	Establishment of body axes in mammals and birds		Development of the tetrapod limb		The saga of the germ line

continued

Table 3.2 *Continued*

Gilbert 1997 (5th edition)		Gilbert 2006 (8th edition)		Gilbert 2010 (9th edition)	
Cellular Interactions During Organ Formation	Proximate tissue interactions: secondary induction		Sex determination	Systems biology: Expanding Developmental Biology to Medicine, Ecology, and Evolution	Medical aspects of developmental biology
	Development of the tetrapod limb		Postembryonic development: metamorphosis, regeneration, and aging		Developmental plasticity and symbiosis
	Cell interactions at a distance: hormones as mediators of development		The saga of the germ line		Developmental mechanisms of evolutionary change
	Sex determination	Ramifications of Developmental Biology	An overview of plant development		
	Environmental regulation of animal development		Medical implications of developmental biology		
	The saga of the germ line		Environmental regulation of animal development		
	Developmental mechanisms of evolutionary change		Developmental mechanisms of evolutionary change		

Table 3.3 Textbook structure for the 2nd and 4th editions of *Principles of Development* (Wolpert et al., 1998, 2010). As with the other two textbooks, the structure is highly conserved and shows a tight correspondence in what are assumed to be the core problems of developmental biology.

Wolpert et al., 1998 (2nd edition)		Wolpert et al., 2010 (4th edition)	
Chapter 1	History and basic concepts	Chapter 1	History and basic concepts
Chapter 2	Model systems	Chapter 2	Development of the *Drosophila* body plan
Chapter 3	Patterning the vertebrate body plan I: axes and germ layers	Chapter 3	Vertebrate development I: life cycles and experimental techniques
Chapter 4	Patterning the vertebrate body plan II: the mesoderm and early nervous system	Chapter 4	Vertebrate development II: axes and germ layers
Chapter 5	Development of the *Drosophila* body plan	Chapter 5	Vertebrate development III: patterning the early nervous system and the somites
Chapter 6	Development of nematodes, sea urchins, ascidians, and slime molds	Chapter 6	Development of nematodes, sea urchins, ascidians, and slime molds
Chapter 7	Plant development	Chapter 7	Plant development
Chapter 8	Morphogenesis: change in form in the early embryo	Chapter 8	Morphogenesis: change in form in the early embryo
Chapter 9	Cell differentiation	Chapter 9	Germs cells, fertilization, and sex
Chapter 10	Organogenesis	Chapter 10	Cell differentiation and stem cells
Chapter 11	Development of the nervous system	Chapter 11	Organogenesis
Chapter 12	Germs cells and sex	Chapter 12	Development of the nervous system
Chapter 13	Regeneration	Chapter 13	Growth and postembryonic development
Chapter 14	Growth and postembryonic development	Chapter 14	Regeneration
Chapter 15	Evolution and development	Chapter 15	Evolution and development

are made' and 'The reliability of development is achieved by a variety of means' (Wolpert et al., 2010: 17, 30). Summary descriptions in what the authors term a 'conceptual framework for the study of development' pull us back to the same clusters of questions seen in other textbook presentations:

> Development is essentially the emergence of organized structures from an initially very simple group of cells. It is convenient to distinguish four main developmental processes, which occur in roughly sequential order in development, although in reality they overlap with, and influence, each other considerably. They are pattern formation, morphogenesis, cell differentiation, and growth (Wolpert et al., 2010: 14; cf. Wolpert, 2011).

Differential gene expression is salient throughout the discussion—'differential gene activity controls development' (Wolpert et al., 1998: 15)—and fits with the recent history of molecularization in studies of ontogeny: 'The molecular metamorphosis of our understanding of embryology has relied on the identification of genes that control development' (Fraser & Harland, 2000: 47). We might attempt to cast this set of principles that includes positional information, lateral inhibition, localization of cytoplasmic determinants, and asymmetric cell division in an organizational role, but it is unclear whether they constitute a governing core since they are not utilized in many developmental explanations and the mechanisms that produce them and other processes (e.g. symmetric cell division) do much of the explanatory work.

Additionally, not all of these mechanisms can be classified as differential gene expression, which makes it more difficult to understand how these principles organize and guide inquiry in developmental biology. For example, lateral inhibition is the phenomenon of structures becoming regularly spaced, such as feathers in the avian integument. Although this can occur in part due to differential gene expression, where cells secrete an inhibitor that blocks cellular neighbors from differentiating, thereby setting up a regular but intermittent cellular pattern, other 'epigenetic' mechanisms can play a role as well. These include physicochemical rules and stochastic processes (Jiang et al., 2004), which are recognized even when differential gene expression is in the foreground: 'As all the key steps in development reflect changes in gene activity, one might be tempted to think of development simply in terms of mechanisms for controlling gene expression. But this would be highly misleading.... To think only in terms of genes is to ignore crucial aspects of cell biology' (Wolpert et al., 1998: 15).

One potential way to sharpen the differential gene expression list is through an appeal to genetic programs:

> [Elements of the genome] contain the sequence-specific code for development; and they determine the particular outcome of developmental processes, and thus the form of the animal produced by every embryo. . . . Development is the execution of the genetic program for construction of a given species of organism (Davidson, 2006: 2, 16).
>
> The genome contains a program of instructions for making an organism—a generative program—that determines where and when different proteins are synthesized and thus controls how cell behave (Wolpert et al., 2010: 29).

But this strategy has empirical and conceptual drawbacks that include an inattention to plasticity and the role of the environment, an ambiguity about the locus of causal agency, and a reliance on metaphors drawn from computer science (Gilbert & Epel, 2009; Keller, 2002; Moss, 2002; Robert, 2004). More importantly, in the present context, it is unhelpful because these appeals to a genetic program do not generalize across the research community (and never did)—they are not pervasive features of the epistemic practice of developmental biology. This is indicated (in part) by their absence from textbooks ('genetic program' doesn't appear in any index, even Wolpert et al., 1998, 2010) and from regular investigative discourse (e.g. see the case study on mammary gland morphogenesis below). Thus, the controversial claim that there is a generative, genetic program for development is not operating as a shared organizing principle to coordinate the welter of molecular detail in developmental biology.

This is a subtle point worth emphasizing. There is not widespread agreement about the existence or explanatory power of a genetic program or blueprint in the way there is for differential gene expression, and we have already observed that the latter is not operating as an organizational principle in developmental biology. This lack of concordance is represented in how they are treated in textbooks and regular investigative venues. In seeking to identify organizational scaffolding for developmental biology, finding commonalities is critical—pervasive features of practice—and the erotetic organizational structure is shared in common. When researchers have vocalized an interest in a general theory for developmental biology by appeal to a genetic program (e.g. Wolpert & Lewis, 1975), these do not achieve consensus within the community and play little to no organizing role in ongoing experimental research. They can provide a theory-informed perspective; theoretical models can be suggestive: 'The problem may be approached by viewing the egg as containing a program for development, and considering the logical nature of the program by treating cells as automata and ignoring the details of molecular mechanisms' (Wolpert & Lewis, 1975: 14). But the concept of a genetic program for ontogeny is not serving the purpose of organizing the knowledge and directing inquiry; it is not a pervasive feature of practice for investigations of ontogeny.

Developmental biology is not organized by a theory, family of models, or some set of core explanatory principles. While it is informed by theoretical knowledge from different domains, developmental biology is organized *erotetically*. These problems (e.g. differentiation, growth, morphogenesis, and pattern formation) are composed of multiple questions themselves, which form an anatomical problem structure for developmental biology in terms of the weighted variables that characterize these problems: abstraction, variety, connectivity, temporality, and spatial composition. What we still lack is a precise understanding of the nature of this erotetic structure (i.e. an ontology of the epistemic units).

Problem agendas as units of scientific organization

Not individual interrogatives

Several of the textbooks frame the core problems of developmental biology as 'questions'. Erotetic *logics* are a natural place to turn for an account of question structure in the sciences (Bromberger, 1992; Hintikka, 2007; Kleiner, 1988). While these approaches contain resources for characterizing particular questions, they are less helpful for explicating the structure of scientific problems already identified within developmental biology. For example, erotetic logics require that the possible answers to a question constitute a mutually exclusive and exhaustive set. But many scientific problems outstrip our ability to formulate exclusive or exhaustive sets of answers (Nickles, 1981), and the stability of the problem agendas found in developmental biology emerged long before some of our best candidate answers were extant (e.g. differential gene expression). Additionally, sometimes the answers offered to particular questions are incompatible but mutually informative as a consequence of different choices about idealization and abstraction. For example, understanding the dynamics of biological populations can be achieved with distinct models that idealize different factors, such as population size, number of alleles in the genetic pool, and the spatial structure among its members.

More generally, an answer to a scientific problem should not be understood as a single proposition. There is a broad mismatch between erotetic *logics* and the nature of scientific problems. A biological problem, such as cell differentiation, does not correspond to a single question ('how do cells differentiate?') like other individual interrogatives ('what did you do last night?'). The problem of cellular differentiation is more of an *agenda*, a composite list of things that need to be addressed or multiple, interrelated questions (Love, 2008a; b). This insight remains if theories are the central unit of analysis because it makes little sense to describe science as a problem solving or theory building activity in response to a single interrogative—a theory is more than an answer to a question.

What we need then is a vocabulary for describing the erotetic structure of problem agendas that is not available from erotetic logics where only individual interrogatives are in view. How do we elaborate on the variables of abstraction, variety, connectivity, temporality, and spatial composition to further comprehend the structure of problems like differentiation, growth, and pattern formation? One thing we should not expect is that this structure is similar across different sciences with different problems. Although there might be commonalities, the structure is likely specific to the science in question ('particular concepts are appropriate for analysing some areas of science rather than others'). Although complex scientific problems are not structured logically or 'well-defined' (Osbeck et al., 2011: Ch. 3), this does not mean that they are unstructured, and their role in guiding long-term investigative programs suggests otherwise. And this structure is something that is epistemologically accessible to the working scientist. Problem agenda anatomy can be delineated further with three dimensions: history, heterogeneity, and hierarchy (Brigandt and Love, 2012). These three dimensions, in combination with weighted values for the five variables already identified (abstraction, variety, connectivity, temporality, and spatial composition), illuminate how problems provide an erotetic organizational framework for coordinating inquiry in developmental biology.

History, heterogeneity, and hierarchy

The first dimension structuring problem agendas is *history*: discussions and debates surrounding longstanding questions yield structural relations and stable individuation (Hattiangadi, 1978, 1979). The stability of problem agendas through historical discussion is an important factor in recognizing the individuality and variety of problems that compose a science. For developmental biology, this hovers in the range of four core domains: regional specification or pattern formation, cellular differentiation, morphogenesis, and growth. Continued investigation into what causes cellular differentiation or morphogenesis over decades, if not centuries, validates the legitimacy of the number and variety of problems, as well as providing benchmarks for empirical progress in these different domains. This is especially evident in disagreements over what explains these phenomena. Researchers agree on the phenomena that need to be explained (shared, multilayered *explanandum* types), but differ (for example) on whether physical rules or genetic factors are more explanatory (Keller, 2002).

History signals different weightings for the values taken by the variables that characterize developmental biology's problem agendas. Molecular biology research over the past 30 years has largely been undertaken on early ontogeny (temporality) because of issues with controlling experimental parameters and observational tractability. Due to therapeutic potential, cellular differentiation has focused on embryonic stem cells. Connectivity is established through repeated interactions among the erotetic domains over time; abnormal growth and particular types of morphogenesis (epithelium to mesenchyme; Figure 3.2) are found juxtaposed in studies of cancer. Differentiation and growth are interweaved in the study of regeneration. Much of this connectivity is implicit in textbook presentations, having grown out of repeated historical associations among these problem agendas. Spatial composition guides research into the origin of different cell lineages, which have been established over long periods of time (e.g. nerve cells are ectodermal derivatives; muscle cells are mesodermal derivatives). Levels of abstraction also are weighted differently through history. A focus on convergent extension, one kind of morphogenesis relevant to gastrulation, was influenced by the introduction of new experimental techniques involving tissue explants (Sive et al., 2007). Epiboly, a different kind of morphogenesis relevant to gastrulation, has received attention over the past two decades because it is exhibited by zebrafish, now one of the most popular model organisms (Rohde & Heisenberg, 2007).

The second dimension of problem agendas is *heterogeneity*: the problem agendas of developmental biology contain different kinds of questions (Laudan, 1977), among which there are specific relationships. Empirical questions ('what regulatory genes control pluripotency?') are answered differently than theoretical questions ('under what conditions does an oscillating wave of gene expression stabilize into segments?'). Questions at one level of organization (e.g. cellular) are addressed differently than questions at other levels (e.g. tissue). This heterogeneity interacts with various weightings of the five variables. Empirical questions at the cellular level of organization can be asked at early stages of ontogeny (temporality) about differentiation or regional specification (variety). Theoretical questions at the tissue level of organization can be asked about the propensity of an epithelial-to-mesenchymal transition in a tumor of a certain size or type (connectivity—morphogenesis and growth).

Abstraction is relevant in separating empirical questions about whether a particular gene makes a difference to a developmental outcome or whether general principles that map genotype to phenotype can be discerned. Spatial composition questions inherently span levels of organization, though particular relationships (e.g. the germ layers involved in the formation of a specific organ) are usually weighted.

The third dimension of structure is *hierarchy*: relationships among definable arrays of questions in problem agendas thought of as parts to the whole (Nickles, 1981). The most obvious of these concerns the variable of abstraction. Questions about blood cell differentiation are part of the larger problem of cellular differentiation. Spatial composition is also germane to hierarchy; understanding the structural characteristics of skin cells is part (but not all) of understanding skin development. But if we broaden our understanding of hierarchy to include functional relationships, temporality is immediately invoked. The dependence of one process (gut formation) on an earlier process (gastrulation) sets up hierarchical relationships between questions of process through time. Once process hierarchies are in view, then connectivity and variety are involved, since functional dependence may occur across phenomena in problem agendas of different kinds. Variety is also relevant to spatial composition and abstraction in order to answer questions about whether different levels of organization behave according to similar principles and at what level of abstraction.

Another feature relevant to hierarchy is generalization. Part of the shift in abstraction with respect to a problem agenda or its component questions can be motivated by the scope of the generalities sought. Moving from regional specification to somite formation involves a reduction in generality; fruit flies have regional specification of many different kinds (e.g. segment formation), but they do not have somites. Thus, a focus on somitogenesis as a more concrete instantiation of pattern formation can involve an adjustment in generalization. But if pattern formation processes are treated at a high enough level of abstraction, it may be possible to comprehend principles that are not reliant on shared molecular constituents, such as in a reaction–diffusion mechanism. Similar considerations about generality apply for the other variables: temporality—gastrulation, an earlier developmental event, occurs across metazoans; neurulation, a later developmental event, occurs only in vertebrates; connectivity—the intersection of questions about differentiation and morphogenesis differ for diploblasts and triploblasts; variety—a generalization for growth in terms of cell cycle regulation contrasted with a generalization for gastrulation in terms of morphogenesis; and, spatial composition—questions about particular organs can differ depending on whether they share the same components in different species (e.g. number of cell types).

These three dimensions of structure are not unique to developmental biology but in conjunction with the five weighted variables generate criteria of explanatory adequacy that provide a template or guide for ontogenetic research (Love, 2008a). The specific problem anatomy *directs* scientific inquiry and indicates how to organize new knowledge; no central theory or family of models is required to achieve this result. The structure of the problem agendas contains information on what kinds of methodological contributions need to be coordinated to generate a deeper understanding of how aspects of development work, and clarify where new results get catalogued and categorized. This structure also suggests how the welter of detail is tamed and where horizons of future research lie, as well as how balanced existing explanatory frameworks are with respect to different problem agendas. A key element of the explanatory appeal of differential gene expression is its involvement across all of the problem agendas, at all developmental stages. It appears to be most (or all) of the answer to many of the heterogeneous questions at different levels of organization studied by developmental biologists, though other factors are championed as well (Forgacs & Newman, 2005).

Case study: mammary gland morphogenesis

Different editions of the major textbooks demonstrate that broad problem domains associated with differentiation, morphogenesis, growth, and pattern formation organize inquiry in developmental biological research. Now that we have an account of the ontology of these units and their structure as problem agendas, it is helpful to consider a concrete example of research from developmental biology. There we should observe a similar kind of organizational pattern in the context of a specific research program. The case I will consider is a study of the phenomenon of branching morphogenesis in the mammary gland using a mouse model and concentrating on the role of a cellular receptor named Neuropilin-2 (Goel et al., 2011).

The central problem agenda relevant to this study is morphogenesis, or the origination of shape and form in ontogeny due to cellular death, migration, movement, growth, and rearrangement. There are approximately eleven distinct types of morphogenesis (see Figure 3.2), of which branching morphogenesis is only one. It is associated with structures such as kidneys, lungs, and blood vessels. There are three different types of branching morphogenesis: clefting, sprouting, and intussusception (Davies, 2005: Ch. 4.6; Figure 3.3). Mammary gland morphogenesis primarily exhibits the phenomenon of sprouting, which can be subdivided into monopodial (side branching from a main trunk) and dipodial (dichotomous branching with no main trunk)

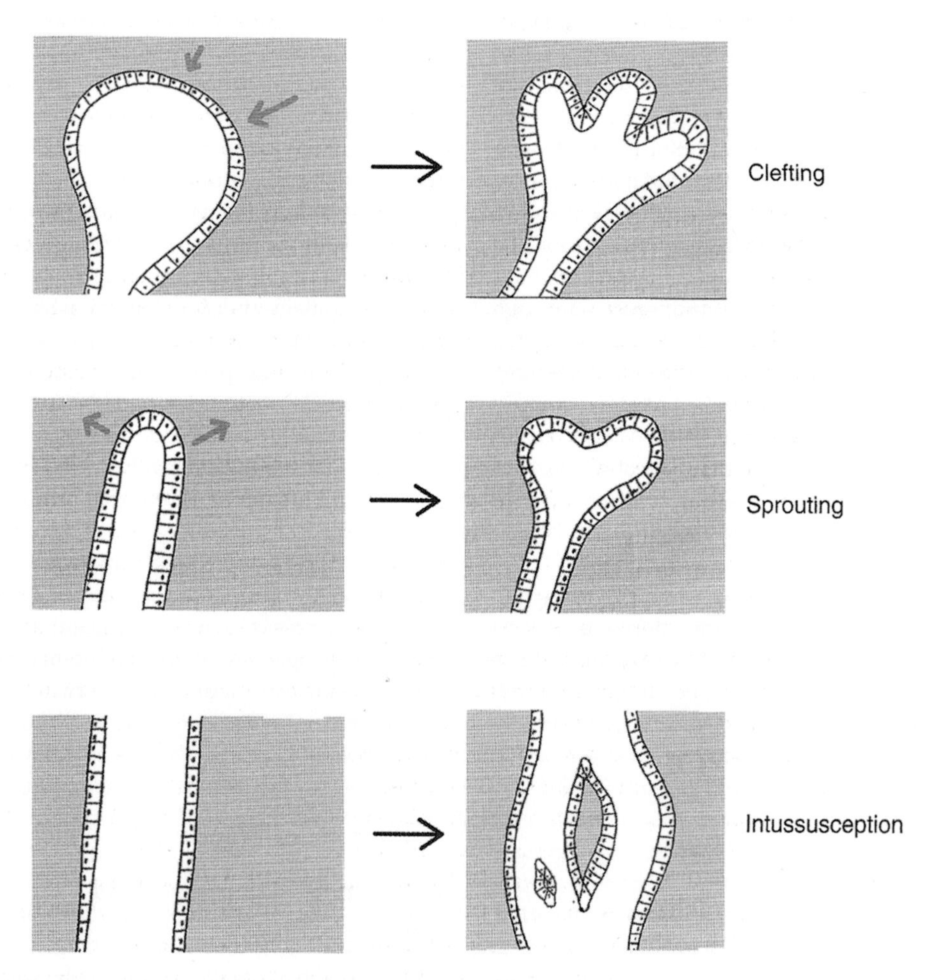

Figure 3.3 Three main types of branching morphogenesis: clefting, sprouting, and intussusception (Davies, 2005: 260). This further illustrates how the variable of abstraction operates in problem agendas (concrete exemplars of a particular kind of morphogenesis), as well as connectivity since the connections with other aspects of development is variable, such as sprouting with mammary gland formation or intussusception with vasculature formation. This material is reproduced with permission of Elsevier.

modes. Obviously, the phenomenon is confined to mammals. This gives us a starting sense of the level of abstraction involved in this research and the associated generality that frames it.

The research group undertook this study due to its connections with cancer and potential applications for humans ('our interest is in the function of the [neuropilins] in mammary gland biology and breast cancer'). Thus, connectivity between the problem agenda of growth and morphogenesis is involved. But the more critical value for the variable of connectivity is at a less abstract level. The neuropilins were first identified as necessary for axon guidance in the development of the nervous system and subsequently found to play a role in the development of blood vessels; both processes involve branching morphogenesis. 'The [neuropilins] have been studied extensively for their involvement in neural development and to some extent vascular development, but their potential contribution to the development of other tissues has not been investigated extensively' (Goel et al., 2011: 2969). Variety is germane in terms of this connectivity being dictated by the role neuropilins play across different instantiations of branching morphogenesis, and spatial composition bears on this connectivity because the morphogenetic transition from epithelium to mesenchyme underlies the metastatic capacities of cancer cells. 'An understanding of the potential role of [neuropilins] in epithelial development and homeostasis would provide mechanistic insight into their role in carcinomas' (Goel et al., 2011: 2969). These researchers had previously demonstrated that the Neuropilin-1 receptor played a role in breast cancer cell migration and survival; connectivity explains the focus on Neuropilin-2 (NRP2) in mammary gland morphogenesis. Additionally, temporality is involved, both at the level of mammary gland morphogenesis occurring later in ontogeny and its recurrent remodelling throughout the life cycle during offspring-bearing events.

The dimensions of history, heterogeneity, and hierarchy also illuminate the structure of this investigation. In terms of history, it is the prior relationships of connectivity among types of branching morphogenesis via the neuropilins that provide guiding rails for the present study. More broadly, the study of mammary gland morphogenesis is anchored in longstanding questions about breast cancer. In terms of heterogeneity, there are a variety of empirical question types at different levels of organization engaged by these investigators. The use of knockout mice enabled a study of the role of NRP2 in mammary gland development in the intact organ: 'the results indicate that NRP2 is required for proper branching morphogenesis of mouse mammary glands and invasion of mammary ducts into fat pads' (Goel et al., 2011: 2971). Mammary gland cell lines permitted a study of how molecular growth factors that bind neuropilins on the cell surface activate or inhibit branching morphogenesis: 'these results substantiate the importance of the microenvironment in response to growth factors and branching morphogenesis' (Goel et al., 2011: 2972). This led to a new form of connectivity because the growth factor tested, vascular endothelial growth factor (VEGF), was already known to be involved in the branching morphogenesis of the vascular system: 'a potential role for VEGF in outgrowth and branching of the mammary gland is novel' (Goel et al., 2011: 2974). In conjunction with other studies of internal cell signaling pathways relevant to mammary gland morphogenesis, the authors established that VEGF binds to NRP2 in order to activate focal adhesion kinase. In terms of hierarchy, the relationships among questions at different levels of abstraction is relevant to how these results bear on cancer. The role of NRP2 is different for different types of morphogenesis: 'NRP2 has a causal role in both [morphogenetic] processes, but they are distinct, and branching morphogenesis of the mammary gland does not appear to involve an EMT [epithelial-to-mesenchymal transition] . . . NRP2 has distinct roles in the EMT and branching morphogenesis' (Goel et al., 2011: 2975) The overall conclusion that ends the paper stresses the connection among questions at a higher level of abstraction, with an eye to generality across variety: 'In a broader context, our data support an emerging hypothesis that directional outgrowth and branching morphogenesis in a variety of tissues are influenced by signals that were identified initially for their role in growth cone guidance' (Goel et al., 2011: 2975).

The significant result derived from zooming in on this study is that we are able to observe in detail how concrete research in developmental biology is organized by erotetic epistemic units and that there is no need for a theory or family of related models to provide its organizational scaffolding. Using the four core problem agendas (differentiation, growth, morphogenesis, and pattern formation), five characterizing variables (abstraction, variety, connectivity, temporality, and spatial composition), and three dimensions (history, heterogeneity, and hierarchy), we can lay bare the reasoning of developmental biologists and comprehend how they are engaged in a science that is theory-informed but erotetically organized.

Conclusion

We now have a comprehensive picture of how problems organize research and guide inquiry in developmental biology. Although there remains much to say about developmental biology's conceptual commitments and evaluative standards, its erotetic organization explains why a central theory or group of theories is not needed to play this role, and why this absence is not a sign of immaturity. Although theoretical knowledge is routinely utilized, there is no 'theory of development' in operation, even in a more relaxed sense of theory structure (e.g. models or diagrams of multiple molecular mechanisms; Darden and Tabery 2005), and the expectation that there must be one is unjustified. The account of erotetic organization uses categories that fit the language of actual research practice, thereby meeting the criterion of epistemic accessibility (Love, 2012). Given that problem agendas are not individual interrogatives, we introduced five variables (abstraction, connectivity, temporality, variety, and spatial composition) and three dimensions (history, heterogeneity, and hierarchy) to characterize the ontology that underlies the organizing capacity of these erotetic units. While it is true that some of the specific labels for the variables and dimensions are not the provenance of practicing researchers, they represent distinctions in operation and discernible from the successful practice itself, evidenced in both textbook presentations and the case study of mammary gland morphogenesis. As noted, the epistemic accessibility criterion does not forbid the existence of theories in developmental biology but indicates that when found, they play very different epistemic roles (e.g. position information models).

In addition to capturing the reasoning structure in developmental biology, an account of erotetic organization for sciences is attractive more generally, especially since some other areas of molecular and cell biology make no appeal to a theory nor rely on groups of theories to coordinate and guide their investigative inquiries. While they often use theoretical knowledge, the structuring of these sciences frequently comes from their problems. And this problem agenda structure that governs the research practices of biologists is present in the context of textbooks, where it is used to communicate the science to students (including the next generation of researchers), as well as regular journal article presentations intended for other active investigators. In addition to identifying some of the defining characteristics and organizational potential of problem agendas, this account of erotetic organization has several other philosophical advantages, which can only be mentioned briefly.

First, it helps in comprehending the stability and transformation of problems in the history of science. Particular sub-questions can disappear, be transformed, or change their relations with other questions within or across problem agendas. Particular explanations or theories can come and go while problem architecture remains. These changes are consonant with researchers seeing themselves as working on the 'same' problem as their predecessors. Second, the presence of exploratory experimentation (Waters, 2007), which is not motivated by theory confirmation or hypothesis testing, becomes more salient within erotetic organization. Diverse but structured problems provide organizing principles to conduct this type of research without an explanatory theory. Third, this account yields a different perspective on scientific progress. Kuhn claimed that, 'the case for cumulative development of science's problems and standards is even harder to make than the case for cumulation of theories' (Kuhn, 1996: 108). But progress can be understood in terms of the refinement, articulation, and specification of questions within and across problem agendas instead of cumulativity. These properties are conducive to making research questions more tractable, which is another form of progress. Multiple problem agendas characterized by abstraction, connectivity, variety, temporality, and spatial composition, as well as the three dimensions of history, heterogeneity, and hierarchy, suggest a more complex space for measuring progress in developmental biology. That the 'problem of development' is more structured than 50 years ago constitutes a genuine aspect of progress, though one not coupled directly to an asymptotic approach to truth. Although theories are often discussed in similar terms—refinement, articulation, and specification—these activities also apply to other epistemic units as well, including problem agendas.

Finally, erotetic organization yields a novel perspective on incommensurability. Most discussions recognize two forms of incommensurability, semantic and methodological, and only with respect to competing theories (Soler et al., 2008). But because some scientific problems lie between disciplines and require diverse explanatory resources to address their heterogeneous questions, another form of incommensurability obtains: no common measure for coordinating who needs to explain what within the context of a complex scientific problem. Erotetic incommensurability, which is not a function of competing theories, is widespread in biological research and is why scientists often worry about the structure of problems and who shoulders the explanatory responsibility.

Each of these philosophical advantages require further articulation but the potential of erotetic organization to deal with diverse conceptual issues in analysing the sciences recommends it as an attractive trajectory for philosophy of science research. It responds to the observation that too much of a focus on theories has distorted our understanding of particular sciences (see, e.g. Waters, 2004) and pushes beyond the truism that problems motivate scientific inquiry by challenging the underlying presumption that sciences must be directed or organized by theories: 'The conscious task before the scientist is always the solution of a problem through the construction of a theory which solves the problem' (Popper, 2002: 301). The present account of erotetic organization in terms of developmental biology's four main problem agendas (pattern formation, morphogenesis, growth, and cell differentiation), characterized by differential weightings for the five variables of abstraction, variety, temporality, connectivity, and spatial composition, and refracted through the three dimensions of history, heterogeneity, and hierarchy, accounts for the structure of complex scientific problems beyond the individual interrogatives scrutinized in erotetic logics. This structure can provide the necessary organizational scaffolding and guidance for a science. Instead of a mark of immaturity or an encouragement to reconstruct the scientific discourse away from its participant's vocabulary, the value of erotetic organization for interpreting developmental biology emerges from a commitment to philosophical analyses exhibiting epistemic accessibility, which yields a robust way to embrace Giere's dictum with scientific theories in mind: 'Science need not be understood in these terms and, indeed, may be better understood in other terms.'

Acknowledgments

I would like to thank Sandro Minelli and Thomas Pradeu for the invitation to contribute and their editorial work on the volume. I am grateful for the feedback from audiences at the Institute for History and Philosophy of Science and Technology, University of Paris, France (September 2010) and the Central Division Meeting of the American Philosophical Association, Chicago, IL (February 2010) where early components of this material were presented. Lucie Laplane, Sandro Minelli, Beckett Sterner, and Thomas Pradeu (especially) provided insightful and critical comments on an earlier version of this chapter despite some sharp disagreements about its conclusions. These comments were instrumental in improving the final outcome. I also am grateful to James Overton for permission to reproduce Figure 3.1.

References

Beatty, J. (1995). The evolutionary contingency thesis. In J.G. Lennox and G. Wolters, eds., *Concepts, Theories, and Rationality in the Biological Sciences*. University of Pittsburgh Press, Pittsburgh, pp. 45–81.

Boyd, R. (1983). On the current status of scientific realism. In R. Boyd, P. Gasper, and J.D. Trout, eds., *The Philosophy of Science*. MIT Press, Cambridge, pp. 195–222.

Brigandt, I., and Love A.C. (2012). Conceptualizing evolutionary novelty: moving beyond definitional debates. *Journal of Experimental Zoology (Molecular and Developmental Evolution)*, **318**B, 417–27.

Bromberger, S. (1992). *On What We Know We Don't Know: Explanation, Theory, Linguistics, and How Questions Shape Them*. University of Chicago Press, Chicago and London.

Darden, L., and J.G. Tabery (2010), Molecular biology. In E.N. Zalta, ed. *Stanford Encyclopedia of Philosophy (Fall 2010 Edition)*: http://plato.stanford.edu/archives/fall2010/entries/molecular-biology/.

Darwin, C. (1959). *The Autobiography of Charles Darwin 1809–1882: With Original Omissions Restored, Edited with Appendix and Notes*, ed. N. Barlow. Harcourt, Brace and Company, New York.

Darwin, C. (1988), *Variation of Animals and Plants under Domestication*, Vol. 1, ed. P.H. Barrett, R.B. Freeman. New York University Press, New York.

Davidson, E.H. (2006). *The Regulatory Genome: Gene Regulatory Networks in Development and Evolution*. Academic Press, San Diego.

Davies, J.A. (2005). *Mechanisms of Morphogenesis: The Creation of Biological Form*. Elsevier Academic Press, San Diego.

Feyerabend, P.K. (1975). *Against Method*. New Left Books, London.

Fraser, S.E., and Harland R.M. (2000). The molecular metamorphosis of experimental embryology. *Cell*, **100**, 41–55.

Forgacs, G., and Newman, S.A. (2005), *Biological Physics of the Developing Embryo*. Cambridge University Press, New York.

Giere, R.N. (1999). *Science Without Laws*. University of Chicago Press, Chicago.

Gilbert, S.F. (1997). *Developmental Biology*, 5th ed. Sinauer, Sunderland.

Gilbert, S.F. (2001). Ecological developmental biology: developmental biology meets the real world. *Developmental Biology*, **233**, 1–12.

Gilbert, S.F. (2006). *Developmental Biology*, 8th ed. Sinauer, Sunderland.

Gilbert, S.F. (2010). *Developmental Biology.*, 9th ed. Sinauer, Sunderland.

Gilbert, S.F., and Epel D. (2009). *Ecological Developmental Biology: Integrating Epigenetics, Medicine, and Evolution*. Sinauer, Sunderland.

Goel, H.L., Bae D., Pursell, B., Gouvin, L.M., Lu, S., and Mercurio, A.M. (2011). Neuropilin-2 promotes branching morphogenesis in the mouse mammary gland. *Development*, **138**, 2969–76.

Griesemer, J.G. 2013. Formalization and the meaning of 'theory' in the inexact biological sciences. *Biological Theory*, **7**, 298–310.

Halvorson, H. (2012). What scientific theories could not be. *Philosophy of Science*, **79**, 183–206.

Hattiangadi, J.N. (1978). The structure of problems, part I. *Philosophy of the Social Sciences*, **8**, 345–65.

Hattiangadi, J.N. (1979). The structure of problems, part II. *Philosophy of the Social Sciences*, **9**, 49–76.

Hintikka, J. (2007). *Socratic Epistemology: Explorations of Knowledge-Seeking by Questioning*. Cambridge University Press, New York.

Jiang, T.X., Widelitz, R.B., Shen, W-M., et al. (2004). Integument pattern formation involves genetic and epigenetic controls: feather arrays simulated by digital hormone models. *International Journal of Developmental Biology*, **48**, 117–36.

Keller, E.F. (2002). *Making Sense of Life: Explaining Biological Development with Models, Metaphors, and Machines*. Harvard University Press, Cambridge.

Kleiner, S.A. (1988). Erotetic logic and scientific inquiry. *Synthese*, **74**, 19–46.

Kuhn, T.S. (1996). *The Structure of Scientific Revolutions*, 3rd ed. University of Chicago Press, Chicago and London.

Laudan, L. (1977). *Progress and its Problems: Towards a Theory of Scientific Growth*. University of California Press, Berkeley and Los Angeles.

Love, A.C. (2008a). Explaining evolutionary innovation and novelty: criteria of explanatory adequacy and epistemological prerequisites. *Philosophy of Science*, **75**, 874–86.

Love, A.C. (2008b). Explaining the ontogeny of form: philosophical issues. In A. Plutynski and S Sarkar, eds., *The Blackwell Companion to Philosophy of Biology*. Blackwell, Malden, pp. 223–47.

Love, A.C. (2010). Rethinking the structure of evolutionary theory for an extended synthesis. In M. Pigliucci and G.B. Müller, eds., *Evolution—The Extended Synthesis*. MIT Press, Cambridge, pp. 403–41.

Love, A.C. (2012). Formal and material theories in philosophy of science: a methodological interpretation. In H.W. de Regt, S. Okasha, and S. Hartmann, eds., *EPSA Philosophy of Science: Amsterdam 2009 (The European Philosophy of Science Association Proceedings, Vol. 1)*. Springer, Berlin, pp. 175–85.

Love, A.C. (2013). Theory is as theory does: Scientific practice and theory structure in biology. *Biological Theory*, **7**, 325–37.

Maienschein, J. (2005), Epigenesis and preformationism. In E.N. Zalta, ed., *Stanford Encyclopedia of Philosophy* (Winter 2005 edition): http://plato.stanford.edu/archives/win2005/entries/epigenesis/.

Minelli, A. (2011a). Animal development, an open-ended segment of life. *Biological Theory*, **6**, 4–15.

Minelli, A. (2011b). A principle of developmental inertia. In B. Halgrímsson and B.K Hall, eds., *Epigenetics: Link-*

ing Genotype and Phenotype in Development and Evolution. University of California Press, San Francisco, pp. 116–33.

Morrison, M. (2007). Where have all the theories gone? *Philosophy of Science*, **74**, 195–228.

Moss, L. (2002). *What Genes Can't Do*. MIT Press, Cambridge.

National Research Council. (2008). 'The role of theory in advancing 21st-century biology: catalyzing transformative research', Committee on Defining and Advancing the Conceptual Basis of Biological Sciences in the 21st Century. http://www.nap.edu/catalog.php?record_id=12026#toc.

Nickles, T. (1981). What is a problem that we may solve it? *Synthese*, **47**, 85–118.

Oppenheim, P., and Putnam H. (1958). The unity of science as a working hypothesis. In H. Feigl, M. Scriven, and G. Maxwell, eds., *Minnesota Studies in the Philosophy of Science*. University of Minnesota Press, Minneapolis, pp. 3–36.

Osbeck, L.M., Nersessian, N.J., Malone, K.R., and Newstetter, W.C. (2011). *Science as Psychology: Sense-Making and Identity in Science Practice*. Cambridge University Press, New York.

Overton, J.A. (2013). 'Explain' in scientific discourse. *Synthese*, **190**, 1383–405.

Popper, K. (2002). *Conjectures and Refutations: The Growth of Scientific Knowledge*, 3rd ed. Routledge, London and New York.

Robert, J.S. (2004). *Embryology, Epigenesis, and Evolution: Taking Development Seriously*. Cambridge University Press, New York.

Rohde, L.A., and Heisenberg C.P. (2007). Zebrafish gastrulation: cell movements, signals, and mechanisms. *International Review of Cytology*, **261**, 159–92.

Rosenberg, A. (1985). *The Structure of Biological Science*. Cambridge: Cambridge University Press.

Rosenberg, A. (2006). *Darwinian Reductionism: Or, How to Stop Worrying and Love Molecular Biology*. University of Chicago Press, Chicago.

Sive, H.L., Grainger, R.M., and Harland, R.M. (2007). *Xenopus laevis* Keller explants. *Cold Spring Harbor Protocols*, **6**, pdb.prot4749.

Slack, J.M.W. (2006). *Essential Developmental Biology*, 2nd ed. Blackwell Publishing, Malden.

Slack, J.M.W. (2013). *Essential Developmental Biology*, 3rd ed. Wiley-Blackwell, Chichester.

Soler, L., Sankey, H., and Hoyningen-Huene, P., eds. (2008). *Rethinking Scientific Change and Theory Comparison: Stabilities, Ruptures, Incommensurabilities?* Springer, Dordrecht.

Sommer, R.J. (2009). The future of evo-devo: model systems and evolutionary theory. *Nature Reviews Genetics*, **10**, 416–22.

Sunderland, M.E. (2010). Regeneration: Thomas Hunt Morgan's window into development. *Journal of the History of Biology*, **43**, 325–61.

Suppe, F., ed. (1977). *The Structure of Scientific Theories*, 2nd ed. University of Illinois Press, Urbana.

Thompson, P. (2007). Formalisations of evolutionary biology. In M. Matthen and C. Stephens, eds., *Philosophy of Biology*. Elsevier, Amsterdam, pp. 485–523.

Waters, C.K. (2004). What was classical genetics? *Studies in the History and Philosophy of Science*, **35**, 783–809.

Waters, C.K. (2007). The nature and context of exploratory experimentation. *History and Philosophy of the Life Sciences*, **29**, 275–84.

Whewell, W. (1989). *Theory of Scientific Method*. Hackett, Indianapolis.

Winther, R.G. (2006). Parts and theories in compositional biology. *Biology and Philosophy*, **21**, 471–99.

Wolpert, L. (2011). *Developmental Biology: A Very Short Introduction*. Oxford University Press, New York and Oxford.

Wolpert, L., Beddington, R., Brockes, J., Jessell, T., Lawrence, P.A., and Meyerowitz, E.M. (1998). *Principles of Development*. Oxford University Press, New York.

Wolpert, L., and Lewis, J.H. (1975). Towards a theory of development. *Federation Proceedings*, **34**, 14–20.

Wolpert, L., Tickle, C., Jessell, T., et al. (2010). *Principles of Development*, 4th ed. Oxford University Press, New York and Oxford.

Woodward, J. (2003). *Making Things Happen: A Theory of Causal Explanation*. Oxford University Press, New York.

CHAPTER 4

On the concept of mechanism in development

Johannes Jaeger and James Sharpe

Introduction: a mechanistic theory of development?

Multicellular organisms come in a tremendous diversity of shapes and sizes. This diversity is not random. There are evident regularities and recurring motifs such as spotted or striped patterns and branching or segmented structures. Some biological forms are clearly more likely to evolve than others. However, our understanding of the mechanisms that shape and constrain organismic form is incomplete. The non-random structure of morphological diversity not only depends on selective pressure and functional adaptation, or physical constraints on the system, although all of these factors are clearly important. It also depends on the complex internal organization of developmental processes (see, for example, Alberch, 1982; Arthur, 2004; Maynard Smith et al., 1985; Salazar-Ciudad, 2006a; Salazar-Ciudad & Riera-Marín, 2013; Wagner, 1988). Only if we understand the principles governing this internal organization will we be able to obtain a sound and deep understanding of the developmental and evolutionary dynamics of biological systems, and hence, the nature and origin of organismic form. This is one of the central challenges in modern biology.

The diversity of morphological traits—and that of the underlying developmental processes producing them—is constrained by the finite amount of dynamic behaviours that can be implemented by genetic, epigenetic, and cellular regulatory systems (Alberch, 1991; Goodwin, 1982; Jaeger & Crombach, 2012; Jaeger et al., 2012; Oster & Alberch, 1982; Thom, 1976). This implies that it should be possible to construct something like a 'periodic table' of developmental principles and processes, an idea first introduced by the Russian geneticist Nikolai Vavilov in the 1920s (Webster & Goodwin, 1996). The metaphor of a periodic table should not be taken too literally. Rather, it suggests the possibility that we can enumerate and classify the limited (or at least finite) number of possible basic developmental processes, and relate them to each other. For instance, limb development in vertebrates can be considered as the combination of various sub-processes: pattern (stripe) formation along the proximal–distal and anterior–posterior axes of the growing limb bud, combined with mechano-chemical changes in tissues that lead to the formation of cartilage and bones (see, for example, Oster & Alberch, 1982; Sheth et al., 2012). Other complex morphogenetic processes can also be understood in terms of similar basic constituents (see, for example, Newman, 2012; Newman & Bhat, 2009; Salazar-Ciudad, 2006b; Salazar-Ciudad et al., 2003). A rigorous classification scheme for these basic building blocks would constitute the foundation for a theory of development that characterizes and explains the regularities and recurring motifs of biological form.

This kind of theory must be stated in mechanistic terms (Machamer et al., 2000). In our view, it should be centred on the concept of a *developmental mechanism*.[1] Purely phenomenological classification schemes, such as those suggested by Davidson (1991), Davidson et al. (1995), Salazar-Ciudad

[1] Terms that are introduced in italics are defined in the glossary.

Towards a Theory of Development. Edited by Alessandro Minelli and Thomas Pradeu

(2006b, 2010), Salazar-Ciudad et al. (2003), and Wray (2000), among others, are valuable steps in the right direction, but do not go far enough, since they are based on correlations representing similarities between observables, instead of causal explanations of how the observed regularities arise.

Classification of organismic forms based on the study of underlying generative mechanisms has a long tradition. It has been attempted, notably, by rational taxonomists such as Étienne Geoffroy Saint-Hilaire, Richard Owen, and Johann Wolfgang von Goethe (reviewed in Webster & Goodwin, 1996). More recent examples include the engineering-inspired approach of D'Arcy Wentworth Thompson, Conrad Hal Waddington's theory of form, and the structuralist research programme proposed by Brian Goodwin (Goodwin, 1982, 1990, 1999; Thompson, 1917; Waddington, 1957, 1975; Webster & Goodwin, 1996). All of these attempts, however, were limited by a lack of empirical evidence and causal understanding of real developmental mechanisms. Today, systems biology is tackling the fundamental problem of biological form again, this time based on rigorous, quantitative experimental characterization of developmental processes. In our chapter, we explore a concept of 'developmental mechanism' which is suitable for this endeavour.

What is a mechanism?

At the most general level, our intention is that the term 'mechanism' should be a causal explanation of how something works.[2] In other words, it should be equally applicable to the question of how a car engine functions, as to how an embryo develops. In this respect, it is useful to contrast this usage of the term with how it is commonly used in physics.

The 'mechanism of gravity', for instance, has been debated for centuries and is still unresolved. In this case, alternative hypotheses propose fundamentally distinct constituents of the system: is gravity mediated by particles? Or by a force field? Or by the curvature of the space–time continuum? The debate centres on what the system is made of—what types of entities are involved.

This is very different from asking how a newly invented engine works. We know that the engine is constructed from a combination of physical components (gears, axles, pistons, belts, etc.), and what these components are made of. Moreover, we can safely assume that all engine parts are obeying Newton's laws of motion. In this context, the appropriate question is not what the engine is constructed of, but rather how it is organized—how the components have been put together. This question is as much about engineering as it is about physics.

The success of molecular biology over the last half-century has put us in a similar position for biological systems. Fifty years ago, we were closer to the gravity analogy: the mystery of multicellular development largely reflected our ignorance of the internal machinery by which cells could make complex and coordinated fate decisions. The central questions were: what is the control system made of? What kind of physical entity could possibly account for the wide range of complex regulatory behaviours we observe?

Since then, we have gained much insight into the chemical nature of biological systems. In particular, it has become abundantly clear that one of the important roles of genes is to implement molecular switches within a cell, as well as a means of intercellular communication. Regions along the DNA responsible for controlling the expression of a gene are activated or repressed by the presence of DNA-binding proteins in the cell (so-called transcription factors). The arrangement of these control sequences results in complex combinatorial control, which is a non-trivial function of the relevant proteins present in a given context. The protein products of many of these genes are transcription factors themselves, which then feed back into the regulatory functions of other genes with the potential of creating an arbitrarily complex internal control circuit (see, for example, Britten & Davidson, 1969; Davidson, 2006; Kauffman, 1993). Some of the genes in this circuit produce secreted proteins, which may diffuse

[2] Our notion of 'mechanism' (and our approach in general) is similar in spirit to the neo-mechanical philosophy of science presented in Machamer et al. (2000). We differ from that approach, however, by our explicit emphasis on dynamics and formulation of our conceptual framework in mathematical terms. We believe that this allows us to avoid some of the conceptual controversies concerning the temporal and spatially distributed nature of systems components that have recently arisen among neo-mechanical philosophers (see McManus, 2012, for a recent review).

through the extracellular matrix and regulate gene expression in neighbouring cells. Since many different secreted proteins exist (tens or hundreds for any given organism), this constitutes a powerful toolkit for communication between cells, thereby extending the concept of the internal regulatory circuit to tissue-level control (see Martinez Arias & Stewart, 2002, a textbook focussing on the fundamental importance of signalling networks in developmental genetics).

Today, gene regulatory and signalling-based networks represent a major part of the explanation of development (see also Morange, this volume), and they form the focus of this chapter. However, it is important to note that there are other essential components of developmental systems, which are increasingly being recognized but have been studied in much less detail. Apart from gene products (RNAs or proteins), and other chemical substances in the cell (such as ions and metabolites), other important factors include mechanical stresses, bioelectrical potentials, pH, or even extra-organismic triggers and factors that causally contribute to proper development (see, for example, Forgacs & Newman, 2005; Oyama, 2000; Oyama et al., 2001; Robert, 2004; Webster & Goodwin, 1996).

Our success in discovering many of the basic regulatory components—the building blocks, or molecular toolkit of biological systems—has transformed the study of developmental mechanisms. As with the example of physics vs engineering above, our current questions deal less with what the components of the system are, and more with how these components are put together into a functional whole. In other words, we are increasingly focussing again on *biological organization* rather than composition.

In addition to what we have discussed so far, it is important to distinguish our use of the term 'mechanism' from how it is often used in biology. Biologists commonly ask: what is the 'molecular mechanism' regulating gene X? By analogy with the design of an engine, this is equivalent to asking what causes the pistons to move. This is indeed a mechanistic question, but one which restricts itself to just a small part of the system. Ultimately, understanding how an explosion of petrol pushes the pistons does not explain the full picture of how the car drives around. Similarly, it is true that activation by the extracellular morphogen Sonic Hedgehog (Shh) is one of the molecular mechanisms explaining the regulation of the Gli3 transcription factor in the developing limb bud (Sasaki et al., 1999). Understanding the detailed and exact molecular nature of the regulation of specific factors is important, and constitutes a great achievement of modern molecular biology. However, it explains very little about how the intricate three-dimensional structure of a limb is generated. Here we are concerned with a larger-scale question—the principles of how tissues are shaped and structured. Again, understanding which individual component regulates which is useful but not enough. We need to put this insight into context. It is the question of how the system achieves interesting, controlled, reliable behaviours, which is at the heart of our concept of mechanism.

Developmental mechanisms: complexity and function

Biology displays a number of obvious differences from other natural sciences, but two particular features stand out as having an impact on our concept of mechanism: firstly the high complexity of biological processes and organisms (when compared to non-living systems), and secondly the deeply embedded concept of 'function'. Here, we will briefly discuss how these features impact the idea of a developmental mechanism, and in particular how they involve the concepts of teleology, hidden variables, and design principles.

The high *complexity* of a biological system can be appreciated from a few simple observations. It is well known from dynamical systems theory that feedback can lead to complex behaviours such as multi-stability and homeostasis (Hirsch et al., 2004; Strogatz, 2000). Since genes can regulate their own expression, this repertoire of interesting behaviours is available to gene regulatory networks. In addition, the regulation of gene expression is usually non-linear, which adds another level of complexity by introducing threshold effects and other drastic changes in state, which are hard (or even impossible) to predict. Finally, and most importantly, the existence of many hundreds of distinct regulatory genes—each capable of positively or negatively regulating any other gene in the system—allows for an almost arbitrary level of organizational diversity and complexity. It is easy to appreciate the astronomical number of possible *regulatory structures* (often called topologies[3] in the literature) for even just a handful of genes if we consider gene networks as directed graphs (Figure 4.1). As more genes are added, the number of possible topologies exhibits a combinatorial explosion. For example, one recent study calculated 39 possible distinct topologies for stripe-forming networks containing two genes, 9710 for three genes, and more than seven million different topologies for networks with four genes (Figure 4.1; Cotterell & Sharpe, 2010). Clearly, the situation appears formidably complicated to understand.

In which way may we hope a theory would help with this situation? What do we wish to understand? At first glance, we might want a theory to address the following question: for any given gene network, what will it do? What will its dynamic behaviour be? This we may call the physics approach. With this type of theory, we could define the equations and parameters of the system, along with the initial and boundary conditions, we then calculate its dynamics. Although the complexity of anything but the simplest networks is beyond the capabilities of current analytical approaches, numerical simulations can give us good estimates of what (reasonably small) networks are theoretically capable of.[4]

Being able to predict the *dynamical repertoire* of any hypothetical network is important and useful. However, this does not constitute the kind of theory we are interested in, for a number of reasons. The first problem arises from the absence of a concept of *function*. A straight mathematical relationship between each model (equations, parameters, plus initial and boundary conditions) and its dynamical repertoire contains no distinction between functional networks and dysfunctional ones—in other words, between successful networks and failed ones. It is simply a complete genotype-to-phenotype map, but with no concept of fitness. Instead our goal is not just a 'universal predictor' capable of explaining how any, and every, possible gene regulatory network works. In the context of biological systems, we need to focus on that subset of possible networks which achieves a given function.

This is where a form of teleology becomes relevant, and potentially problematic (see also Neander, 1991; Wright, 1973). The dynamics of randomly sampled gene regulatory networks are not 'wrong' because they contravene any physical law, and thus could not possibly exist. Instead, they are 'wrong' because they do not achieve a given biological function (Figure 4.1). This function depends on context and must be defined up front. It is completely independent of the particular networks capable of implementing it.

Such a goal-oriented definition is sometimes considered to sit uneasily with the purposeless laws of scientific causality. However, in essence, it simply represents one way to define the concept of biological function. Let us illustrate this with the following analogy: wings within a certain range of shapes allow a bird to fly. Outside this range, they will fail to achieve this function.[5] It is clearly unreasonable to make claims such as: birds evolved a certain wing shape 'in order to fly', or bird wings were 'designed for flight'. Nevertheless, it is equally clear that we can objectively identify those features which are shared between all successful wing shapes. If engineers were attempting to design such a system, they would claim to be seeking the *design principles* for functional flying wings. These principles simply define why some organizational features function correctly in a given context, and why others fail. In an equivalent manner, we can define the function of a gene regulatory network: it must produce a specific outcome (e.g. dynamical expression patterns

[3] The word 'topology' is used in two quite different ways in our argument. In the context of gene networks, it stands for the regulatory structure or organization of a network, i.e. the set of regulatory interactions (activation or repression) between genes. Further below, 'topology' is used in its much more abstract mathematical sense, describing the basic geometrical arrangement of features in a space, e.g. the number, relative placement, and connectedness of attractors and their basins in phase space. To avoid confusion, we will use 'regulatory structure' for the former and will reserve the use of 'topology' for the latter meaning.

[4] Note: confirming whether a model matches a given real biological system is a different and much harder question. But this is not the immediate concern of a theory of development.

[5] The fact that wings may serve purposes other than flying (e.g. penguins' flippers) is important for understanding wing evolution, but is not directly relevant for our more limited task of classifying mechanisms that achieve a predefined function (i.e. flight).

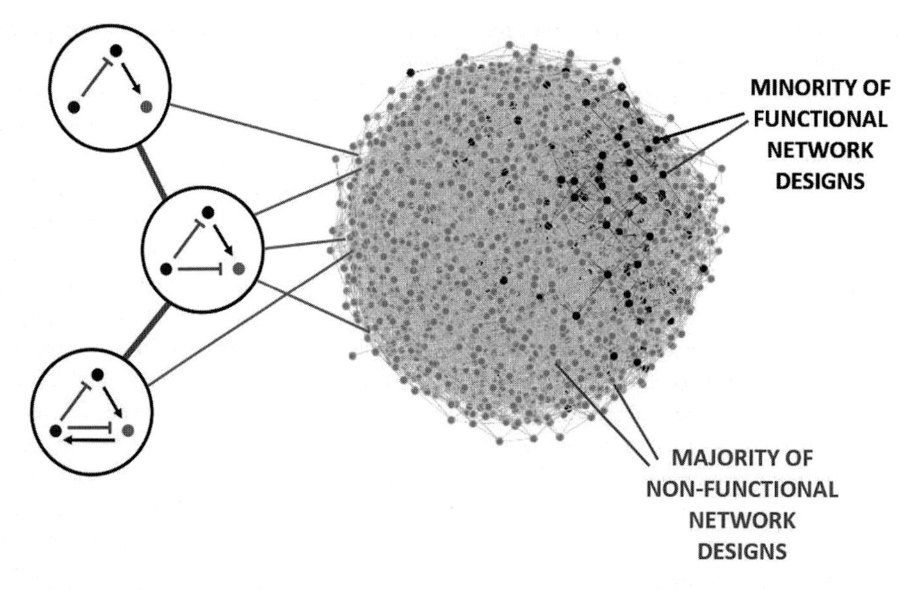

Figure 4.1 Graph showing the mutational relationships of all possible regulatory structures for three-gene networks. In this three-dimensional graph, each specific instance of the network is shown as a node. Directed graphs representing the regulatory structures of three of these networks are shown on the left. Pairs of nodes are directly connected to each other if they display only one regulatory change between them (i.e. the addition or removal of a regulatory link between genes, or a change in regulatory sign from activation to repression in one specific link). In this example, nearly 10 000 network structures are shown (Cotterell & Sharpe, 2010). Of these, just over 450 are capable of producing a predefined phenotype (within a given modelling context): the creation of a stripe of gene expression within a one-dimensional array of cells. It is thus possible to map out the distinction between functional (black) and non-functional (grey) network structures. The successful design principles may therefore be studied as an objective endeavour, independently of any hypothetical 'designer'.

that result in a stripe of gene expression) which corresponds to a given set of design principles. Given a clear definition of the successful behaviour, these design principles exist independently of the engineers' desire to build it. Design principles do not imply any hypothetical 'designer'. Or in other words: although the act of designing implies a purpose, the existence of design principles does not.[6]

If we manage to define the function of an organ or a gene regulatory network—and thus the range of possible successful designs—independently of how the outcome is actually achieved during development, then it follows by definition that some of the variables of the developmental process may be *hidden variables*. In other words, such hidden variables are not strictly necessary for the realization of the outcome (the selected phenotype). They will be crucial for the system to function, but are not directly assessed in determining the fitness of the system. To give a concrete example: the Shh morphogen may be essential for normal limb development, but the fitness of the limb for running, once it has been built, will not depend on the state of Shh per se. If an alternative set of genes could build an identical limb structure, it would be equally fit.

This leads to an intriguing realization: defining the function of a biological system does not fully determine the underlying developmental mechanism. In fact, this is an intrinsic aspect of distinguishing function from structure. It is precisely this distinction which allows that a whole collection of different generative processes may be compatible with the observed phenotype. In fact, such collections of possible mechanisms have been mapped and characterized for the specific functions of biochemical adaptation and stripe formation (Cotterell & Sharpe, 2010; Ma et al., 2009).

[6] This type of goal-orientedness—where the purpose lies in a function that has an evolutionary origin—has been termed *teleonomy* to distinguish it from teleology, where purpose usually implies a designer. The word 'teleonomy', however, remains controversial since it is not obvious that it represents a true difference to teleology, or simply describes the same phenomenon in different—more scientifically acceptable—terms (Neander, 1991).

This observation helps us to define more precisely what we desire from our theory of development. Given that many different processes may exist that achieve a specific function, how can we decide which ones among them genuinely represent different mechanisms? On one hand, many small changes to the regulatory structure of a system will not affect the causal explanations of how it works in any significant way. On the other, some small regulatory changes do alter how the system works—often unexpectedly and drastically—even if they may not alter the biological function. A mechanistic theory of development should give us a rational basis for deciding whether two different networks are minor variants of each other, or whether they represent fundamentally different ways of producing functionally equivalent output.

Developmental mechanisms: two illustrative examples

So far, our discussion has been very general and abstract. To make it more concrete, we introduce two examples which illustrate how alternative mechanisms can explain the same phenotype. In the first case, the two mechanisms differ in a manner which is straightforward to grasp intuitively. The second example then highlights how the question may be approached in a less obvious case, when the mechanistic similarities or differences between alternative designs are much less clear.

The classical example: digit patterning in vertebrates

A classic case of alternative mechanistic explanations for the same phenotypic outcome can be found in the literature on digit formation in the vertebrate limb bud. It provides an excellent illustration of the concept of developmental mechanism because it is easy to see how fundamentally different the two proposed regulatory processes are.

The earliest signs of an emerging digit pattern during embryogenesis are two-dimensional stripes of gene expression by early markers of chondrogenesis (Healy et al., 1999). Since these stripes are more or less parallel to each other along the proximal–distal axis of the limb, the pattern can be simplified to a one-dimensional arrangement of expression peaks and troughs along the anterior–posterior axis (Figure 4.2).

One hypothesis to explain this pattern is based on the concept of positional information—first introduced by Lewis Wolpert (1968, 1969). This mechanism assumes that every cell in the developing limb field acquires a unique positional value—in other words, a unique coordinate in space. In early proposals, positional information was encoded by a spatial gradient of a diffusible molecule (a morphogen), which was synthesized asymmetrically, and released from the posterior side of the field (Saunders & Gasseling, 1968; Wolpert, 1969). Cells within the limb bud would thus experience a different concentration of morphogen at every position along the anterior–posterior axis. In a paper from 1975—intriguingly titled 'Towards a theory of development' (Wolpert & Lewis, 1975)—Wolpert proposed that for each cell 'this positional information is then interpreted in terms of muscle, cartilage, loose connective tissue, and so on'. Since the pattern of digits displays the specific feature of being repetitive, with the same tissue-type specified at different positions, Wolpert clarifies that 'different [positional] specifications can lead to the same mode of differentiation' (Wolpert & Lewis, 1975). Herein lies precisely the key feature of this hypothetical mechanism: although all digits involve the same modes of cellular differentiation (i.e. the same cell types), each digit is specified by a different positional value. Indeed, to specify an alternating pattern of two different cell fates would require at least six different threshold responses to produce three digits (Figure 4.2). This has clear consequences for the behaviour of the system to perturbations. If the limb field is enlarged (e.g. by stimulating higher growth rates) the number of digits should remain the same, and—assuming the gradient scales with the diameter of the field—the resulting digits should be thicker spanning the entire enlarged tissue (Wolpert, 1969). Essentially, the whole digital pattern will have scaled up (Figure 4.2).

The alternative mechanism proposed to explain the same pattern is a Turing reaction–diffusion system (Newman & Frisch, 1979; Turing, 1952). Unlike the global, hierarchical, and feed-forward mechanism of positional information, this is a local, self-organizing, and feedback-driven system. Two (or more) molecular species interact such that they spontaneously produce a pattern of stripes. Pattern

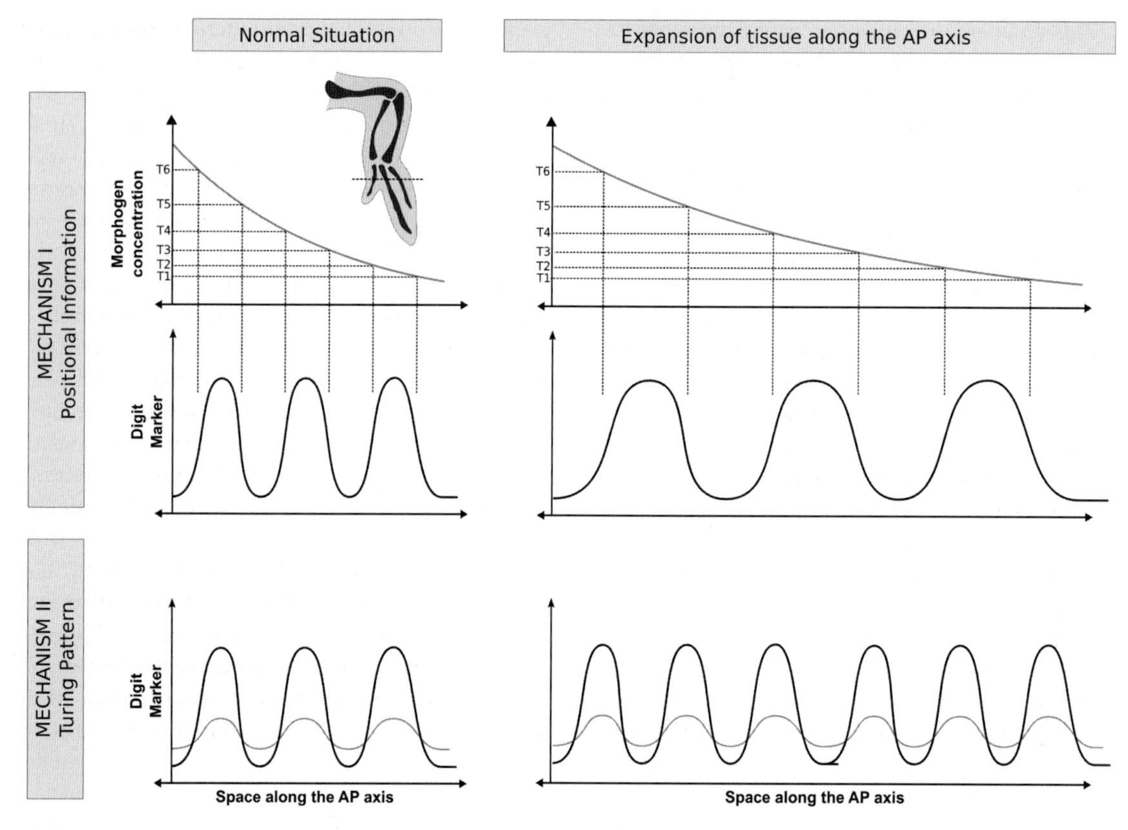

Figure 4.2 How two fundamentally different mechanisms can produce the same spatial pattern. Graphs show striped pre-patterns of gene expression underlying the formation of digits along the anterior–posterior (AP) axis of the vertebrate limb (see line in the inset crossing a schematic drawing of a chick limb). Mechanism I is based on positional information driven by a morphogen gradient (Wolpert, 1969). For a digit-marker gene to respond by making a three-digit pattern (i.e. a three-stripe pattern), at least six independent threshold responses are necessary (T1 to T6). By contrast, mechanism II—a Turing system (Turing, 1952)—is able to spontaneously create a periodic pattern without reference to a global coordinate system—purely through local self-organization (grey line indicates the second component of the system, which is itself distributed in a stripe-like manner, rather than a global gradient). Although the two mechanisms can produce the same pattern (left-hand panels), their response to a manipulation of the tissue reveals how different they are. If the tissue is stretched, mechanism I cannot produce extra digits (although it may produce wider digits if the gradient scales with the tissue). However, mechanism II can only make extra digits—it cannot make wider ones. In the case of vertebrate digit patterning, there is increasing evidence that mechanism II applies (see main text for details).

width is not controlled by independent factors for each stripe. Instead the system has an inherent periodic wavelength, which is controlled by a few global parameters. Although it can produce the same pattern as the morphogen gradient, this mechanism is fundamentally different. For example, if the size of the tissue is increased, the number of digits will also increase to fill the available space (Figure 4.2). Recent evidence shows that this is indeed what happens in polydactylous mutants, which have an enlarged limb bud (Sheth et al., 2012). This behaviour is impossible for a mechanism based on positional information, as each digit is specified individually in that context. In this case, it is not only straightforward to intuitively recognize the distinct nature of the two proposed mechanisms, but it is also easy to come up with specific experiments to distinguish the two possibilities.

The value of discussing this example is to highlight two central, but contrasting points: on one hand, it is evident that the two proposed mechanisms, which can achieve the same pattern, operate in a fundamentally different way. This is especially clear when considering the manipulative experiments described above. On the other hand, they could both be built out of the same basic building blocks—collections of genes and signalling molecules wired together into regulatory networks. The distinct dynamics arise from the fact that these same types of components are wired together in fundamentally different ways.

The test case: more than one way to make a stripe

Both pattern-forming processes in the previous example—positional information and Turing patterning—are capable of producing the same outcome but operate on such radically distinct principles that we readily accept that they represent two different mechanisms. However, often it is not so straightforward to discern the alternatives. Can our conceptual framework be generalized to such situations? In other words, can we find rigorous criteria that allow us to find qualitative differences between more similar mechanisms?

A first step towards addressing this question is to search for examples where many different designs appear to achieve the same function. For this purpose, it is useful to reduce the problem to one of the simplest examples that still displays such mechanistic diversity. Our test case is the collection of simple three-gene networks shown in Figure 4.1. These circuits are able to produce an expression stripe—two non-expressing territories surrounding a domain of high expression—in a simple, one-dimensional tissue (Cotterell & Sharpe, 2010). In each case, the tissue is composed of a row of cells, and each cell contains the same gene regulatory network. Each gene product can regulate any of the three genes of the network (in a manner defined by its regulatory structure), and gene products can diffuse into the neighbouring cell, thereby influencing the regulation of genes in a cell-non-autonomous, spatial manner.

By an exhaustive search across all possible network topologies, Cotterell and Sharpe (2010) were able to find that a handful of distinct regulatory structures were able to produce the desired stripe phenotype (Figure 4.3). Analysis of these networks revealed that they appeared to operate in

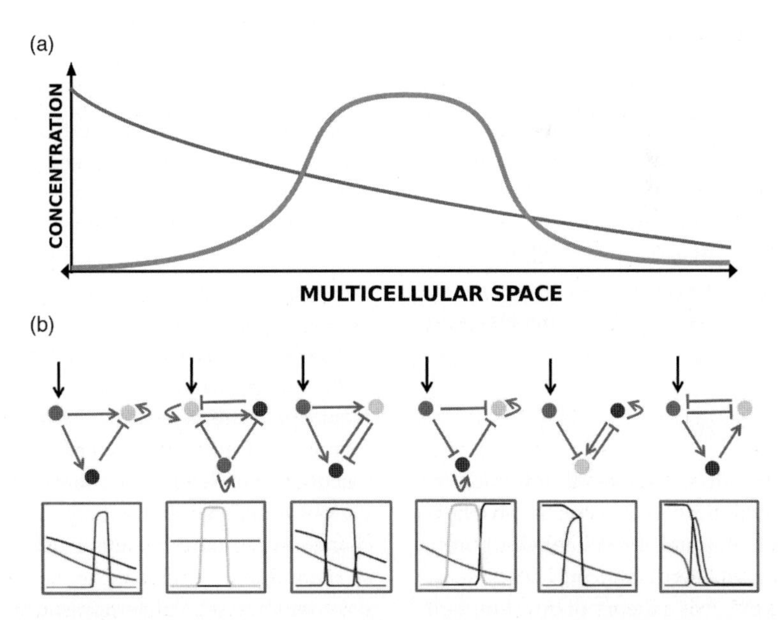

Figure 4.3 Exploring three-gene gene regulatory networks capable of forming a single-stripe pattern. (a) The function to be achieved was defined as: forming a stripe of expression bounded by zones of low expression on both sides (thick line). The network can achieve this pattern by interpreting a morphogen gradient (thin line) which provides positional information. (b) Six different network designs were found that all successfully create a stripe of expression of the output gene. Graphs showing the final expression patterns for all three genes are shown below each network design. In all cases the output gene displays a stripe near the centre, but the other genes are variable, as they correspond to hidden variables. Their diversity of patterns corresponds to the diversity of the underlying mechanisms.

fundamentally different ways. Some were quite intuitive, while others were difficult to interpret. One approach to explore whether these are significantly distinct from each other is to consider all possible regulatory structures as part of one larger *parameter space*. In this context, every design (i.e. every parameter combination) is represented as a single point in a vast multidimensional parameter space whose dimensions are defined by the type and strength of each regulatory interaction. Exploration of this parameter space reveals that different regulatory structures form separate clusters. In other words, one could not smoothly change one structure into another without passing through non-functional parameter regions (i.e. networks which would not preserve the desired phenotype).

In summary, the study described above revealed that many different network designs can achieve the same phenotype. The differences between them can be attributed to 'hidden variables'—those genes which contribute to the pattern, but do not constitute the stripe-pattern readout. Numerical analysis demonstrates that they operate in distinguishable ways, and therefore can be classified as distinct mechanisms. However, as discussed above, one goal for our theory is to provide some degree of rigour in deciding whether two networks represent genuinely different mechanisms, or are just variants on the same theme. In other words, we require not only numerical and statistical evidence on the separation of mechanisms, but want to provide a causal, mechanistic explanation. We will now proceed to discuss which type of theoretical framework may be productive in putting this task onto a firm footing.

What kind of theory?

So far, we have introduced a general definition for what developmental mechanisms are, and we have discussed why it is important to establish rigorous criteria by which to distinguish them. These are important first steps towards a theory of development. However, they are not sufficient. As mentioned in 'Introduction: a mechanistic theory of development?', a theory of development should enable us to systematically classify developmental processes. To achieve this, we must learn how different mechanisms relate to each other. This requires an appropriate conceptual framework.

Several such frameworks from engineering, computer science, physics, and mathematics have been suggested as basis for a theory of development. Information theory, for example, is widely used to study causal flow in regulatory networks (see Tkačik & Walczak, 2011, for a recent review). This approach is rooted in the concept of Shannon entropy, which measures the amount of information that can be transmitted through a communication channel. It treats development as such a channel, conveying information from egg to adult, and has been successfully applied to the study of pattern formation, especially aspects of positional information and patterning precision (e.g. Bialek & Setayeshgar, 2005; Dubuis et al., 2013; Gregor et al., 2007; Tkačik et al., 2008).

However, information theoretic approaches are limited in a very fundamental way by the fact that Shannon information cannot increase while being transmitted. This implies that the overall amount of information within an organism must remain constant (or decrease) from egg to adult, which amounts to a kind of implicit preformationism. Such a notion is difficult to maintain considering the evident increase in complexity during embryogenesis (see Apter & Wolpert, 1965; Oyama, 2000; Thom, 1976; Wolpert & Lewis, 1975, for detailed critical discussions).

As an alternative, it has been suggested that the theory of computation is the appropriate framework to study development (see, for example, Apter & Wolpert, 1965; Britten & Davidson, 1969; Istrail et al., 2007; Wolpert & Lewis, 1975). In this view, the genome of the egg is seen as encoding a genetic or developmental programme, which determines embryological trajectories. Cells are treated as automata (finite-state machines) whose internal state is represented by various functional levels of regulatory factors, which change in response to signals from surrounding tissues and the external environment. Development is governed by a set of deterministic instructions implementing programme flow and decision points (e.g. through the equivalent of 'if . . . then' statements). In this way, the genome instructs the organism on how to grow while avoiding the problem of preformation. The nucleus does not contain an image of the adult

variables does not change. In our two-dimensional example in Figure 4.4b, nullclines are lines whose trajectories are either horizontal (if factor Y does not change) or vertical (if X remains the same). Fixed points occur where nullclines intersect, and we can determine whether a fixed point is an attractor or a saddle based on the direction of the trajectories crossing each nullcline in the vicinity of the intersection (Figure 4.4b).

In addition to state variables, which represent the components (nodes) of a regulatory network (e.g. genes and their associated gene product concentrations), we also need to define *system parameters*, which determine the initial and boundary conditions (e.g. morphogen concentration in the stripe-forming example above) as well as the interactions among regulatory factors (the edges or vertices of the network defining activation or repression between genes). If we consider a time-window in development that is short enough, these parameters usually stay constant for the duration of the process under study, such that we may consider many of these parameters to be defined by the genotype of the organism. Systems whose structure does not explicitly depend on time are called *autonomous systems* (Strogatz, 2000). Their phase portrait remains constant over time.

The dynamical repertoire of an autonomous system is defined entirely by the fixed geometrical arrangement of attractor basins in its phase space. If multiple attractors are present, the system will converge to the steady state associated with the basin in which its initial conditions lie. For instance, in Figure 4.4a, the system will converge to the high-X attractor as long as we start with a slightly higher concentration of X compared to Y. As those conditions change (e.g. due to altered maternal contributions or environmental conditions) the system can exhibit a transition to a different attractor state. In our example of the toggle switch, it can suddenly change to a high-Y state once we cross the separatrix (Figure 4.4a,b). Which different attractors can be reached, depends entirely on which basins are in the neighbourhood of the original initial conditions in phase space (Figure 4.4c). Some basins are inaccessible since they lay far away from the current state. Others may not be connected at all—meaning they do not share a common separatrix—which implies that transitions between them cannot be direct, but must go through an intermediate dynamical regime (Figure 4.4c). This implies that some phenotypic transitions are impossible to achieve directly, and can only occur through an intermediate phenotypic state. In this way, phase space geometry determines the dynamical behaviours the system can exhibit under any given circumstances. If this geometry is known, it is possible to derive probabilities for specific dynamical behaviours and their developmental or evolutionary transitions.

A complication of the situation is introduced by the fact that most biological systems of sufficient spatio-temporal extent and complexity have parameters and initial or boundary conditions that are time-variable. For example, maternal gradients play an important role in the early development of animals such as the fruit fly *Drosophila melanogaster* (reviewed in Jaeger, 2011; St Johnston & Nüsslein-Volhard, 1992). These gradients are implemented as initial and boundary conditions in models of pattern formation, and therefore represent parameters of the system. Since their concentrations change over time (sometimes drastically, as in the case of maternal Hunchback (Hb) or Caudal (Cad) protein; see Surkova et al., 2008), the parameters of the system also change during the duration of development. Such systems where parameter values depend on time are called *non-autonomous systems*.

Under some circumstances, non-autonomy can be ignored, or the system can be simplified to make it autonomous (see, for example, Manu et al., 2009b). In most cases, however, considering the non-autonomous nature of the system is essential for our understanding. This is because such systems show a much richer range of behaviours than autonomous ones, due to the fact that the structure of their phase portrait changes over time. The model of vulval induction in *C. elegans* by Corson and Siggia (2012), for example, implements the effect of inductive signals through a change in system parameters that represent the regulatory landscape of the cell. In this case, cell fate decisions are affected due to the transient distortion of the phase portrait during developmental periods when cells are competent to respond to the induction.

Non-autonomous systems not only move along trajectories through phase space, but also move through parameter space over time. As a consequence, parameter changes lead to a constant rearrangement of the geometry of phase space. Again, the geometry of phase space is crucial for understanding the dynamical repertoire of the system. Many of these phase space changes will be of no effect, since they affect regions of the phase portrait that lie far from the current state of the system. A new potential cell state may be enabled by a signal, but may not be reachable depending on the current state of competence in the responding cell. Moreover, parameter changes often affect the structure of phase space in a smooth and continuous manner, which does not alter the trajectory of the system, and its associated attractor, in any significant manner. In these cases, the local structure of phase space does not undergo any drastic changes, and the system will still converge to a similar state as before. This property of phase space underlies most of the robustness of development (Thom, 1976; A. Wagner, 2005), since most dynamical systems (even very complex ones) have a limited number of basins of attraction, and most changes in parameter values will not lead to qualitative changes in the output, or phenotype of the system.

In some less frequent cases, however, there will be more significant effects. New attractor regimes can come into being, or old ones can be annihilated or change properties (e.g. from stable to unstable) through a process called *bifurcation* (Hirsch et al., 2004; Kuznetsov, 2004; Strogatz, 2000). In the case of the toggle switch model, introducing auto-activation leads to a bifurcation that creates a third stable state in which low levels of both factor X and Y coexist. Once the current state of the system crosses a separatrix as its underlying phase space changes, sudden and drastic transitions occur (Jaeger & Crombach, 2012; Jaeger et al., 2012). These rare but important events are called *critical transitions*.[9] They occur in developmental and morphogenetic processes, which break symmetry or drastically alter the geometry of a trait in a relatively short period of developmental time. Examples of such symmetry-breaking processes occur in Turing systems,[10] where increasingly complex striped or spotted patterns can be created from uniform initial conditions if system parameters such as diffusion rates or the spatial range of patterning change over time due to the growth of the tissue (reviewed in Meinhardt, 1982; Murray, 2002).

Interestingly, not only attractors and their basins, but also transitions driven by bifurcation and attractor switches can exhibit robustness if they retain their basic geometric features over a large range of parameter values (i.e. genotypes). To study such phenomena, it is useful to combine phase and parameter axes into a higher-dimensional *configuration space* (Thom, 1976). In configuration space, some axes are defined by state variables, while others represent system parameters. Assuming there is only one state variable, and two control parameters, equilibrium states can be drawn as a surface in a three-dimensional configuration space. In this example, configuration space is subdivided into a horizontal parameter plane and an associated phase space above it (representing the single state variable of the system). We show an example of such a diagram, the cusp surface, in Figure 4.4d. In this representation, bifurcations show as creases, folds, and other discontinuities in the equilibrium surface. Assuming rapid convergence of the system to its steady state, its trajectory through developmental time can be traced along this surface. The system moves smoothly as parameters change through time until it reaches a discontinuity, where it will suddenly change its position, even if parameter values are only altered by small amounts (Figure 4.4d). This marks a dramatic shift to a new steady state, which can represent a symmetry-breaking event (such as those occurring in the growing Turing systems described above), or any other sudden morphogenetic or pattern-forming transition. Analogous to the case of autonomous systems, the nature of the process, its dynamical transitions, and its response to variation, will be determined by the geometrical arrangement of equilibrium surfaces and bifurcation events in configuration space.

[9] They have also been named *catastrophes* (Thom, 1976), a term which is not commonly used anymore and has so many colloquial connotations that we prefer to use 'critical transitions'.

[10] Note that Turing systems need not be non-autonomous to create patterns. Here we just review a particular non-autonomous instance of this type of mechanism that has been used to explain the formation of increasingly complex patterns in growing tissues (see Murray, 2002, for detailed discussion).

Geometrical similarities imply similarities in dynamical behaviour, and hence functional output of a regulatory network.

Towards a geometrical theory of developmental mechanisms

The geometrical picture of phase space or configuration space that we have drawn above introduces a powerful way to establish equivalence or differences between developmental mechanisms. Equivalent developmental mechanisms are governed by geometrical features of phase or configuration space that are *topologically equivalent* (Thom, 1976). *Topological equivalence* means that mechanisms share the same arrangement and connectedness of attractors and their bifurcations, leading to families or classes of trajectories and critical transitions across configuration space which share the same qualitative properties. Such a system will also be robust, since its qualitative behaviour will be insensitive to most small changes in parameter values.

This notion of robustness and equivalence through topological equivalence can be made mathematically rigorous (Kuznetsov, 2004; Thom, 1976). Here, we chose a more pragmatic approach since it is challenging (and often impossible) to precisely measure or infer the topology of configuration space. Instead, we can identify certain properties of a pattern-forming system that are characteristic of the underlying geometry of their attractors and bifurcations. Let us illustrate this with a few examples.

Figure 4.5 shows two different network structures capable of creating the single-stripe phenotype (Munteanu et al., 2014). They both contain three genes, and in both cases there is regulatory input by a global morphogen gradient (Figure 4.5a). The output gene (shown in black in Figure 4.5a) forms the stripe. Structurally, the networks have much in common—they both contain a combination of positive and negative links, and they both have a positive feedback loop between the two downstream regulatory targets. However, in the first case (Figure 4.5b) the positive feedback is constructed from mutual inhibition, while in the second case (Figure 4.5c), the two target genes mutually activate each other. Given all these similarities (including their ability to produce the same phenotype) on what basis shall we decide if they are different in some fundamental way, or just variants on the same theme?

Munteanu and colleagues have used phase space analysis to examine exactly how each design achieves the bi-phasic response of the output gene to differing levels of morphogen inputs (Munteanu et al., 2014). The system can be usefully analysed with respect to the two target genes only (X and Y in Figure 4.5b,c), because these networks have no feedback on to the morphogen. Since there is no diffusion of gene products between neighbouring cells, the possible dynamics of each cell can be represented by a two-dimensional phase portrait. For cells at different positions within the tissue, the morphogen exhibits higher or lower values. This directly determines the arrangement of features within the phase portrait—especially the shapes and distributions of nullclines.

In the case of mutual inhibition (Figure 4.5b), as we shift from cells with high morphogen input to cells with low input, one of the nullclines gradually moves downwards, while the other moves to the left. As mentioned above, it is the intersections of the nullclines which define the attractors present in each cell. In the cells experiencing high morphogen levels (Figure 4.5b, left) only one attractor exists, with low levels of gene product X, so the cell moves directly into this state. However, as we move to medium morphogen levels (Figure 4.5b, middle) one of the nullclines moves downwards and intersects the other, causing a bifurcation which results in the creation of a new stable attractor. With the correct system parameters the newly formed separatrix is positioned such that the initial condition of the cell (at the origin of the graph) falls within the basin of attraction of the new steady state, so the cell moves into the high-X state. Further reduction of the morphogen levels leads to further shifting of the nullclines such that at a certain point the X nullcline 'escapes' one of its intersections with the Y nullcline, and again only one stable attractor remains (Figure 4.5b, right). This mutual-inhibition mechanism can therefore be summarized as displaying two bifurcations in phase space along the length of the tissue: one for the boundary switching gene X from low to high expression, and one for the boundary switching X from high to low. Both cases involve a change in the number of attractors.

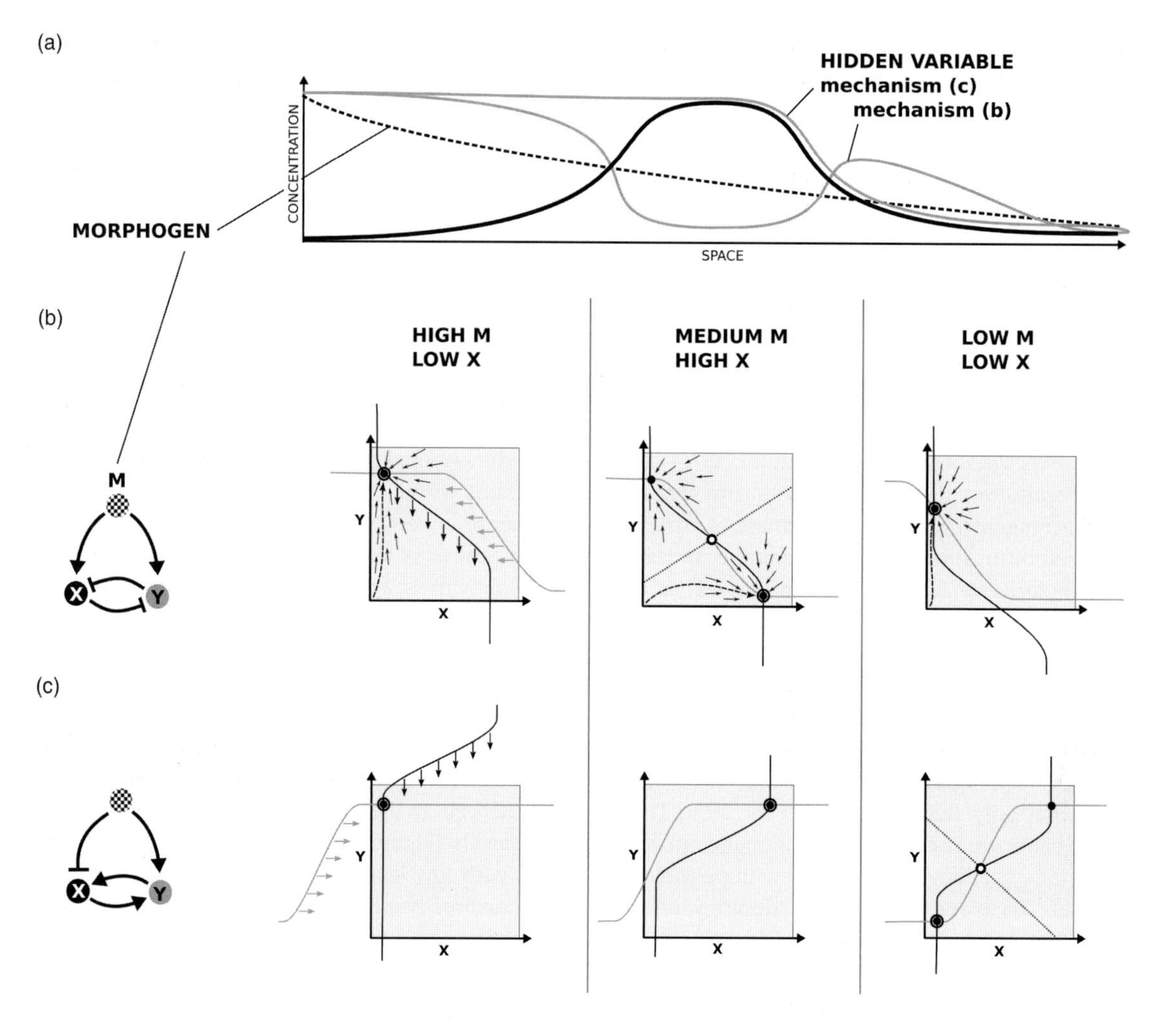

Figure 4.5 How phase space may be used to differentiate mechanisms from each other. (a) Expression patterns: we consider here a mechanism which interprets a monotonic morphogen gradient (M, dashed line) to produce a single stripe of expression of a downstream gene (X, black line). The other gene (Y, grey lines, representing the two alternative circuits described below) is not constrained and therefore constitutes a hidden variable of the system. Here we illustrate two distinct network mechanisms which both produce a stripe. Analysis of their phase spaces highlights how distinct their dynamics are. (b) A mutual-inhibition mechanism. At high morphogen levels (left side of graph in panel (a)) the nullclines of X and Y only intersect at one position, which has a low level of X. As M decreases, the nullclines shift across the phase space (rows of bold black and grey arrows). At medium levels of M, the realignment of nullclines leads to three intersections, which creates a new stable attractor at a high level of X. At lower levels of M, only one intersection exists, which again corresponds to low levels of X. (c) A mutual-activation mechanism displays distinct dynamical behaviour. The shift from high to intermediate levels of X does not involve creation of a new attractor, but rather a shift of the existing one. Only at low levels of M, a new attractor is created, which leads to low levels of X. The fundamental differences in the geometry of phase portraits between (b) and (c) provides an appropriate basis to define them as representing different developmental mechanisms.

If we now provide a similar type of analysis for how the mutual-activation network works (Figure 4.5c), we find a qualitatively different sequence of phase space changes. The first position (at high morphogen concentration) is superficially similar to the mutual-inhibition case above: only one attractor exists, with high levels of Y and low levels of X (Figure 4.5c, left). Again the nullclines shift as the morphogen level decreases (although not in the same direction as before). However, before a bifurcation can occur, the shape of the X nullcline (which shifts to the left) pushes the attractor upwards along the X-axis, so that the attractor comes to lie at a new position with high levels of X (Figure 4.5c, middle). The first boundary of the output stripe is therefore defined by attractor movement—not attractor switching or selection as seen above for mutual inhibition. Further reduction in morphogen levels (Figure 4.5c, right) causes the Y nullcline to shift upwards sufficiently to cause it to intersect with the X nullcline, which creates a new attractor by bifurcation. This new attractor is low in X and high in Y.

The important conclusion we can draw from the graphical characterization of phase space geometry is that the two alternative network topologies lead to qualitatively different modes of dynamics—a conclusion which is hard (if not impossible) to reach without this approach. For this reason, our example provides an illustrative suggestion that a theory of development based on dynamical systems theory will satisfy the major objective we have defined above: it will provide rigorous and measurable criteria by which to determine if two different ways of achieving a given phenotype genuinely represent different developmental mechanisms, rather than just minor variations on a theme.

Conclusions

In this chapter, we propose a definition of 'developmental mechanism' in terms of dynamical systems theory, which could serve as the conceptual foundation of a future theory of development. A developmental mechanism is determined by the particular geometry of its phase portrait—the specific arrangement and connectedness of its attractors, saddle points, basins—and its bifurcations. In other words, equivalent mechanisms share an equivalent topology of their phase and configuration spaces.

This definition suggests a rational classification scheme for developmental mechanisms based on the similarity of their respective phase space geometries. Such a classification scheme is not intended to contradict, but rather to complement phylogenetic classification. In contrast to phylogenetic reconstruction, we are not focussed on the elucidation of common descent. This is clearly illustrated by the fact that convergently evolved developmental processes will be deemed closely related to each other in our classification scheme. But this is not the point: we are not trying to reconstruct what actually happened during the evolutionary history of any particular organism. Rather, rational classification of developmental mechanisms will yield insights into the design principles of pattern-forming regulatory networks performing a defined function. These design principles, in turn, enable us to understand which sort of mechanism can happen at all, and which type of pattern-forming process is more likely to arise under a given selection regime for a specific function. In other words, our classification scheme provides a map of the possible (and probable): a substrate on which evolution by natural selection can work.

Our argument is based on ideas and concepts which go back to the rational taxonomists of the 18th and 19th Century, and to more recent proponents of a theory of biological form, such as D'Arcy Thompson, Waddington, Thom, Goodwin, Oster, and Alberch (see 'Introduction: a mechanistic theory of development?'). We review and extend these ideas in the light of recent progress in the fields of dynamical systems theory and systems biology. On one hand, increased computing power and conceptual advances in dealing with autonomous and non-autonomous complex non-linear systems provide the foundation for systematic numerical exploration of regulatory network dynamics. On the other, quantitative and high-throughput data sets on gene expression now allow us to test highly detailed and intricate models using suitable experimental evidence. This enables us, for the first time in history, to combine experimental and computational approaches for empirical and rigorous quantitative studies of phase and configuration space geometry and its associated dynamical behaviours.

We have illustrated our outline for a possible theory of development with a number of examples. We are aware that these examples remain preliminary. They are only intended to provide a proof of principle for the proposed approach at this point. In the future, we must move beyond the current state of affairs, and apply our conceptual framework much more generally and widely. Such a research programme suggests new lines of investigation, and opens unique new opportunities for understanding development, which go beyond earlier attempts at arguing for a dynamical systems view of development and evolution. It is now possible to measure or infer features of phase and configuration space—such as attractors and their associated basins and bifurcations—in a growing number of real developmental systems (Jaeger & Crombach, 2012). Such empirical investigations—combined with systematic mapping of pattern-generating processes for specific functions (as in Cotterell & Sharpe, 2010)—provide a powerful entry point for an exploration of what we could call the natural history of phase and configuration space: the systematic quantitative study and rational classification of types of developmental mechanisms. Only such systematic classification, based on data from many systems, will reveal whether there is indeed a periodic table for pattern formation, whether we can achieve a general theory of development, whether there are regularities, rules, or even laws to be discovered. Although we cannot guarantee the success of such an endeavour, one thing is clear: finding and characterizing such laws would fill a central gap in our current understanding of development and evolution, a gap which constitutes one of the major unresolved issues in modern biology. This prospect, in our opinion, more than justifies the challenges, risks, and the considerable effort required to carry out such a pioneering and ambitious new line of research.

Acknowledgments

This book chapter would not have been possible without numerous stimulating discussions and the encouragement we have received from many of our lab members and colleagues over the past few years. Nick Monk provided essential contributions to the initial development of our conceptual framework for dynamical systems. Many of the examples we use to illustrate our concepts are

Box 4.1 Glossary

attractor	A point or bounded subregion in phase space to which trajectories converge over time.
(non-)autonomous system	An autonomous system is a dynamical system whose *parameters* do not change over time. A non-autonomous system is a dynamical system whose parameters do change over time.
basin of attraction	The region containing all those trajectories in phase space which converge to a specific *attractor*.
bifurcation	A discontinuous and drastic change in the behaviour of a dynamical system—involving the creation or annihilation of *attractors*—caused by a smooth change in *parameter* values.
catastrophe	See *critical transition.*
complexity	Difficult to define rigorously and precisely. Here, we use this term to describe the large number of possible combinatorial arrangements of regulators and the highly non-linear nature of interactions in gene regulatory networks (and biological regulatory systems in general).
configuration space	An abstract space whose dimensions are defined by both *state variables* and *system parameters* of the system. Each point in configuration space represents a specific state of the system, given a specific set of parameters.
critical transition	A discontinuous and drastic change in the behaviour of a dynamical system that is either due to a *bifurcation*, or an *attractor* switch, caused by the changing geometry of *phase space* in response to changes in the *parameters* of the system.
design principle	Specific conditions concerning regulatory interactions that must be met for a *developmental mechanism* to perform its given *function*. Implies no 'designer' of any kind.
developmental mechanism	Regulatory process which shows a specific geometrical arrangement and connectedness of attractors and their bifurcations. Equivalent developmental mechanisms share *topologically equivalent* geometries of phase or configuration space.
dynamical repertoire	The set of outputs (dynamical behaviours, or phenotypes) a system is able to produce. Defined by the *phase portrait* (*attractors*, *saddles*, and *basins*) of the system.

continued

Box 4.1 *Continued*

flow	The flow of a dynamical system consists of the totality of its trajectories. It defines the *dynamical repertoire* of the system.
function	We define function as a specific task that must be performed by a biological system to make it suitable for survival and reproduction in its specific context.
genotype space	An abstract space whose points are the set of all possible genotypes of a system.
genotype-phenotype map	A function that maps genotypes to their associated phenotypes. This function is degenerate due to *phenotypic plasticity* and *robustness* of biological systems.
hidden variable	A *state variable* of the system that is necessary to produce a given *function*, but does not directly contribute to it. Different *mechanisms* performing the same function will contain differences in the expression and regulation of their hidden variables.
morphospace	*Phenotypic space* as applied to morphological characters only.
nullcline	The set of points in phase space at which one of the *state variables* of the system does not change.
organization (biological)	See *structure*.
parameter space	An abstract space whose dimensions are defined by the *parameters* of the system. In the context of gene regulatory networks, parameter space is often considered equivalent to *genotype space*.
phase space	An abstract space whose dimensions are defined by the *state variables* of the system. Contains *attractors* and their associated *basins*.
phase portrait	Consists of the specific arrangement of *attractors*, *saddles*, and *basins* in *phase space* that characterizes a particular dynamical system (e.g. a gene network with a specific *regulatory structure*).
phenotype space	An abstract space whose points are the set of all possible phenotypes of a system. In our context, a phenotype can consist of patterns of gene expression.
phenotypic plasticity	A property of biological systems where the same genotype produces different phenotypes, depending on environmental conditions.
robustness	A property of biological systems where different genotypes produce the same phenotype despite molecular, genetic, or environmental fluctuations or perturbations.
saddle point	A fixed point of a dynamical system located on a *separatrix*. Saddle points are unstable, since only trajectories within the separatrix converge towards the saddle, while all others *diverge* to their respective attractors.
separatrix	The boundary between two basins of attraction. Can contain *saddle points*.
state space	See *phase space*.
state variable	Variable representing components of a dynamical system that change over short timescales (e.g. gene product concentrations).
structure (regulatory)	The set of interactions between components of a regulatory network, including information on strength and kind—activating or repressing. Usually depicted by a directed graph.
structural stability	See *robustness*.
system parameters	Determine the strength and type of interactions between *state variables*, as well as other system properties. Do not change over time (autonomous systems), or at a slower rate/less frequently than *state variables* (non-autonomous systems).
topological equivalence	Abstract geometrical concept which implies an equivalence of the arrangement of features in some space (in our case: *attractors* and their *basins* in *phase space*). Equivalent geometries are homeomorphic since they can be transformed into each other without disrupting any contact points (*separatrices*) between basins.
teleonomy	A form of teleology where the purpose inherent in a system (the biological *function* in our context) has an evolutionary origin, rather than being created by a designer.

based on modelling work by Andreea Munteanu, James Cotterell, Luciano Marcon, and David Irons. Furthermore, we would like to acknowledge Stas Shvartsman's continuing encouragement and constructive criticism of our approach. Kolja Becker, Bàrbara Negre, Berta Verd, Hilde Janssens and other members of our groups provided suggestions that helped to clarify the argument. Finally, we acknowledge critical feedback from Giuseppe Fusco, as well as the editors of this book, Alessandro Minelli and Thomas Pradeu, which greatly helped to improve the structure, rigour, and scope of our chapter.

References

Alberch, P. (1982). Developmental constraints in evolutionary processes. In J.T. Bonner, ed., *Evolution and Development*. Springer, Heidelberg, pp. 313–32.

Alberch, P. (1991). From genes to phenotype: dynamical systems and evolvability. *Genetica*, **84**, 5–11.

Alon, U. (2006). *An Introduction to Systems Biology: Design Principles of Biological Circuits*. Chapman and Hall/CRC, London.

Apter, M.J., and Wolpert, L. (1965). Cybernetics and development I. Information theory. *Journal of Theoretical Biology*, **8**, 244–57.

Arthur, W. (2004). *Biased Embryos and Evolution*. Cambridge University Press, Cambridge.

Bialek, W., and Setayeshgar, S. (2005). Physical limits to biochemical signaling. *Proceedings of the National Academy of Sciences USA*, **102**, 10040–5.

Bolouri, H. (2008). *Computational Modeling of Gene Regulatory Networks—A Primer*. Imperial College Press, London.

Britten, R.J., and Davidson, E.H. (1969). Gene regulation for higher cells: a theory. *Science*, **165**, 349–57.

Corson, F., and Siggia, E.D. (2012). Geometry, epistasis, and developmental patterning. *Proceedings of the National Academy of Sciences USA*, **109**, 5568–75.

Cotterell, J., and Sharpe, J. (2010). An atlas of gene regulatory networks reveals multiple three-gene mechanisms for interpreting morphogen gradients. *Molecular Systems Biology*, **6**, 425.

Davidson, E.H. (1991). Spatial mechanisms of gene regulation in metazoan embryos. *Development*, **113**, 1–26.

Davidson, E.H. (2006). *The Regulatory Genome: Gene Regulatory Networks in Development and Evolution*, Academic Press, Burlington, MA.

Davidson, E.H., Peterson, K.J., and Cameron, R.A. (1995). Origin of bilaterian body plans: evolution of developmental regulatory mechanisms. *Science*, **270**, 1319–25.

Dawkins, R. (1989). The evolution of evolvability. In C. Langton, ed., *Artificial Life*. Addison-Wesley, Redwood City, pp. 201–20.

De Jong, H. (2002). Modeling and simulation of genetic regulatory systems: a literature review. *Journal of Computational Biology*, **9**, 67–103.

Dubuis, J.O., Samanta, R., and Gregor, T. (2013). Accurate measurements of dynamics and reproducibility in small genetic networks. *Molecular Systems Biology*, **9**, 639.

Fakhouri, W.D., Ay, A., Sayal, R., et al. (2010). Deciphering a transcriptional regulatory code: modeling short-range repression in the *Drosophila* embryo. *Molecular Systems Biology*, **6**, 341.

Félix, M.-A. (2012). Evolution in developmental phenotype space. *Current Opinion in Genetics and Development*, **22**, 593–99.

Forgacs, G., and Newman, S.A. (2005). *Biological Physics of the Developing Embryo*. Cambridge University Press, Cambridge.

François, P., and Siggia, E.D. (2012). Phenotypic models of evolution and development: geometry as destiny. *Current Opinion in Genetics and Development*, **22**, 627–33.

Fusco G. (2001). How many processes are responsible for phenotypic evolution? *Evolution and Development*, **3**, 279–86.

Fusco G., and Minelli A. (2010). Phenotypic plasticity in development and evolution: facts and concepts. *Philosophical Transactions of the Royal Society of London B*, **365**, 547–56.

Gilbert, S.F., and Epel, D. (2009). *Ecological Developmental Biology: Integrating Epigenetics, Medicine, and Evolution*. Sinauer Associates, Cambridge.

Goodwin, B.C. (1982). Development and evolution. *Journal of Theoretical Biology*, **97**, 43–55.

Goodwin, B.C. (1990). Structuralism in biology. *Science Progress*, **74**, 227–44.

Goodwin, B.C. (1999). D'Arcy Thompson and the problem of biological form. In A.J. Chaplain, G.D. Sing and J.C. McLachlan, eds., *On Growth and Form: Spatio-temporal Pattern Formation in Biology*. Wiley and Sons, Chichester, pp. 395–402.

Gregor, T., Tank, D.W., Wieschaus, E.F., et al. (2007). Probing the limits to positional information. *Cell*, **130**, 153–64.

Grieneisen, V., Xu, J., Marée, A.F.M., et al. (2007). Auxin transport is sufficient to generate a maximum and gradient guiding root growth. *Nature*, **449**, 1008–13.

Healy, C., Uwanogho, D., and Sharpe, P.T. (1999). Regulation and role of Sox9 in cartilage formation. *Develomental Dynamics*, **215**, 69–78.

Hirsch, M.W., Smale, S., and Devaney, R.L. (2004). *Differential Equations, Dynamical Systems, and an Introduction to Chaos*. Elsevier, Amsterdam.

Hoyos, E., Kim, K., Milloz, J., et al. (2011). Quantitative variation in autocrine signaling and pathway crosstalk in the *Caenorhabditis* vulval network. *Current Biology*, **21**, 527–38.

Istrail, S., Ben-Tabou De-Leon, S., and Davidson, E.H. (2007). The regulatory genome and the computer. *Developmental Biology*, **310**, 187–95.

Jaeger, J. (2011). The gap gene network. *Cellular and Molecular Life Sciences*, **68**, 243–74.

Jaeger, J., and Crombach, A. (2012). Life's attractors: understanding developmental systems through reverse engineering and *in silico* evolution. In O. Soyer, ed., *Evolutionary Systems Biology*. Springer, Berlin, pp. 93–119.

Jaeger, J., Irons, D., and Monk, N. (2012). The inheritance of process: a dynamical systems approach. *Journal of Experimental Zoology: Molecular and Developmental Evolution*, **318B**, 591–612.

Jaeger, J., Surkova, S., Blagov, M., et al. (2004). Dynamic control of positional information in the early *Drosophila* embryo. *Nature*, **430**, 368–71.

Janssens, H., Hou, S., Jaeger, J., et al. (2006). Quantitative and predictive model of transcriptional control of the *Drosophila melanogaster even-skipped* gene. *Nature Genetics*, **38**, 1159–65.

Karlebach, G., and Shamir, R. (2008). Modelling and analysis of gene regulatory networks. *Nature Reviews Genetics*, **9**, 770–80.

Kauffman, S.A. (1993). *The Origins of Order: Self Organization and Selection in Evolution*. Oxford University Press, Oxford.

Kazemian, M., Blatti, C., Richards, A., et al. (2010). Quantitative analysis of the *Drosophila* segmentation regulatory network using pattern generating potentials. *PLoS Biology*, **8**, e1000456.

Kim, A.-R., Martinez, C., Ionides, J., et al. (2013). Rearrangements of 2.5 kilobases of noncoding DANN from the *Drosophila even-skipped* locus define predictive rules of genomic *cis*-regulatory logic. *PLoS Genetics*, **9**, e1003243.

Klipp, E., Liebermeister, W., Wierling, C., et al. (2009). *Systems Biology: A Textbook*. Wiley, Weinheim.

Kuchen, E.E., Fox, S., Barbier de Reuille, P., et al. (2012). Generation of leaf shape through early patterns of growth and tissue polarity. *Science*, **335**, 1092–6.

Kuznetsov, Y. (2004). *Elements of Applied Bifurcation Theory*. Springer, New York.

Ma, W., Trusina, A., El-Samad, H., et al. (2009). Defining network topologies that can achieve biochemical adaptation. *Cell*, **138**, 760–73.

Machamer, P., Darden, L., and Craver, C.F. (2000). Thinking about mechanisms. *Philosophy of Science*, **67**, 1–25.

Mangan, S., and Alon, U. (2003). Structure and function of the feed-forward loop network motif. *Proceedings of the National Academy of Sciences USA*, **100**, 11980–85.

Manu, Surkova, S., Spirov, A.V., et al. (2009a). Canalization of gene expression in the *Drosophila* blastoderm by gap gene cross regulation. *PLoS Biology*, **7**, e1000049.

Manu, Surkova, S., Spirov, A.V., et al. (2009b). Canalization of gene expression and domain shifts in the *Drosophila* blastoderm by dynamical attractors. *PLoS Computational Biology*, **5**, e1000303.

Martinez Arias, A., and Stewart, A. (2002). *Molecular Principles of Animal Development*. Oxford University Press, Oxford.

Maynard Smith, J., Burian, R., Kauffman, S., et al. (1985). Developmental constraints and evolution. *Quarterly Review of Biology*, **60**, 265–87.

McGhee, G. (2007). *The Geometry of Evolution: Adaptive Landscapes and Theoretical Morphospaces*. Cambridge University Press, Cambridge.

McManus, F. (2012). Development and mechanistic explanation. *Studies in History and Philosophy of Biological and Biomedical Sciences*, **43**, 532–41.

Meinhardt, H. (1982). *Models of Biological Pattern Formation*, London, Academic Press.

Munteanu, A., Cotterell, J., Solé, R.V., Sharpe, J. (2014). Positive feedback discriminates between alternative designs of morphogen interpretation in the case of the single-stripe pattern. *Scientific Reports. In Press.*

Murray, J.D. (2002). *Mathematical Biology*, Berlin, Springer.

Neander, K. (1991). The teleological notion of 'function'. *Australasian Journal of Philosophy*, **69**, 454–68.

Newman, S.A. (2012). Physico-genetic determinants in the evolution of development. *Science*, **338**, 217–19.

Newman, S.A., and Bhat, R. (2009). Dynamical patterning modules: a 'pattern language' for development and evolution of multicellular form. *International Journal of Developmental Biology*, **53**, 693–705.

Newman, S.A., and Frisch, H.L. (1979). Dynamics of skeletal pattern formation in developing chick limb. *Science*, **205**, 662–8.

Oster, G., and Alberch, P. (1982). Evolution and bifurcation of developmental programs. *Evolution*, **36**, 444–59.

Oyama, S. (2000). *The Ontogeny of Information: Developmental Systems and Evolution*, 2nd ed. Duke University Press, Durham, NC.

Oyama, S., Griffiths, P.E., and Gray, R.D., eds. (2001). *Cycles of Contingency: Developmental Systems and Evolution*. MIT Press, Cambridge.

Pigliucci, M. (2008). Is evolvability evolvable? *Nature Reviews Genetics*, **9**, 75–82.

Pigliucci, M. (2010). Genotype-phenotype mapping and the end of the 'genes as blueprint' metaphor. *Philosophical Transactions of the Royal Society of London B*, **365**, 557–66.

Robert, J.S. (2004). *Embryology, Epigenesis, and Evolution: Taking Development Seriously*. Cambridge University Press, Cambridge.

Salazar-Ciudad, I. (2006a). Developmental constraints vs. variational properties: how pattern formation can help to understand evolution and development. *Journal of Experimental Zoology: Molecular and Developmental Evolution*, **306B**, 107–25.

Salazar-Ciudad, I. (2006b). On the origins of morphological disparity and its diverse developmental bases. *BioEssays*, **28**, 1112–22.

Salazar-Ciudad, I. (2010). Morphological evolution and embryonic developmental diversity in metazoa. *Development*, **137**, 531–39.

Salazar-Ciudad, I., and Jernvall, J. (2010). A computational model of teeth and the developmental origins of morphological variation. *Nature*, **464**, 583–6.

Salazar-Ciudad, I., Jernvall, J., and Newman, S.A. (2003). Mechanisms of pattern formation in development and evolution. *Development*, **130**, 2027–37.

Salazar-Ciudad, I., and Riera-Marín, M. (2013). Adaptive dynamics under development-based genotype-phenotype maps. *Nature*, **497**, 361–4.

Sasaki, H., Nishizaki, Y., Hui, C.-C., et al. (1999). Regulation of Gli2 and Gli3 activities by an amino-terminal repression domain: implication of Gli2 and Gli3 as primary mediators of Shh signaling. *Development*, **126**, 3915–24.

Saunders, J., and Gasseling, M. (1968). Ectodermal-mesenchymal interactions in the origin of limb symmetry. In R. Fleischmajer and R.E. Billingham, eds., *Epithelial-Mesenchymal Interactions*. Williams and Wilkins, Baltimore, pp. 78–97.

Saunders, P.T. (1980). *An Introduction to Catastrophe Theory*. Cambridge University Press, Cambridge.

Schlichting, C.D., and Pigliucci, M. (1998). *Phenotypic Evolution: A Reaction Norm Perspective*. Sinauer Associates, Sunderland.

Sheth, R., Marcon, L., Bastida, M.F., et al. (2012). *Hox* genes regulate digit patterning by controlling the wavelength of a Turing-type mechanism. *Science*, **338**, 1476–80.

Siegal, M. L., Promislow, D.E.L., and Bergman, A. (2007). Functional and evolutionary inference in gene networks: does topology matter? *Genetica*, **129**, 83–103.

St Johnston, D., and Nüsslein-Volhard, C. (1992). The origin of pattern and polarity in the *Drosophila* embryo. *Cell*, **68**, 201–19.

Strogatz, S.H. (2000). *Nonlinear Dynamics and Chaos: With Applications to Physics, Biology, Chemistry and Engineering*. Perseus Books, New York.

Surkova, S., Kosman, D., Kozlov, K., et al. (2008). Characterization of the *Drosophila* segment determination morphome. *Developmental Biology*, **313**, 844–62.

Thom, R. (1976). *Structural Stability and Morphogenesis*. Benjamin, Reading.

Thomas, R., and D'Ari, R. (1990). *Biological Feedback*. CRC Press, Boca Raton.

Thompson, D.W. (1917). *On Growth and Form*. Cambridge University Press, Cambridge.

Tkačik, G., Callan, C.G., and Bialek, W. (2008). Information flow and optimization in transcriptional regulation. *Proceedings of the National Academy of Sciences USA*, **105**, 12265–70.

Tkačik, G., and Walczak, A.M. (2011). Information transmission in genetic regulatory networks: a review. *Journal of Physics: Condensed Matter*, **23**, 153102.

Turing, A.M. (1952). The chemical basis of morphogenesis. *Philosophical Transactions of the Royal Society of LondonB*, **237**, 37–72.

Waddington, C.H. (1957). *The Strategy of the Genes*. Allen & Unwin, London.

Waddington, C.H. (1975). *The Evolution of an Evolutionist*. Cornell University Press, Ithaca.

Wagner, A. (2005). *Robustness and Evolvability in Living Systems*. Princeton University Press, Princeton.

Wagner, A. (2011). *The Origins of Evolutionary Innovations: A Theory of Transformative Change in Living Systems*. Oxford University Press, Oxford.

Wagner, G.P. (1988). The significance of developmental constraints for phenotypic evolution by natural selection. In G. De Jong, ed., *Population Genetics and Evolution*. Springer, Berlin, pp. 222–9.

Webster, G., and Goodwin, B.C. (1996). *Form and Transformation: Generative and Relational Principles in Biology*. Cambridge University Press, Cambridge.

West-Eberhard, M.J. (2003). *Developmental Plasticity and Evolution*. Oxford University Press, Oxford.

Wolpert, L. (1968). The French Flag problem: a contribution to the discussion on pattern development and regulation. In C.H. Waddington, ed., *Towards a Theoretical Biology*, Vol 2. Edinburgh University Press, Edinburgh, pp. 125–33.

Wolpert, L. (1969). Positional information and the spatial pattern of cellular differentiation. *Journal of Theoretical Biology*, **25**, 1–47.

Wolpert, L., and Lewis, J.H. (1975). Towards a theory of development. *Federation Proceedings*, **34**, 14–20.

Wray, G.A. (2000). The evolution of embryonic patterning mechanisms in animals. *Seminars in Cell and Developmental Biology*, **11**, 385–93.

Wright, L. (1973). Functions. *The Philosophical Review*, **82**, 139–68.

CHAPTER 5

The epistemological resilience of the concept of morphogenetic field

Davide Vecchi and Isaac Hernández

Dirty boots in metaphysical swamps

The concept of morphogenetic field (hereafter MF) was the central theoretical concept of the *Gestaltungsgesetze* research programme of experimental embryology in the first part of the 20th century. It was introduced by Boveri (1910) and Gurwitsch (1910) in order to account for the phenomenon of coordination of cellular activities across a tissue. Starting from 1922 the latter called them *Embryonalen Felder* (i.e. embryonic fields). Harrison (1918), Weiss (1926), Needham (1931), Huxley and de Beer (1934), and Spemann (1938), among others, defended versions of the concept (Gilbert et al., 1996). The analogy between the concept of morphogenetic *field* (referring to the putative biological property of groups of cells in developmental contexts) and the concept of electromagnetic *field* (referring to the epistemically accessible physical properties of groups of electrically charged physical particles) was probably meant to give prima facie respectability to the former. Mangold's and Spemann's (1924) transplantation experiments are a typical example of the kind of research propelled by *Gestaltungsgesetze*. Spemann (1935) labelled the MF *Organisationfeld* (i.e. organizer). The organizer was thought to be analogous to a physical 'force' external to the cells that makes them differentiate and assume the individual characters appropriate within the whole. This determining 'force' is not within the cell but is rather a holistic property of the transplanted tissue. So, despite its acknowledged physical nature, the MF concept challenged the prevailing mechanistic and preformationist explanations of the phenomena of morphogenesis. The concept of MF was ultimately related to organicist (Haraway, 1976) and even vitalistic metaphysical positions (Hamburger, 1999).

Gilbert et al. (1996) argued that the emergence of developmental genetics is to be considered the main cause of the demise of the MF concept, for the reason that the gene and the MF were perceived as being concepts in competition. The difficult relationships between *Gestaltungsgesetze* and genetics meant, as Lawrence and Levine (2006: R236) recently argued, that 'embryology drifted off into metaphysical swamps, while genetics explored the dry savannahs of statistics'. The central theoretical concept of *Gestaltungsgesetze* remained indefinite. In order to explicate it, not very rigorous metaphorical language was often used (e.g. analogies with Newtonian forces or electromagnetic fields). Furthermore, the concept was used in different ways by different scientists in order to refer to different aspects of complex causal processes. Embryonic fields, organizers and morphogenetic gradients all referred to phenomena of developmental regulation. They were all postulated on empirical grounds in order to conceptualize specific morphogenetic dynamics. They all possessed a physical substrate and a biochemical nature that remained to be investigated and that was largely experimentally inaccessible. They were generally considered as areas of biological information that was not genetic in origin (Gilbert et al., 1996). They were, in brief, concepts referring to material and physical entities (or processes) that contained part of the organizational

Towards a Theory of Development. Edited by Alessandro Minelli and Thomas Pradeu

information necessary for the regulation of morphogenesis. But, apart from this general definition, it is difficult to extrapolate the commonalities between the varieties of field concepts used. Haraway (1976: 55) notes that developmental biologists disagreed on the issue concerning the 'material status' of the MF. While for Harrison, Needham, and Weiss the substratum was clearly biochemical, Gurwitsch was not so clear, while Spemann considered a purely biochemical definition insufficient (Hamburger, 1999). Such conceptual idiosyncrasy contributed to the general perception that the concept of MF was elusive and mystical and that the *Gestaltungsgesetze* research programme was degenerative (in Lakatos' sense). The MF concept was gradually seen as the embryologists' phlogiston.

Lawrence and Levine (2006) have also argued that developmental biology must supersede, in the light of the available molecular understanding, the historical dichotomies that have plagued its practice. They specifically name four philosophical disputes: preformation vs epigenesis; vitalism vs mechanicism; thinking of embryonic cells as preprogrammed vs thinking of them as totipotent; and mosaicism vs regulation. They argue that the conceptual distinctions at the basis of these dichotomies are today indefensible trivializations. We agree with Lawrence and Levine that, for instance, the dispute between preformationism and epigenesis has evolved and that, as a consequence, it is difficult to find a contemporary developmental philosophy that does not share some of the classical features of both positions. No experimental embryologist would today reject the significance of developmental genetics or refuse to characterize developmental processes in molecular terms. Preformationism and epigenesis might be thought of as *metaphysical research programmes* (an expression, intended without derogatory meaning, used by Popper (1992) to characterize Darwinism). They have no essences and change through time like biological species (Hull, 1988). However, accepting the strategy of molecularization does not solve in one stroke all the fundamentally metaphysical issues at the core of developmental biology. For this reason, we believe that the metaphysical swamps in which *Gestaltungsgesetze* supposedly muddled itself are still very squelchy in the molecular age.

Accordingly, in the next two sections the paper will tackle two metaphysical issues. In the second section the focus will be on developmental information. *Gestaltungsgesetze* implicitly defined the MF as an area of non-genetic embryological *information*. However, if the additional information does not physically reside in the nuclear material of the cell, where is it localized? We will consider this issue by analysing the nature of the concept of positional information originally proposed by Wolpert (1969). We will show that the concept of positional information was gradually 'geneticized' by Wolpert, concomitantly with his endorsement of the radical research programme of 'computational embryology'. We will argue that Wolpert's version of the science of morphogenesis (based on informational preformationism and the heuristics of computing the embryo) makes the concept of MF redundant.

In the following section the focus will be on three causal issues that need to be investigated when trying to understand the nature of morphogenesis: (i) How does the formation of a pattern—of a tissue—of an organ—start? (ii) What is the nature of the physical processes governing intercellular information exchange? (iii) What is the nature of the interactions between the components of the developmental system? The various models briefly presented in this contribution provide different answers to these questions because they are based on different metaphysical commitments concerning the nature of developmental causation as well as the nature of the MF. Indeed, we suggest that genuine progress in developmental biology is parasitic on the achievement of more clarity concerning the metaphysical issues we identify. Therefore, we suggest that, contrary to what Lawrence and Levine argue, developmental biologists, like all other scientists, still need to get their boots dirty in metaphysical swamps.

Developmental information

One of the assumptions of *Gestaltungsgesetze* is that the kind of biological information apportioned by the field is not genetic. It would be absurd to think otherwise because undifferentiated cells all have the same genes (what has been called the paradox of cellular differentiation, cf. Sapp, 2003: 181).

Morphogenesis for *Gestaltungsgesetze* was not a degenerative but an integrative process in informational terms. The task was therefore to account for the nature of the additional information necessary for developmental regulation. Does the additional information physically reside in the cytoplasm, in the internal environment of the developing embryo, or in its external environment? Is it localized somewhere in the first place or is it dispersed? In order to focus on these issues, we will concentrate on the concept of positional information introduced by Lewis Wolpert.

Return to Weismann

The rise of developmental genetics was eventually accompanied by the use of informational language in biology propelled by Crick (1958). The history of information talk in developmental biology is complex. In the book *What is Life?* Schrödinger (1944) famously suggested that development is regulated by a codescript structure. If there is a code, then there must exist, at least metaphorically speaking, 'information'. However, despite this implicit acknowledgment, Schrödinger does not explicitly use the term 'information'. Crick and Watson were probably first to introduce the term in genetics: 'it . . . seems likely that the precise sequence of the bases is the code which carries the genetical information' (Watson & Crick, 1953: 964). Crick (1958) then made this theoretical usage common and philosophically solid. The thesis that DNA contains the instructions to 'code' for proteins and mature phenotypes instead of contributing to 'causing' them is linked to a preformationist view of development, a reductionist view of science, a localized view of causation, and the thesis of the primacy of endogenous factors in biology. Sarkar (2005: Ch. 14) argues that Waddington was the first proponent of a synthesis of developmental biology and genetics based on the idea that genes are the primary source of information. Waddington, according to Sarkar, thought that regulation at the genetic level can explain tissue differentiation and morphogenesis: all these processes are basically a matter of switching genes on and off. Waddington's ideas reflected an informational and genetic *zeitgeist*. In roughly the same period Turing (1952) ventured on the developmental biology scene by making an important contribution (see 'A clash of causal ideologies'), namely, introducing the concept of morphogen, an atomistic concept that became a competitor of the holistic concept of MF. Around the 1970s two other relevant contributions appeared in the literature on morphogenesis. First, Wolpert (1969) proposed the concept of *positional information* (henceforth PI) in order to conceptualize the process of pattern formation. Second, Crick (1970) suggested that diffusion is the physical process at the basis of intercellular coordination during morphogenesis (see 'A sinking model?').

The idea of PI is that cells have a positional value that is related to a cell's position in a developing system and that cells 'interpret' their positional value and differentiate in specific position-dependent ways. Regulation can be viewed in terms of the process of interpretation of PI. Initially the concept of PI provided an alternative to explanations couched in terms of genes. The phenomenon of spatial organization (i.e. that cells in a developing embryo seem to 'understand' where they are and 'know' how to differentiate in a coordinated and seemingly teleological manner) was generally considered as an insurmountable obstacle for accounts of morphogenesis based on the expression of genetic information. So, given that genes were not considered as carrying spatial information, given that a vitalistic explanation referring to the cognitive capacities of cells was out of the question, and given the popularity of information talk in biology, Wolpert introduced the concept of PI in order to refer to the spatial information required by developing cells. The concept was initially sufficiently vague as to remain interpretable in many physical ways, for instance as the information conveyed by a specific concentration of a diffusing chemical substance or as the information communicated by cells to other cells. Coupled with the French flag model—a representational device that enables to understand how genetically identical cells, by responding to different concentrations of diffusing chemicals in the embryonic context, end up assuming different identities symbolised by the characteristic colours of the flag—the concept of PI provided a theoretical framework to conceptualize the process of spatial regulation. Indeed, Wolpert thought that his framework was general enough to be unifying and supplying a new meaning to traditional embryological concepts such as gradient, induction, and field. In synthesis, according to Wolpert, the MF of the *Gestaltungsgesetze* was a variant of positional field.

But the PI concept was problematic from the outset. PI was based on the ability of cells to *estimate* their position within the embryo, that is, on the ability of cells to *interpret* PI. But nothing was known about such interpretive powers. The French flag model just said that cells *estimate* the concentration of a morphogen, *compute* the information provided by the gradient, and then change their fate. The anthropomorphic and information-processing language used by Wolpert should be considered metaphorical. Wolpert's model of morphogenesis constituted a theoretical model of the process of morphogenesis (Wolpert preferred to speak of pattern formation) that not only lacked an explanation of how cells *estimated* their position, but that was additionally naturally coupled to the simplest gradient model based on the diffusion of one morphogen, considered, with few exceptions (see 'A sinking model?'), too simplistic. The idea that the complexity of the variegated processes of pattern formation could be captured with models consisting of the diffusion of just one morphogen was considered a trivialization by Turing himself, who, certainly not a developmental biologist, only considered models with interacting morphogens as biologically realistic. Given these limitations, why did PI become such an important concept in developmental biology? Fox Keller (2002: Ch. 6) reckons that it was because, eventually, the concept of PI was interpreted in terms of genes.

The transformation of the concept happened, according to Fox Keller, between 1971 and 1975, and was an outcome of its original malleability. In 1971 Wolpert rejected the ideas that the genetic material provides the description of the adult and that 'if we understood cytodifferentiation or molecular differentiation then pattern would be explicable' (Wolpert, 1971: 184). This view was in line with the first proposal of 1969, particularly with the idea that it is the egg (and not solely its genetic material) that contains the instructions of a 'programme for development'. This constitutes an embryonic view of preformation, not a genetic one. The egg contains spatial information, and cells are able to estimate this not purely genetic information. But by 1975, in a paper co-authored with Lewis, Wolpert seems to have changed his mind, embracing the strong research programme of computational embryology: 'A theory of development would effectively enable one to compute the adult organism from the genetic information in the egg. The problem may be approached by viewing the egg as containing a program for development, and considering the logical nature of the program by treating cells as automata and ignoring the details of molecular mechanisms' (Wolpert and Lewis 1975, : 14). Computational embryology is founded on the hypothesis that the egg possesses enough information to make an adult. In fact, Wolpert and Lewis argue, arguments that posit that such information mysteriously increases during development are 'not valid'. What needs to be added is simply that the egg does not contain a complete description of the adult but rather a genetic programme for making it (Wolpert & Lewis, 1975: 14). In fact, genes control cells' behaviour and account for their crucial developmental properties such as memory, competence and interpretation of PI: 'The system of genes controlling a cell provides it with a memory, and governs its behavior according to past as well as present circumstances' (Wolpert & Lewis, 1975: 19). Gene networks with around 100 genes might be complex enough, they argue, to generate complex structures like, for instance, the vertebrate limb. The methodological advantage of considering the programme in terms of gene control networks is that it can be characterized in logical terms. Understanding development in logical terms might have been a very appealing feature in an era in which cybernetics and computer sciences were in full explosion.

The crucial assumptions at the heart of computational embryology are, first, that the egg contains internally all the developmental resources it needs to build an organism; second, that these resources are genetic; and third, that they control cells' behaviour. While the first and second theses were part of Wolpert's theoretical framework from the start (i.e. 1969), the third was added later. Fox Keller's interpretation of Wolpert's developmental philosophy stresses the rupture between the original concept of PI and the programme of computational embryology, whereas, as a matter of fact, there are many continuities between the two. Despite acknowledging such continuities, we argue that the genuine rupture is in assuming that genes control cells' behaviour, making cells causally redundant. Since 1975, Wolpert's model of morphogenesis effectively reduced the process of spatial regulation to the process of differential gene expression. The concept of developmental programme was subsumed under the category of genetic programme, and the concept of PI was subsumed under the category of genetic information. In the end, the problem of spatial regulation was subsumed under the problem of cellular differentiation, a view originally proposed by Weismann (1893).

Computing the embryo

In computational embryology all the major features that characterize classic preformationism can be identified: the egg contains a genetic programme (i.e. the latent preformed structure) for development that is complete and operates deterministically (i.e. it is an algorithm potentially allowing the computation of the adult organism, discarding the informational contribution of all non-genetic factors). The 1970s were possibly an over-optimistic period dominated by the misuse of information language (Kay 2000) and the assumed impending vindication of genetic reductionism. Wolpert's developmental philosophy is seasoned with such over-optimism. But this attitude was unjustified in the face of the conceptual void left by Wolpert's model of morphogenesis. In fact, the nature of the process by means of which cells interpret PI was totally unknown. Of course, Wolpert realized that this was the most problematic aspect of the conceptual framework he proposed: 'the least attractive feature of positional information . . . is that it places a great burden on the process of interpretation' (Wolpert, 1989: 3). If one considers cells as automata, as Wolpert and Lewis (1975) did, a clear problem emerges: for cells to be automata, a plan—a software—is required. Wolpert's (1989: 8) solution to the problem of how cells manage to build a pattern given specific positional values was to postulate such a master plan: 'A formal solution is that each cell contains a complete specification of the behaviour of every cell in every position. There must be a complete list of the cells that will differentiate as type A, and another list of those that form type B and so on.' This radical and 'formal' solution is consistent with a form of extreme preformationism: all 'lists' postulated by Wolpert are presumably localized in DNA: 'there is some evidence that there may be a listing of the type postulated for the sensory bristles in *Drosophila*. There are 11 sensory bristles on the thorax of the fly which can be removed, often in pairs, by a series of mutations in the *achaete-scute* complex' (Wolpert, 1989: 8). It is also deterministic given that cells just follow automatically the rules set by the master plan. In effect, the concept of interpretation seems to be eliminated in favour of an automatism.

Wolpert's original aim was to supply a unifying framework in which it could be possible to translate in a 'rigorous' and 'precise' language idiosyncratic concepts such as fields and gradients. However, the concept of MF received a blow with Wolpert's proposal. The interpretation of the concept of MF as a kind of *positional field* lacks many of the characteristics of the original concept, particularly its holistic and non-genetic character. We argue that the 'geneticization' of the concept of PI, coupled with the over-optimistic expectations of the programme of computational embryology, effectively makes the concept of MF redundant for the reason that all the relevant embryological information necessary for all kinds of regulative developmental processes is thought to be specified in the genes. In its strongest form, the research programme of computing the embryo is based on the methodological simplification of considering position to be totally specified by DNA, because each cell contains a list of rules determining its behaviour. The original concept of PI contained two variables: the informational one and the interpretive one. However, by making interpretation equivalent to the processing of the genetic information encoded in the series of preformed lists with which every cell is supposedly endowed from the start, the interpretation variable is reduced to the information one. Thus, it could also be argued that, instead of providing a way to translate the concept of MF in biochemical terms, Wolpert's geneticized version of PI makes the concept of MF redundant. Thus, in their geneticized versions, the concepts of PI and positional field are rather the ultimate realization of informational preformationism in developmental biology (Mahner & Bunge, 1997:

Section 8.2.3.1). The nuclear DNA contains all the necessary information—the 'programme' that will underwrite a successful and normal developmental process. The cytoplasmic and embryonic environments are just negligible variables from this preformationist perspective. As Wolpert (1991) surmised, paraphrasing William Harvey: *ex DNA omnia*.

An important corollary of the view Wolpert endorses from 1975—with its emphasis on the programme of computational embryology and the geneticized version of the concept of PI—is that cells' responses become causally irrelevant, that is, redundant causal intermediaries between the fundamental generative process (i.e. input) of gene expression and the fundamental effect (i.e. output) of morphogenesis. This point has also been highlighted by Rosenberg (1997), who argues that the aim of computational embryology is to understand how development can be fully explained through the identification of the relevant molecules and their rules of interactions. The basic idea of computational embryology is therefore to compute embryological output given only the macromolecular input relative to the components (i.e. nucleic acids and proteins) of the embryo, abstracting from the cell's behavioural responses. As Wolpert (1994: 572) explained, computing the embryo 'is a formidable task, for it implies that . . . it may be necessary to compute the behavior of all the constituent cells. It may, however, be feasible if a level of complexity of description of cell behavior can be chosen that is adequate to account for development but that does not require each cell's detailed behavior to be taken into account.' The level of description chosen by geneticizing the concept of PI is to ignore the interpretive variable and the nature of cellular responses. In Rosenberg's (1997: 450) clear words: 'Wolpert's thesis must be understood as claiming that the function which renders the embryo computable will take us from macromolecules to organisms without having to pass through the way-station of a complete description of all the constituent cells.'

It must be acknowledged that, in its strongest form, Wolpert's conceptual framework is capable of yielding a set of coherent answers to some of the questions concerning the nature of information and causation set in the introduction: PI is ultimately assumed to be genetic information, actualized in the putative genetically determined lists all cells possess; cells are automata that compute rather than interpret PI; genetic switching is the triggering event that could be considered the proximate cause of the process of pattern formation; and genes are the prime movers of all developmental processes. At the same time, it must be highlighted that Wolpert's framework is just a *formal conceptualization* of the morphogenetic process; that is, an idealized, over-optimistic simplification. Note that Wolpert and Lewis wanted to emphasize 'the logical nature of the [developmental] program by treating cells as automata and ignoring the details of molecular mechanisms'. As Wolpert acknowledges today (Kerszberg & Wolpert, 2007), his conceptual framework does not provide a satisfactory model for understanding morphogenesis for at least two reasons: diffusion of morphogens is not the only physical process of information transmission between cells, and cells are not programmed automata. Computational embryology is based on the methodological assumption that the molecular interactions between cells are 'logically' secondary, even though they are, without doubt, biologically primary for most experimentally prone developmental biologists; and, second, in Wolpert's model cells remained 'black boxed', with no attempt to answer the crucial question of how they were actually interpreting PI. In both cases, we suspect that the rationale substantiating such a position was that cellular processing and communication were until very recently beyond the reach of molecular methods. Wolpert (1994) in fact considered cellular complexity to be much greater than the embryo's. Accordingly, genes assume the fundamental causal role in computational embryology, while cells simply act as causal proxies, dispensable because they represent an irrelevant intermediate hierarchical level between molecular input and organismal output. In its most extreme, geneticized formulation, the metaphor of PI masquerades as a preformationist and genetic reductionist bias that privileges the causal importance of the endogenous and rejects the role of context-dependence in developmental processes (Laubichler & Wagner, 2001). Informational preformationism commits what could be called the endogenous fallacy, that is, the fallacy of considering gene sequences as determinants rather than as co-determinants of developmental processes (Gilbert & Sarkar, 2000).

We conclude our analysis in this section by suggesting that the basic limit of Wolpert's theory of morphogenesis might reside in its theoretical nature. The tension between proposing theoretical models and engaging in experimental work is well known. Meinhardt (Gordon & Beloussov, 2006: 110) states that the usual reception of his models was the following: 'First they were regarded as unrealistic or misleading: "cannot be". More or less abruptly this changed later into: "that is trivial, how else should it be?"' Analogously, the two major assumptions at the core of computational embryology (i.e. the 'geneticization' of the concept of PI, and the heuristic strategy of ignoring the causal role of extremely complex cellular responses) can be perceived as unrealistic as well as misleading. We also argued that they effectively made the concept of MF redundant.

Another way to spell out our suggestion is the following. Wolpert's theory of morphogenesis is partly couched in informational terms. Exploiting metaphorical informational terminology might be perceived as a way to avoid dealing with crucial causal issues. Accordingly, it could be argued that the model of computing the embryo was not accompanied by a serious attempt to delve into the causal issues we have identified in the introduction. The reason is straightforward in Wolpert's case: DNA is the prime mover of development. If DNA is the only causal agent, all the other causal questions we identified become uninteresting. In the following section we will show that the recent interpretations of the concept of MF we illustrate not only eschew information language but primarily deal with issues of causation.

Developmental causation

In the introduction we proposed three causal questions:

(i) How does the formation of a pattern, a tissue, or an organ start? More specifically, what is the triggering event of specific morphogenetic processes?
(ii) What is the nature of the physical processes governing intercellular information exchange during morphogenesis?
(iii) What is the nature of the interactions between the components of the developmental system?

In this section we will try to give answers to these questions by partially analysing some historically relevant and contemporary interpretations of the process of morphogenesis. We will first analyse the causation involved in the source-sink model, using it as a kind of conceptual scapegoat and default model. We will then start to introduce, not in chronological order, conceptual complexity to this default model by considering contributions that led to the postulation of more complex causal interactions.

A sinking model?

Crick (1970: 420) proposed that diffusion might be the causal mechanism that could explain the nature of the 'embryonic fields' of the developmental literature. His source-sink model was theoretically simple: 'At one end of a line of cells one postulates a source—a cell that produces a chemical (which I shall call a morphogen) and maintains it at a constant level. At the other end the extreme cell acts as a sink: that is, it destroys the molecule, holding the concentration at that point to a fixed low level. The morphogen can diffuse from one cell to another along the line of cells.' The great advantage of the source-sink model is that the postulated causal process is familiar: the diffusion of a substance is an example of mechanistic causation via contact. The alternative proposed by some *Gestaltungsgesetze* theorists never seemed to be so easy to grasp (i.e. as it involved the downward causation exerted on the differentiating cells by a physically unlocalizable entity such as the MF). Crick substituted a holistic concept whose causal influence is difficult to conceptualize mechanically (i.e. the MF) but with clear reference (the organizational effect of the MF was experimentally observable) with an easily conceptualizable mechanistic concept with indefinite reference (i.e. diffusion of hypothetical chemical substances). After estimating the value of the diffusion constant, of the permeability of cellular membranes and of the time needed to set up a gradient,

Crick concluded that diffusion could work if the tissue width is around 30–70 cells, depending on the size of the cell. Embryonic fields of a millimetre or less are possible, but larger ones are not. The results are 'striking': 'Even allowing for the very approximate nature of the calculations, the agreement with the figures given in the quotation at the start of the article is striking' (Crick, 1970: 421). The figures referred to come from Wolpert's first paper on PI (Wolpert, 1969), in which the latter suggests that most embryonic fields seem to involve distances of less than 100 cells. It is interesting to note that, at the time, Crick was one of the few supporters of Wolpert's French flag model (Richardson, 2009: 661).

After illustrating the basics of the model, let us now analyse the nature of the answers it provides to the causal questions we posed above. Crick's answer to our first causal question was that the triggering event should be identified with the time when the source (i.e. the founder cell of the field) starts secreting the morphogen. But Crick did not try to answer the question of why that specific cell starts secreting the morphogen at that specific time. Crick gave a clear answer to the second causal question: diffusion is the central physical process of intercellular communication. The third causal question was implicitly answered by assuming that all the causal interactions are molecular and mechanical, determined by the source-produced morphogens diffusing through the extracellular matrix and affecting cells' behaviour.

Crick's model can be criticized for three reasons: the first concerns the explanatory power of diffusion theory in biology; the second concerns the nature of the cellular response to morphogenetic inputs; and the third regards the universality of the process proposed.

Starting with the first, it should be noted that diffusion theory had been strongly criticized in biology since the 1930s. Agutter et al. (2000: 103) contend that the long-term success of diffusion theory in biology was due to the simplified and idealized nature of the causal framework it provided. However, it must be highlighted that Agutter et al.'s analysis mainly refers to intracellular rather than intercellular diffusion. This implies that the putative epistemological limits of diffusion theory can be considered deficiencies of Crick's model only to a partial extent.

Passing to the second and third criticisms, it must be remembered that Crick proposed his model emphasizing that it was important 'to make two reservations'. The first caveat concerned an already known problem: 'Even when the gradient has been set up the cell has to recognize it' (Crick, 1970: 421). This is the problem concerning the way cells process and interpret the morphogenetic input. Crick did not provide many clues on how to solve this conceptual and empirical problem, even though one could argue the following. His model is heavily based on Wolpert's insights on PI. In 'Developmental information' we saw that eventually Wolpert's conception of PI relied on the strategy of ignoring cellular complexity. This strategy is compatible with one important simplification Agutter et al. (2000) see in diffusion theory as applied to biology: cells are considered passive entities, but passive entities cannot be the actors of regulative processes like those observed in physiology and development for the reason that they require 'sensors, effectors, information processing and feedback mechanisms' (Agutter et al., 2000: 85). The way in which the information conveyed by morphogens is processed and communicated by cells could not be answered in 1970. It would therefore by preposterous to criticize Crick for this reason. However, neglecting the complexity of cellular responses was arguably a common strategy in developmental biology in the period in which Crick's model was proposed.

The third criticism is related to the other reservation introduced by Crick: 'There may be special cases, involving setting up gradients quickly over large distances (of the order of several centimetres) which may require other mechanisms' (Crick, 1970: 421). This reservation concerned the limitations, but just in 'special' cases, of the process of morphogen diffusion in reaching the sink when the sink is very distant. In such cases Crick thought that additional processes of cellular signalling might be involved. The issue at this level was partly empirical (i.e. was the evidence in favour of Wolpert's hypothesis that most fields consist of maximum 100 cells reliable enough?) and partly methodological (i.e. what was the theoretical rationale for deeming that a model so simple could be universal and causally primary?).

In retrospect, it should be stressed that Crick's model has been recently confirmed experimentally

(Yu et al., 2009): it is henceforth not a sinking model. However, such experimental vindication must not overshadow the fact that the model is based both on a simplified account of cellular processing and intercellular signalling, and that these limitations might have implications concerning its putative universality and primacy. Furthermore, it could be argued that the idea of process universality and primacy itself is anachronistic in the light of current experimental evidence. First, Crick's model was proposed in order to account particularly for the dynamics of early embryogenesis, when the number of cells is low and the fields to be organized relatively small: in this context diffusion might work, but in the context of later stages of development it might not. Furthermore, and more importantly philosophically speaking, today a variety of ways are known by means of which groups of cells can differentiate and form specific tissues. Minelli (2009: 118–119), for instance, identifies four such ways: apart from inheriting the epigenetic state of their mother cells, cells could process the morphogenetic information from neighbours, engage in self-sorting, and participate in what Minelli calls the 'chorus effect', that is, 'the sustained, two-way exchange of chemical or electrical inputs between neighbouring cells'. Note that in Crick's model only the first two processes are taken into account. And also note that it is the latter in particular that evokes the traditional concept of MF (see 'The renaissance of a concept in two forms').

A clash of causal ideologies

Turing (1952) not only introduced the concept of morphogen in the literature, but he arguably proposed the first physically plausible model of morphogenesis (Fox Keller, 2002: 100). And he achieved this result by means of an extremely idealized mathematical model based on a number of biologically powerful simplifications. One of these simplifications concerned cells. In Crick's and Wolpert's models, morphogens diffuse and cells interpret their information. But in Turing's the details of the interpretive process remain irrelevant because cells are treated as 'idealized geometrical points', analogous to Leibniz's monads. The outcome is that the diffusion of morphogens is assumed to happen in an embryonic context that is continuous and totally homogeneous. This is an unrealistic simplification, but it bore an unexpected fruit.

It must be emphasized that, even if this simplification had been part of Turing's original proposal, it is certainly not part of the most popular biological version of his reaction–diffusion model (Meinhardt & Gierer, 2000). Naturally, in the local activation–lateral inhibition model proposed by Meinhardt and Gierer, the reactors are cells. There is therefore a huge biological difference between the nature of the physical interactions that can be imagined between self-synthesizing chemical substances (i.e Turing's actual model) and the nature of the physical interactions between chemical substances mediated by complex information-processing organisms like cells (i.e. all contemporary models based on Turing's mathematical insights). Also note that Wolpert and Lewis (1975:16) observed that Turing totally ignored the crucial process by means of which cells interpret new morphogenetic signals in the light of their developmental history. This is certainly true given that in Turing cells are totally black boxed. It is also true that computational embryology makes cellular activity totally dependent on genes, making cells cybernetic automatons (see 'Computing the embryo').

Nonetheless, the simplification bore an unexpected fruit in the sense that Turing's model offered a creative solution to our first causal question. When it is asked 'What is the triggering event of X, where X is a specific morphogenetic process (i.e. the formation of a pattern, of a body segment, of a leg)?', a variety of answers are possible. Within such variety a first rough dichotomy can be recognized: either a causal agent or initiating event is identified, or it is not. In the first case the answer to the question is 'classical': the attribution of causal responsibility to localizable entities or precipitating events is conceptually necessary in order to explain the emergence of the effect. In the second case the attribution of causal responsibility is not to some specific causal agent or event, but rather to the instability of the initially stable interaction dynamics; in this case 'causes' are distributed and unlocalizable. These two conceptualizations of causation represent a clash of causal ideologies (Fox Keller, 2002). Which one is preferable in the case of morphogenesis?

The great disadvantage of the first conceptualization is that it leads to infinite regress, apart from being compatible with incompatible biophilosophies. In the first respect, assuming there exists a triggering event implies the existence of an agent causing that event. We end up with the Aristotelian concept of the prime mover, where the prime mover has been traditionally associated either with an intelligent founder cell or with DNA. Delbrück (1971: 55), for example, argues that: '"unmoved mover" perfectly describes DNA: it acts, creates form and development, and is not changed in the process'. In the second respect, assuming the existence of a prime mover does not tell us anything about its nature. A variety of answers are available: it could be the MF, or DNA, or the cell. These three answers could not be more different, as they have been respectively historically linked to three different biophilosophies: the MF to organicism, DNA to genetic reductionism, and the intelligent founder cell to vitalism.

The great disadvantage of the second conceptualization of causation is that it looks just like a partial explanation. Turing claims that it is a random disturbance that causes the instability of the homogenous equilibrium in the embryonic environment which, as a consequence, causes the emergence of the pattern or morphological structure. This process is spontaneous and lacks the central control through agency. Or does it? At least, it remains always compatible with the postulation of an agent as the creator of the random disturbance. Furthermore, given that the order, form, and organization Turing tries to explain arise de novo in each generation spontaneously thanks to the amplification of random fluctuations and positive feedback processes, one might still wonder what is the evolutionary origin of this precise and robust sequence of events in the first place. But in Turing's model the random disturbance that unsettles the homogeneity of the embryonic system is not temporally and spatially programmed. The localization of the random disturbance is not predictable: many cells could happen to be the source. It could be thus said that in Turing's model causation is distributed and organization decentralized over the components of the system. The robustness of the morphogenetic process and its capability to resist perturbations is based on this distributed logic. Nonetheless, the random disturbance has causal efficacy as it precipitates a self-organizing process: Turing's analysis is appealing precisely because 'it offered a way out of the infinite regress into which thinking about the development of biological structures often falls. That is, it did not presuppose the existence of a prior pattern, or difference, out of which the observed structure could form. Instead, it offered a mechanism of self-organization in which structure could emerge spontaneously from homogeneity.' (Fox Keller, 1983: 515–516). In brief, Turing provided a powerful conceptualization of how structure can arise from undifferentiated material due to self-organization and without central planning and preformationist input. Thanks to Turing, self-organization became a respectable alternative biophilosophy.

We suggest that it is this ontological property of Turing's model that makes it epistemically persistent and theoretically suggestive. Turing's greatest achievement in biology might be considered the fact that he proposed a viable new conceptualization of one important aspect of the process of morphogenesis (i.e. the nature of the event triggering morphogen formation) by abstracting away from the details relative to the nature of genes and cells. Following Turing's line of reasoning, there might be no need to conceptualize DNA as the prime mover of developmental processes, nor any need to think in terms of programmed or intelligent founder cells. Thus, Turing's model provides a serious alternative to the preformationist, genetic reductionist, and vitalist biophilosophies that have traditionally dominated developmental research.

At the same time, it might also be added that Turing's model remains naturally compatible with all these philosophical positions. For instance, in Meinhardt's and Gierer's model, genes are particularly important as the producers of the signalling molecules (i.e. morphogens), and this might be interpreted as compatible with the attribution of a primary causal role to genes in development. But it would be stretching the point too far, since, as Meinhardt reminds us, 'Equally important is the interpretation of these signals' (Gordon & Beloussov, 2006: 7), and henceforth cellular behaviour. Turing's model is therefore philosophically neutral despite being causally loaded.

The renaissance of a concept in two forms

All the conceptual models we have brought so far to the attention of the reader try to capture the *essence* of morphogenesis. Turing's reaction–diffusion model, Crick's source-sink model, and Wolpert's computational embryology are all instances of theoretical contributions that adopt powerful and inevitably partially misleading simplified auxiliary hypotheses in order to describe the 'logic' of morphogenesis. It is well known that mathematical and theoretical biologists have a penchant for simplicity, symmetry, and other aesthetical virtues. For instance, Meinhart argues: 'signalling between cells requires the production and secretion of ligands, their reception by another cell and a signalling cascade to the nucleus. The assumption of a simple diffusion, however, is in most cases a sufficiently good approximation' (Gordon & Beloussov, 2006: 8) In a recent review of Turing's model, Kondo et al. (2010: 1616) claim that 'the logic of pattern formation can be understood with simple models, and by adapting this logic to complex biological phenomena, it becomes easier to extract the essence of the underlying mechanisms'. One of the basic features of Turing's model, at least in the form generating the so-called 'Turing pattern' (Kondo et al., 2010: 1618) is that the morphogens interact. It is because of such interaction that the system becomes self-regulating and the embryo organizes itself. Crick's source-sink model (the causal model at the basis of Wolpert's account of morphogenesis) can be considered as the simplest form of a reaction–diffusion model in which the reaction term is removed (Kondo et al., 2010: 1617). This interpretation provides a straightforward way to capture one basic difference between Crick's classic source-sink model and Turing's. While in the source-sink model a morphogen diffusing from the specific location known as the source forms a gradient that provides PI to the cells in the embryo, thus producing a pattern/form programmed and determined from the outset, in Turing's model the gradient is rather the result of the interactions between morphogens which produce a pattern/form that is dependent on the feedback relationships between the reacting elements. In the actual model proposed by Turing, the reacting elements were two self-synthesizing morphogens diffusing across a continuous embryonic context. As already pointed out, the model is therefore very idealized. But the 'logic' of the model remains intact whenever a non-linear wave is maintained by the dynamic equilibrium of the interactions between its constitutive elements. The real biological system could consist of more morphogens interacting, its constituents elements could be molecules, cells, circuits of cellular signals, or even other organisms (see Nyholm and McFall-Ngai, this volume), and the physical processes of biochemical information transmission could be diffusion, chemotaxis, 'chorus effect', or other mechanisms of cell signalling. Again, the 'logic' of the Turing model remains intact despite these biological refinements. After Turing, what was needed was more knowledge concerning the physical realization (i.e. molecular components and physico-chemical processes of interaction between them) of the developmental system he hypothesised. In the last 60 years developmental biology has advanced tremendously in this sense.

It has been suggested that a developmental system instantiates a Turing model if it includes a network with two self-regulating feedbacks (one enhancing and one inhibitory). De Robertis (2009) thinks that exactly a biochemical system of this kind regulates the development of dorsal–ventral tissue differentiation in the embryos of all bilateral animals. The developmental system De Robertis postulates obeys the logic of the Turing model; but what is most interesting is that such a system supplies a biochemical and molecular description of the classic MF concept, specifically a vindication of Spemann's organizer. De Robertis started to unravel the molecular nature of the organizer in 1991, and in the last 20 years he has advanced considerably in this task. The revived concept of organizer is defined as the reciprocal transcriptional control between two signalling centres localized in the ventral and dorsal sides of the developing embryo where, for each action on the dorsal side, there is a reaction on the ventral one: 'Self-regulation occurs because transcription of ventral genes is induced by BMP [bone morphogenetic protein] while transcription of dorsal genes is repressed by BMP signals.' (De Robertis, 2009: 925)

De Robertis's model is interesting because it provides an alternative answer to our second causal question. Turing considered only diffusion as the process of transmission of biochemical information. This is because, in Turing's model, cells are black boxes, as they effectively were, so we argued, in Wolpert's computational embryology. But the formation of morphogen gradients over long distances (apparently an existing phenomenon) is a process that cannot be explained through diffusion dynamics alone. De Robertis and colleagues have established that field formation requires intercellular information signalling and cell–cell interactions. The basic point is that the simplified source-sink model with diffusion of biochemical information is abandoned in favour of a much more complex model in which a variety of proteins and molecules are involved in the process of intercellular signalling (see also Newman, this volume).

In the molecular era conceptualizing cellular communication becomes necessary in order to solve the puzzle of morphogenesis. Informational interactions between cells can be conceptualized in different ways. One traditional way is, of course, to avoid the problem by treating cells as black boxes (Turing). Another is to consider molecule-mediated interactions. In this latter case causation is reducible to interaction by contact (for instance mediated by signalling molecules). This is a standard view of causation that relies on mechanically understood physical processes and molecular interactions. This causal view is at the basis of the MF envisaged by De Robertis's model. But, are there any other conceptually viable alternatives to this model? As a matter of fact, Newman and Bhat (2009) propose that cells in an embryonic context act in certain circumstances as biochemical oscillators. This is relevant because, as they (Newman & Bhat 2009: 699) remind us, a 'key property of oscillators is their ability to become synchronized'. They then propose that the synchrony of biochemical oscillators is a possible physical process explaining the 'coordination of cell state . . . over a broad tissue domain, a phenomenon described in the older embryological literature as a "morphogenetic field"'. Minelli (2009: 119) labels this phenomenon a 'chorus effect' (see 'A sinking model?') stating that 'a developmental system that is suggestive of this mechanism is the insect wing disc, within which small clusters of cells within the disc progress through the cell cycle together, tightly coupled to their neighbours, irrespective of their clonal origin', with reference to Johnston and Gallant (2002).

Returning to Newman and Bhat, what is interesting about this alternative interpretation of the concept of MF is the fact that it relies on a physical process (i.e. synchrony) whereby the physical interactions between the components of the system are seemingly not by contact but at a distance. Given the bad reputation action at a distance has in some physical quarters, a similar reaction to such claims might be expected in the biological community. However, the phenomenon of synchronized oscillation is well established in biology (e.g. Garcia-Ojalvo et al., 2004) and was not suggested in the first place by Newman and Bhat in its biological form. Nonetheless, they were the first, to our knowledge, to associate this process with the MF of the literature. Newman and Bhat believe that the explanation they provide is mechanistic. We would contend that it could be considered mechanistic if it were possible to identify the way in which the coordination effect they describe (i.e. between the oscillatory behaviour of the cells belonging to the field) were to be explained as instantiations of causation via contact through the transmission of some physical entities (e.g. molecules, chemical stimuli, electrical stimuli). But in this latter case cellular synchrony would not be genuine causation at a distance. In any case, the novelty of this approach, independently of whether cellular synchrony is a mechanical process or not, is its reliance on a clear alternative to diffusion dynamics. If it is indeed true, as it seems, that embryonic developmental processes sometimes require longer cellular distances than the maximum 70 cells calculated by Crick (1970), then cellular synchrony provides a suitable alternative to the various processes at the basis of De Robertis' model.

Ontologically speaking, the MF in this latter model is the effect of the synchronic oscillatory behaviour of the cells involved. Newman and Bhat (2009) ascribe this effect to a specific cause, what they call a developmental patterning module or DPM. DPMs are defined as a category of the toolkit genes supposedly constituting the pre-metazoan genome and that specify for characteristic molecules which, in turn, mobilize certain specific physical processes and, finally, produce specific effects (e.g. field formation) when applied on specific biological substrates (e.g. the cells of a vertebrate embryo). They label as OSC (reference to oscillation) the DPM that causes the MF. OSC chemically consists of the interaction of characteristic molecules (e.g. Wnt, Notch and Hes); is physically realized by means of synchronous biochemical oscillation; and produces a MF when applied to the cells of a vertebrate embryo. Newman's and Bhat's model apparently mixes the self-organizing and systemic properties already found in Turing's and De Robertis's models with the allure of the action at a distance that preserves the holistic character of the original MF concept. Its novelty resides in giving an alternative answer to the second causal question we posed, an answer that is different from that provided by De Robertis but compatible with it.

Conclusion

In 'Developmental causation' the focus was on three causal questions. The aim was to show that understanding proximate causation in developmental biology requires exploring the metaphysical swamps Lawrence and Levine urge developmental biologists to avoid. In the conclusion of this contribution, our only constructive task will be to identify many of these open metaphysical questions while summing up our preceding analysis.

In the case of the first causal question, we identified a clash of causal ideologies: one could either assume a causative agent or accept that what causes the precipitating event is a random instability. In the latter case we appeal to self-organization: one cell just happens to start secreting a morphogen, or starts oscillating, etc. On the other hand, if we choose the first causal ideology, we need to postulate a specific agency. But what kind of agency? At this point, other metaphysical disputes emerge: the agent could be DNA as the creator of form, the MF considered as a holistic entity, or even the intelligent founder cell. The last response might be considered ludicrous unless we accept that cells are agents. The histories of embryology and developmental biology show that many attempts to understand regulation were based on the strategy of black-boxing cells, effectively nullifying their behaviour. To take an extreme case, Wolpert's solution to the puzzle of how cells interpret morphogenetic information was to assume they are programmed. At the same time Wolpert constantly uses terms like 'record', 'compare', 'compute', and 'estimate' in order to describe the cells' abilities to interpret PI. The use of this teleological language is assumed to be largely metaphorical. But what happens if cells turn out to be processing and sensing organisms? First, cells would be considered cognitive, rational, possibly intentional agents (Shapiro, 2011). Secondly, the appeal to teleological behaviour on the part of cells, whose elimination presumably represents one of the central achievements of molecular biology, would be considered suspicious. *Pace* Lawrence and Levine, there remains a world of difference between treating cells as programmed machines and treating cells as interpreting agents. One could consider this contrast as identical to the classic debate between mechanicism and vitalism, but it would be partly missing the point as these philosophical positions have evolved. After all, some biologists have coined the oxymoron *molecular vitalism* (Kirschner *et al.* 2000).

The second causal question was traditionally answered by considering diffusion as the central physical process in morphogenesis. Molecular biology has enabled the study of cellular information processing and signalling. The answer to the second causal question therefore is also inextricably linked to issues concerning the nature of the cell. One outstanding issue concerns the nature of the physical processes necessary in order to comprehend cellular signalling, and whether such processes can be understood molecularly and mechanically. At the end of 'A sinking model?' we indicated that a variety of ways are recognized through which cells can process environmental information in order to coordinate phenogenesis (cf. Minelli, 2009: 118–119). It seems to us that the process of synchronized oscillation on the part of cells is, causally speaking, particularly interesting as it seemingly contravenes some of the most coveted intuitions concerning the mechanistic view of physical causation.

The third causal question concerned the kinds of interactions involved in the process of developmental regulation. It seems to us that De Robertis as well as Newman and Bhat eschew appeal to causal interactions such as holistic, top-down, or downward causation. In both models all the interactions between the components of the system constituting the MF are molecular (e.g. protein–protein and protein–cell), ultimately mechanistic, and therefore reducible to the components' level.

We suggest that a reductionist position labelled *molecular reductionism* is compatible with the two contemporary models we illustrated. Molecular reductionism includes as a special case genetic reductionism because it allows all molecular interactions, and not only those that pertain to the replication and transfer of genetic information. A similar view to ours, ontologically speaking, has been endorsed by Rosenberg (1997), who defended the reductionism inherent in computational embryology. However, we disagree with Rosenberg in a variety of respects. Rosenberg makes two additional claims with which we deeply disagree. The first is that understanding cellular behaviour is epistemologically irrelevant for a comprehensive theory of development (see 'Computing the embryo'), and the second is that all the genuine generalizations of developmental biology will eventually make use solely of molecular concepts. Contrary to what argued by Rosenberg, we would only like to note that progress in developmental biology has been parasitic on increased knowledge concerning the mediators of molecular interactions, namely cells. It is simply not true that cells can be black boxed and treated as an irrelevant intermediate hierarchical level between the molecular and the organismal. Indeed, this whole contribution was founded on the realization—we believe historically grounded—that ignoring cells' behaviour has been increasingly seen as a fundamentally misleading simplification in developmental biology. Furthermore, there are many good philosophical reasons to reject the epistemological legitimacy of purely molecular generalizations in developmental biology. The most general is that to assume that the molecular is the most fundamental ontological level in biology, and therefore the most fundamental level of biological analysis, are two controversial theses probably without sound metaphysical grounding. The second reason stems from empirical considerations: the same morphogenetic function can be molecularly realized in different ways, suggesting that morphogenetic function is developmentally more stable than its molecular basis (Laubichler & Wagner, 2001).

At the same time, the epistemological deficiencies of Rosenberg's reductionism have no straightforward ontological repercussions in our opinion. In particular, we do not think that the obvious fact that developmental biology needs non-molecular concepts implies the causal primacy of function vis-à-vis its molecular realization or that it vindicates the attribution of causal powers to non-molecular properties and entities. Laubichler and Wagner (2001: 58) disagree, criticizing Rosenberg's molecular reductionism, and denying that the fact that all developmental entities (e.g. morphological phenotypes, etc.) can be characterized in molecular terms 'is opposed to the possibility that functionally characterized biological objects can have an autonomous reality'. Among such objects Laubichler and Wagner would include MFs. The nearest to an argument for this claim is the suggestion that the same molecular basis can produce different developmental outcomes in different developmental contexts. If this is the case, then, the argument seems to imply, the difference maker in causal terms is the developmental context (cellular and organismal) itself. But, we ask, are the developmental contexts identical at the molecular level?

The supposed centrality of the concept of MF for a theory of development depends on providing a satisfactory answer to two questions concerning their ontological nature. The first is: what kind of things are fields? Are they entities or processes? Causes or effects? Gilbert et al. (1996) define MFs as heritable developmental causal processes (i.e. homologous developmental pathways). De Robertis defines the MF interchangeably as a self-regulating system, a biochemical pathway, a positional information network and an extracellular biochemical network of interacting proteins. Laubichler and Wagner (2001: 63) seem to consider the MF as the holistic developmental context in which macromolecules interact in order to produce specific morphogenetic outcomes. Newman and Bhat (2009) describe the MF as the effect of the action of physical processes of synchronous biochemical oscillation on cells. It might be argued that the same kind of conceptual idiosyncrasy characterizing the historical interpretations of the field concept is represented in its contemporary interpretations. All these definitions have a sound molecular basis, but they seemingly radically diverge ontologically, either considering the MF as a cause (Gilbert et al., Laubichler and Wagner) or reducing it to a mere effect (Newman and Bhat, De Robertis).

The second question concerns the evolutionary origin of the MF, an issue we did not even try to touch. At this level questions of proximate and ultimate causation (Mayr, 1996) mingle. Gilbert et al. (1996), De Robertis (2009), and Newman and Bhat (2009) all agree that fields are evolutionarily central, even though they might disagree about the way the MF is inherited: how is the robustness of the morphogenetic process achieved de novo at each generation unless it is genetically inherited? And do the 'teleonomic' features of the morphogenetic process cry for an explanation in terms of genes, programmes and agency? We would only like to report a lurking confusion concerning these partly metaphysical issues.

The concept of MF clearly shows some kind of epistemological resilience. Its renaissance provides a test case for evaluating epistemological issues concerning the nature of scientific change. Is the path of developmental biology progressive, at least in the sense of overcoming traditional dichotomies and metaphysical traps? *Pace* Lawrence and Levine, in this essay we have tried to show that, despite the undeniable advances of molecular biology, deep-rooted metaphysical questions remain without a clear answer. But this would not be particularly surprising if, as Feyerabend (1975) argued, the only way in which science can progress is by guaranteeing conceptual pluralism. It might be that the 'epistemological persistence' (Chang, 2011) of the concept of MF vindicates the function of conceptual pluralism in developmental biology. Its renaissance should therefore be welcomed.

Acknowledgments

We would like to thank audiences at the Taller de Magíster en Filosofía de las Ciencias at the Universidad de Santiago (Chile), at the XIII Jornadas Chuaqui Kettlun en Filosofía de la Ciencia in Santiago (Chile), and at the XXIII Jornadas de Epistemología e Historia de la Ciencia in Cordoba (Argentina) for providing a conducive context in which to present and discuss our preliminary work. We also thank Stuart Newman and both editors for very helpful comments and suggestions. We must also acknowledge the financial support of the Fondo Nacional de Desarrollo Científico y Tecnológico de Chile (grant number 11110409).

References

Agutter, P.S., Malone, P.C., and Wheatley, D.N. (2000). Diffusion theory in biology: a relic of mechanistic materialism. *Journal of the History of Biology*, **33**, 71–111.

Boveri, T. (1910). Die Potenzen der *Ascaris*-Blastomeren bei abgeänderter Furchung, zugleich ein Beitrag zur Frage qualitativ-ungleicher Chromosomen-Teilung. *Festschrift für Richard Hertwig, vol. 3*. Gustav Fischer, Jena.

Chang, H. (2011). The persistence of epistemic objects through scientific change. *Erkenntniss*, **75**, 413–29.

Crick, F. (1958). On protein synthesis. *Symposia of the Society for Experimental Biology*, **12**, 138–63.

Crick, F. (1970). Diffusion in embryogenesis. *Nature*, **225**, 420–2.

De Robertis, E. (2009). Spemann's organizer and the self-regulation of embryonic fields. *Mechanisms of Development*, **126**, 925–41.

Delbrück, M. (1971). Aristotle-totle-totle. In J. Monod and E. Borek, eds. *Of Microbes and Life*. Columbia University Press, New York, pp. 50–5.

Feyerabend, P. (1975). *Against Method*. New Left Books, London.

Fox Keller, E. (1983). The force of the pacemaker concept in theories of aggregation in cellular slime mold. *Perspectives in Biology and Medicine*, **26**, 515–21,

Fox Keller, E. (2002). *Making Sense of Life: Explaining Biological Development with Models, Metaphors and Machines*. Harvard University Press, Cambridge.

Garcia-Ojalvo, J., Elowitz, M.B., and Strogatz, S.H. (2004). Modeling a synthetic multicellular clock: repressilators coupled by quorum sensing. *Proceedings of the National Academy of Sciences USA*, **101**, 10955–60.

Gilbert, S.F., Opitz, J.M., and Raff, R.A. (1996). Resynthesizing evolutionary and developmental biology. *Developmental Biology*, **173**, 357–72.

Gilbert, S.F., and Sarkar, S. (2000). Embracing complexity: organicism for the 21st century. *Developmental Dynamics*, **219**, 1–9.

Gordon, R., and Beloussov, L. (2006) From observations to paradigms; the importance of theories and models. An interview with Hans Meinhardt. *International Journal of Developmental Biology*, **50**, 103–11.

Gurwitsch, V.A. (1910). Über Determination, Normierung und Zufall in der Ontogenese. *W. Roux' Archiv für Entwicklungsmechanik*, **30**, 133–93.

Hamburger, V. (1999). Hans Spemann on vitalism in biology. *Journal of the History of Biology*, **32**, 231–43.

Haraway, D.J. (1976). *Crystals, Fabrics, and Fields: Metaphors of Organicism in Twentieth-Century Developmental Biology*. Yale University Press, New Haven.

Harrison, R.G. (1918). Experiments on the development of the forelimb of *Amblystoma*, a self-differentiating equipotential system. *Journal of Experimental Zoology*, **25**, 413–61.

Hull, D.L. (1988). *Science as a Process*. University of Chicago Press, Chicago.

Huxley, J., and de Beer, G.R. (1934). *The Elements of Experimental Embryology*. Cambridge University Press, Cambridge.

Johnston, L.A., and Gallant, P. (2002). Control of growth and organ size in *Drosophila*. *BioEssays*, **24**, 54–64.

Kay, L.E.. (2000). *Who Wrote the Book of Life? A History of the Genetic Code*. Stanford University Press, Stanford.

Kerszberg, M., and Wolpert, L. (2007). Specifying positional information in the embryo: looking beyond morphogens. *Cell*, **130**, 205–9.

Kirschner, M., Gerhart, J., and Mitchison, M. (2000). Molecular 'vitalism'. *Cell*, **100**, 79–88.

Kondo, S., and Miura, T. (2010). Reaction-diffusion model as a framework for understanding biological pattern formation. *Science*, **329**, 1616–20.

Laubichler, M., and Wagner, G.P. (2001). How molecular is molecular developmental biology? A reply to Alex Rosenberg's reductionism redux: computing the embryo. *Biology and Philosophy*, **16**, 53–68.

Lawrence, P.A., and Levine, M. (2006). Mosaic and regulative development: two faces of one coin. *Current Biology*, **16**, R236–9.

Mahner, M., and Bunge, M. (1997). *Foundations of Biophilosophy*. Springer, Berlin-New York.

Mangold, H., and Spemann, H. (1924). Über Induktion von Embryonalanlagen durch Implantation artfremder Organisatoren. *Archiv fur Mikroskopische Anatomie und Entwicklungsmechanik*, **100**, 599–638.

Mayr, E. (1996). The autonomy of biology. *Quarterly Review of Biology*, **71**, 97–106.

Meinhardt, H., and Gierer, A. (2000). Pattern formation by local self-activation and lateral inhibition. *Bioessays*, **22**, 753–60.

Minelli, A. (2009). *Perspectives in Animal Phylogeny and Evolution*. Oxford University Press, Oxford.

Needham, J. (1931). *Chemical Embryology*. Cambridge University Press, Cambridge.

Newman, S.A., and Bhat, R. (2009). Dynamical patterning modules. *International Journal of Developmental Biology*, **53**, 693–705.

Popper, K.R. (1992). *Unended Quest: An Intellectual Autobiography*. Routledge, London.

Richardson, M.K. (2009). Diffusible grandients are out—an interview with Lewis Wolpert. *International Journal of Developmental Biology*, **53**, 659–62.

Rosenberg, A. (1997). Reductionism redux: computing the embryo. *Biology and Philosophy*, **12**, 445–70.

Sapp, J. (2003). *Genesis: The Evolution of Biology*. Oxford University Press, New York.

Sarkar, S. (2005). *Molecular Models of Life*, MIT Press, Cambridge.

Schrödinger, E.(1944). *What is Life?* Cambridge University Press, Cambridge.

Shapiro, J. (2011). *Evolution. A View from the 21st Century*. FT Press Science, Upper Saddle River.

Spemann, H. (1935). *The Organizer-Effect in Embryonic Development*. Nobel Lecture. http://www.nobelprize.org/nobel_prizes/medicine/laureates/1935/spemann-lecture.html

Spemann, H. (1938). *Embryonic Development and Induction*. Yale University Press, New Haven.

Turing, A.M. (1952). The chemical basis of morphogenesis. *Philosophical Transactions of the Royal Society of London B*, **237**, 37–72.

Watson J., and Crick, F. (1953). Genetical implications of the structure of deoxyribonucleic acid, *Nature*, **171**, 964–7.

Weismann, A. (1893). *The Germ Plasm*. Charles Scribners, New York.

Weiss, P. (1926). Morphodynamik. Ein Einblick in die Gesezte der organischen Gestaltung an Hand von experimentellen Ergebnissen. (*Abhandlungen zur theoretischen Biologie*, 23). Borntraeger, Berlin.

Wolpert, L. (1969). Positional information and the spatial pattern of cellular differentiation. *Journal of Theoretical Biology*, **25**, 1–47.

Wolpert, L. (1971). Positional information and pattern formation. *Current Topics in Developmental Biology*, **6**, 183–224.

Wolpert, L. (1989). Positional information revisited. *Development*, **1989 Supplement**, 3–12.

Wolpert, L. (1991). *The Triumph of the Embryo*. Oxford University Press, Oxford.

Wolpert, L. (1994). Do we understand development? *Science*, **266**, 571–2.

Wolpert, L. and Lewis, L.J. (1975). Towards a theory of development. *Federation Proceedings*, **34**, 14–20.

Yu, S.R., Burkhardt, M., Nowak, M., et al. (2009). Fgf8 morphogen gradient forms by a source-sink mechanism with freely diffusing molecules. *Nature*, **461**, 533–6.

CHAPTER 6

Physico-genetics of morphogenesis: the hybrid nature of developmental mechanisms

Stuart A. Newman

Introduction

Like all parcels of matter, embryos are subject to physical forces and effects relevant to their composition and scale (Forgacs & Newman, 2005). They thus bear the morphological signatures of their material nature. In the case of the metazoans (multicellular animals), the subject of this chapter (see Hernández-Hernández et al., 2012 for a parallel treatment of multicellular plants), these motifs, which we have called 'generic' (i.e. common to living and certain forms of non-living matter; Newman & Comper, 1990), include cavities, immiscible layers, segments, tubes, and appendages.

The basis for many of the morphogenetic properties of animal embryos is the liquid-like behaviour of their tissues (Amack & Manning, 2012; Forgacs & Newman, 2005). This phenomenon is not seen at the level of individual cells, which have stiff, submembrane, actin-rich cortices and navigate only by virtue of active reorganization of their internal cytoskeletons. In developing animal tissues and organ primordia, in contrast, the cellular subunits at early stages of development (i.e. before the elaboration of mature junctions or extracellular matrices (ECMs)), typically translocate independently of one another (Amack & Manning, 2012), endowing the collectivity with the formal character of a viscous liquid. The tissues of plant embryos are not 'liquids' in this sense, since their cells have solid walls that require breakdown and resynthesis during remodelling.

In addition, the cells of embryonic tissues (in common with all cells) sustain metabolic and biosynthetic processes and transport ions and other charged molecules. And because their cells also take up and store energy-rich compounds, developing tissues are chemically and electrically 'excitable', meaning that they are capable of generating a response disproportionate to the stimulus that provoked it (Levin, 2012; Levine & Ben-Jacob, 2004). Animal tissues are also capable of producing ECMs of various compositions and consistencies which, along with the cytoplasm of their constituent cells, endow them with a *viscoelastic* character. Because spring-like energy can be stored in such materials, some embryonic tissues are also mechanically excitable (Beloussov, 1998).

None of these features uniquely distinguishes animal tissues from non-living materials. 'Soft matter' refers to condensed viscoelastic materials such as colloids, liquid crystals, foams, and gels (de Gennes, 1992). Such materials have a range of exotic properties and laws of behaviour only discerned late in the 20th century and which are distinct from those of classically studied solids, liquids, and gases. Similarly, the category of *excitable media* extends beyond the cell- and tissue-biological examples mentioned above to include a broad category of chemical, electrical, and mechanical systems—oscillating chemical reactions, fire-susceptible forests—largely unexplored before the last century. While both categories are found outside of biology, it is rare to find non-living materials that are simultaneously soft matter and excitable media. Early stage animal embryos and organ primordia fit this description, however, and in this physical sense are inherently *hybrid*.

Towards a Theory of Development. Edited by Alessandro Minelli and Thomas Pradeu

The hybridity of developmental systems goes beyond the unusual coalescence of different categories of matter they represent, however. Because they simultaneously have chemical, mechanical, and electrical properties, they behave in a *multiscale* fashion, continually and simultaneously changing their composition and organizational properties on several temporal and spatial scales. While the morphology and spatiotemporal compositional patterns of such forms of matter are in principle understandable in terms of the generic mechanisms of mesoscale physics, in practice the hybridity of the systems makes this a difficult task.

One aspect of chemically and mechanically excitable soft matter that actually simplifies the physicogenetic analysis of developing systems is the fact that such materials exhibit preferred, even inevitable, morphologies (Ball, 2009; Newman & Comper 1990). While the degree to which this can provide insight into development pertains more to the origination of morphological motifs than to their present-day realization (Newman, 2012; Müller & Newman, 1999, 2000), it can serve as a starting point for a general theory of development. In particular, the physicalist perspective[1] suggests that once multicellularity had been achieved, the emergence of distinct body plans likely occurred with much less genetic change and at a faster pace than would be predicted by gradualistic models of evolution by natural selection (Newman et al., 2006).

In order to base a theory of development on the properties of soft, excitable media, however, it is essential to deal with the 'complexification' of developmental processes over the course of their evolution (Newman, 2011b). This historical dimension is an important and ultimately non-generic source of the hybridity of such processes. Specifically, while the morphological motifs of present-day organisms can reasonably be tied to the stereotypical outcomes of generic physical processes that acted on ancient cell aggregates, development of present-day embryos is much more than the playing out of generic physical effects. Indeed, deniers of evolution have capitalized on difficulties in understanding the profound intricacy of biological development in attempts to build a case that such complexity is 'irreducible'. A major objective of this paper is to show, on the contrary, how starting from the rough-hewn morphological templates that generic physical determination provided at the dawn of multicellular organization, the evolution of highly integrated, precise, and 'overdetermined' (i.e. characterized by redundant mechanisms supporting the same outcome) developmental programs is the expected result of entirely naturalistic processes.

At least three different factors contribute to complexification of developmental systems:

(i) The cellular nature of living matter makes it respond differently from non-living matter (however active or excitable) to generic forces, and thus to embody 'biological' versions of generic physical processes. From the outset of metazoan multicellularity then, while the physically driven organization of primitive cell clusters produced many stereotypical outcomes characteristic of such determinants, the underlying causation also had non-generic (i.e. biologically unique) aspects.

(ii) While generic physical effects produce stereotypical morphological outcomes inherent to the matter they act on, such outcomes are more reproducible from one instance to the next if prespecified *initial* and *boundary* conditions are in place before the physical determinants act. The evolutionary advent of eggs, the shapes and interior molecular patterning of which are determined before cleavage, ensured that when embryos reach the multicellular stage (when phylum-associated generative forces come into play; Newman & Bhat, 2008), they do so in a 'prepared' fashion with respect to initial and boundary conditions (Newman, 2011a).

(iii) While *generic* physical determinants (both with and without egg-stage specification) acted on ancient cell clusters to produce what became characteristic morphological motifs of animal embryos, *genetic* change over subsequent evolution profoundly altered the manner in

[1] By advocating a physicalist position I am not endorsing an ontological 'physicalism' that asserts that all phenomena, however macroscopic, are reducible to physical phenomena at a more microscopic level (Stoljar, 2009). Rather, I support a pluralistic materialism in which novel effects and regularities, not necessarily predictable from lower level phenomena, emerge at different scales (Chalmers, 2006).

which these forms were achieved. In particular, primitive physics-based quasi-developmental mechanisms were transformed into more sophisticated gene network-dependent processes of increasing reliability. These complex mechanisms, unlike any seen in the non-living world (due to their evolved organization, to be clear, not any vital principle), integrated and stabilized the simpler generic ones, though often with little or no departure from the ancient morphological templates.

In the following sections, examples of each of these complexification processes will be described. A concluding section will discuss the implications of the interplay and relationship among generic and non-generic processes for a theory of development and its evolution. As will become apparent, the hybrid nature of developmental systems, both in the multiscale nature of their operation at all phases of their evolution, and in the ever-increasing imprint of their history over time, forecloses the possibility of a unitary theory or set of laws of the kind applicable to monoscale physical systems (e.g. classical or quantum mechanics, thermodynamics). Nonetheless, the possibility of disentangling the levels of determination in such systems reasonably constitutes them as theoretical objects.

Multicellular systems as loci of 'bio-generic' physical processes

Tissue multilayering

Many embryonic tissues behave like liquids as a result of their constituent cells being independently mobile while remaining collectively cohesive (Foty et al., 1994; Steinberg, 2007). Newly formed embryos are thus spherical by default (like free liquid droplets) and topologically solid (lacking any interior spaces). Distinct tissues may also exhibit liquid-like immiscibility: if two cell populations with different affinities are present in a single aggregate, they will sort out into distinct layers, much as oil and water, shaken together, will spontaneously undergo phase separation (Foty et al., 1996; Steinberg & Takeichi, 1994).

For many years, such cell sorting and multilayering effects were widely believed to be due solely to differential adhesion strengths between the cell types. In the words of its main proponent, 'the "differential adhesion hypothesis" (DAH) explains these liquid-like tissue behaviours as consequences of the generation of tissue surface and interfacial tensions arising from the adhesion energies between motile cells' (Steinberg, 2007). In this interpretation differences in the amount of a given homophilic cell adhesion molecule (CAM; P-cadherin, for example), between different populations of cells could account for sorting and multilayering. The DAH hypothesis gained experimental confirmation when mouse L cells, a mesenchymal cell type that normally does not express any CAMs, were transfected with P-cadherin cDNA and populations expressing the adhesion molecule in substantially differing amounts were selected. When the two cell populations were intermixed, they sorted out and segregated to a sphere-within-a-sphere configuration, with the cell population expressing more CAM becoming internalized. When the two cell populations were instead first formed into separate aggregates which were subsequently allowed to fuse, the cell population expressing more CAM was enveloped by the lower-expressing population, ending up in the same configuration as the sorting experiments (Steinberg & Takeichi, 1994).

The differential adhesion mechanism was a classic example of a generic physical process—differential affinity between subunits of a liquid—being used to account for the behaviour of embryonic tissues. The physical basis of the proposed mechanism was quantitative differences in adhesive strength between like and unlike cells, attributed primarily to differences in levels of expression of a shared CAM. Qualitative CAM differences between cell types could also play a role in this picture insofar as the capacity of the proteins to mediate homotypic and heterotypic cell adhesion could be measured on a common quantitative scale (Steinberg, 2007). In this fashion, the mechanism depended on a functionally relevant analogy between cell–cell interactions and the interactions between molecules in binary liquids, in which phase separation can occur when molecules of one type bond to each other more strongly than they do molecules of the other type.

The major difficulty with this model, however, is that cell–cell attachment strength is not strictly a function of the relative affinities of CAMs. Recognizing that in addition to the number of CAMs, the dynamics of the cytoskeleton contributes to cell–cell adhesion, Brodland (2002) proposed the more general 'differential interfacial tension hypothesis' (DITH), which took into account both the density and affinity of external molecular contacts, which seal the membranes of adjacent cells to each other, and the tension on the membrane in the cells' interiors (cortical tension), which pulls the cell surfaces away from one another. While the preferential association of like vs unlike cells based on the binding of their CAMs is analogous to the preferential association of like molecules in a binary liquid, there is no functional analogy to the role of intracellular cortical tension in the interaction of molecules in a non-living liquid.

In fact, both differential adhesion and cortical tension contribute to interfacial tension between cells and tissues, and it is *interfacial tension*, not either of its components alone, that determines the relative position of the tissues and the curvature of the boundary between them, in a fused aggregate in vitro or the primordia of a developing embryo (Brodland, 2002). This implies that in some systems cortical tension, rather than differential adhesion, will be the decisive factor in sorting and boundary formation in developing systems. This was confirmed for the formation of distinct germ layers during gastrulation in zebrafish (Krieg et al., 2008) and *Xenopus* (Ninomiya et al., 2012) embryos. In vitro, differential adhesion was indeed sufficient to drive the sorting of freshly isolated cells of the *Xenopus* embryo which had been modified to overexpress or underexpress a common cadherin, confirming the earlier cell-line studies of Steinberg and Takeichi (1994), though it was unable to do so in the intact embryo. This implies that regulatory interactions in the in vivo context may oppose the spontaneous action of the purely generic process (Ninomiya et al., 2012).

Even in those cases when cortical tension is the main factor determining multi-tissue organization, adhesion per se can never be dispensed with. Specifically, adhesion is necessary to mechanically couple the surfaces of adjacent cells so that cortical tension can exert its morphogenetic effects (Maître et al., 2012).

To summarize, the generic physics of liquid phase separation predicts the relative positioning of tissues in simple cell clusters and in embryos. The basis for the interfacial tension between adjacent tissue layers (the magnitude of which will determine whether the tissues will mix or not, and if not, what the curvature of the boundary will be), is different (though for understandable mechanistic reasons) for 'cellular liquids' than for non-living liquids.[2] Nonetheless, the components of interfacial tension generation—CAMs and organized cortical cytoskeletons—must have been present in the earliest protometazoan cell aggregates (Abedin & King, 2008; Baines, 2009).

Other cellular contributions to 'generic' effects

There are several other ubiquitous properties of living cells that, like cortical tension, contribute to generic-appearing morphological outcomes. In some cases, these properties also provide biological bases for physically quantifiable properties. For example, in non-biological liquids, polar molecules (defined by charge or shape asymmetry) will self-assemble, arranging themselves into micelles or liquid crystals. In a similar fashion (though on a larger scale), if cells in a tissue mass have structurally (Rodríguez-Fraticelli et al., 2012; Wansleeben & Meijlink, 2011) or mechanically (Amack & Manning, 2012) polar properties, sorting effects can cause the mass to become hollow (analogously to a micelle) or elongated, via cell intercalation leading to 'convergent extension' (Keller et al., 2008) (analogously to an oriented domain in a liquid crystal), thus deviating from the topologically solid and spherical defaults of aggregates of non-polar cells.

Universal properties of cells that are entirely independent of those they share with liquid subunits also contribute to their response to generic physical effects. For example, cells' ability to change their biochemical states in a discontinuous fashion is based on the multi-stability of their gene regulatory networks (Guantes & Poyatos, 2008; Kaneko, 2006). Multi-stable dynamical systems are not uniquely biological, but the thousands of components specified by the genomes of animal (and plant) cells,

[2] The bubble subunits of closed-cell foams also exhibit surface tension, but they are not independently and randomly mobile like the molecular subunits of liquids or the cells of embryonic tissues. Thus, while liquids provide generic models for many tissue properties, foams do not.

the countless interactions among them, and the exquisite responsiveness of cells to their external environments, make living cells highly unusual examples of such systems, though with similar generic properties. The differentiated cells that arise from such multi-stability, which may differ in contractility, electrical excitability, stiffness, or adhesivity, represent local compositional differences in a cell aggregate or embryo that can lead to heterogeneous responses to tissue-scale physical effects.

One additional example will serve to show the scope of these generic developmental phenomena. The mathematician A. M. Turing six decades ago presented a formal model of a process in which coupled chemical reactions with different diffusivities and certain positive and negative feedback relationships among their products spontaneously produce patterned (rather than uniform) distributions of the chemicals involved (Turing, 1952). Turing's theoretical mechanism has been confirmed to operate in chemical systems (Castets et al., 1990), and similar reaction–diffusion models have been shown to be applicable to pattern formation in other non-living systems, such as ripples on sand dunes.

The excitable, multi-stable nature of cells, however, guarantees that they have much more complex response characteristics than any purely chemical reactions. In addition, the mechanisms of transport of developmentally active molecules ('morphogens') across tissue domains can be considerably more complex than the physical diffusion contemplated by Turing (Lander, 2007).

The recasting of the Turing chemical reaction–diffusion mechanism in terms of biological 'local autoactivation–lateral inhibition' (LALI; Meinhardt & Gierer, 2000) systems preserves the logic of Turing's analysis while accommodating it to biological reality. LALI mechanisms, like tissue boundary formation by interfacial tension described above, are cases in which generic physical processes are recognizably operative in their tissue-based versions, and can usefully be termed biological-generic or 'bio-generic' processes, for which dynamical patterning modules (see below) are dissociable basic components. LALI and related 'cell reactor'-morphogen mechanisms have been proposed and experimentally supported for left–right symmetry breaking in the vertebrate embryo (Müller et al., 2012), the distribution of hair follicles on mammalian skin (Sick et al., 2006), and the number and arrangement of skeletal elements in the vertebrate limb (Hentschel et al., 2004; Sheth et al., 2012; Zhu et al., 2010; Glimm et al., forthcoming), among other developmental phenomena.

Role of the egg in stabilizing the morphological phenotype

Minelli (2011) usefully notes in a recent essay that 'development does not necessarily start from an egg.' He goes on to say:

> The fact that the cellular progeny of the egg goes on diverging, morphologically and functionally, to eventually include the whole range of cell types we recognize in an adult animal, is likely to suggest that the egg represents an undifferentiated and perhaps primitive cell. Such a conclusion, however, would be ill-advised. With its enormous size, its yolk content, its specialized envelopes, the egg is one of the most specialized cell types evolved in the animal lineage (Minelli, 2011; 5).

Acknowledging both the dispensability of the egg stage for many forms of development, as well as the dramatic specializations that often characterize this stage, we have suggested that events at the egg stage are indeed of lesser consequence for body plan organization compared to the major morphogenetic and patterning events that are initiated subsequently in the clusters of cells derived from the egg (Newman, 2011a). These mid-development clusters are referred to as the 'morphogenetic stage' of an organism's ontogeny, and depending on the organism, correspond to the blastula, blastoderm, or inner cell mass. In organisms not, or not necessarily, derived from an egg, such as placozoans, sponges, and hydroids, the morphogenetic stage can correspond to multicellular propagules or buds (Hammel et al., 2009).

The morphogenetic stage is when the embryo takes on the character of a viscoelastic excitable medium. At this stage, physical forces, effects, and processes, predominantly the generic ones described above, are mobilized by products of a subset of the highly conserved developmental-genetic 'toolkit' (Carroll et al., 2004; Newman & Bhat, 2008, 2009). These 'interaction toolkit' molecules—cadherins, Hedgehog, Bone morphogenetic protein, Wnt, Notch, collagen—in association with the physical processes they respectively harness,[3] constitute a set of determinants of pattern and form: 'dynamical patterning modules' (DPMs) (Newman & Bhat, 2008, 2009). Most of the DPM-enabling molecules served single-cell activities in one or more unicellular opisthokont[4] ancestors of the Metazoa before being recruited into developmental roles with the emergence of multicellularity (King et al., 2008; Shalchian-Tabrizi et al., 2008).

The inherently multicellular DPMs are employed combinatorially to mediate the formation of aggregates, layers, lumens, folds, tubes, segments, and appendages (Newman & Bhat, 2008, 2009; Figure 6.1). Although some DPMs are common to all the animal phyla by virtue of a shared, ancient set of interaction toolkit genes, different phyla are distinguished by additional DPMs (Newman, 2011a). Indeed, the increase in body plan complexity in moving from sponges and placozoans to eumetazoan diploblasts and triplobasts can be attributed to the expansion in the repertoire of novel DPMs (Newman, 2012). In keeping with the abrupt acquisition of phylotypic genetic 'signatures' (Rokas et al., 2005), the appearance of the animal phyla in the late Precambrian-early Cambrian was compressed in time (Conway Morris, 2006; Shen et al., 2008). It has thus been suggested that the complement of DPMs is what defines phylotype identity in animals (Newman, 2011a), with a similar principle pertaining to the radiation of the multicellular plants (Hernández-Hernández et al., 2012).

Within the various phyla, variability of form is associated with evolutionary modifications in egg size and shape, and most specifically, in the intra-egg cytoplasmic heterogeneities that are implemented both before (i.e. during oogenesis) and after fertilization. Oocytes and pre- and post-fertilized eggs become internally non-uniform by two kinds of processes. In the first, common to most animal taxa (but not eutherian mammals), cytoplasmic determinants ('ooplasms') are incorporated into distinct regions of the egg during oogenesis. The second, 'egg-patterning processes' (EPPs), are physical and physicochemical effects, such as transient bursts and spatiotemporal oscillations of calcium ions, followed or accompanied by reorganization of cytoplasmic factors, induced by sperm entry or parthenogenetic activation (Newman, 2009, 2011a; Figure 6.2). Significantly, the cytoplasmic heterogeneities generated by either or both modes, though often associated with recognizable polarities and landmarks of the adult stage, do not correspond to maps or blueprints of either the embryos or the subsequently developed organism (see, for example, Freeman, 1999). They must have different functions.

While the vertebrates are a classic example of conservation of general body plan characters despite extreme disparity of egg morphology (Richardson, 1999), the indirect relationship between ooplasmic and somatic organization is also seen vividly in the nematodes (Schulze & Schierenberg, 2011). Although this phylum may contain up to one million species, apart from size and a very small number of species with diversely divergent morphology (Zullini, 2012), they all have a very similar anatomy. Notwithstanding this morphological conservation, the deployment of EPPs differs dramatically within this group. The egg of the nematode *Caenorhabditis elegans*, for example, is unpolarized before it is fertilized. Upon sperm entry the egg's cortical cytoplasm becomes reorganized, resulting in an asymmetric distribution of various factors before the first cleavage. This polarity, which is required for the establishment of the anterior–posterior axis during embryogenesis, depends on sperm-contributed microtubules (Rohrschneider & Nance, 2009). Thus, while the sperm does not attach at a preferred site on the egg, its entry point defines the future posterior pole (Munro et al., 2004; Tsai & Ahringer, 2007).

The means by which anterior–posterior symmetry is broken, and even the developmental stage at which it occurs, can be extremely different in different nematode species whose final forms are nonetheless nearly identical. In the egg of *Bursaphelenchus xylophilus*, a nematode anatomically indistinguishable from *C. elegans*, the sperm entry point becomes the future *anterior* pole of the embryo, and the pattern of cortical flow and its relation to the sperm microtubules are entirely different from those in the latter species (Hasegawa et al., 2004). In another nematode, *Romanomermis culicivorax*, the

[3] 'Harness' is used in the loose sense of providing necessary conditions for the action of a particular physical process in a cellular mass (see Newman, 2011c).

[4] Opisthokonts comprise animals, fungi, and related single-celled organisms.

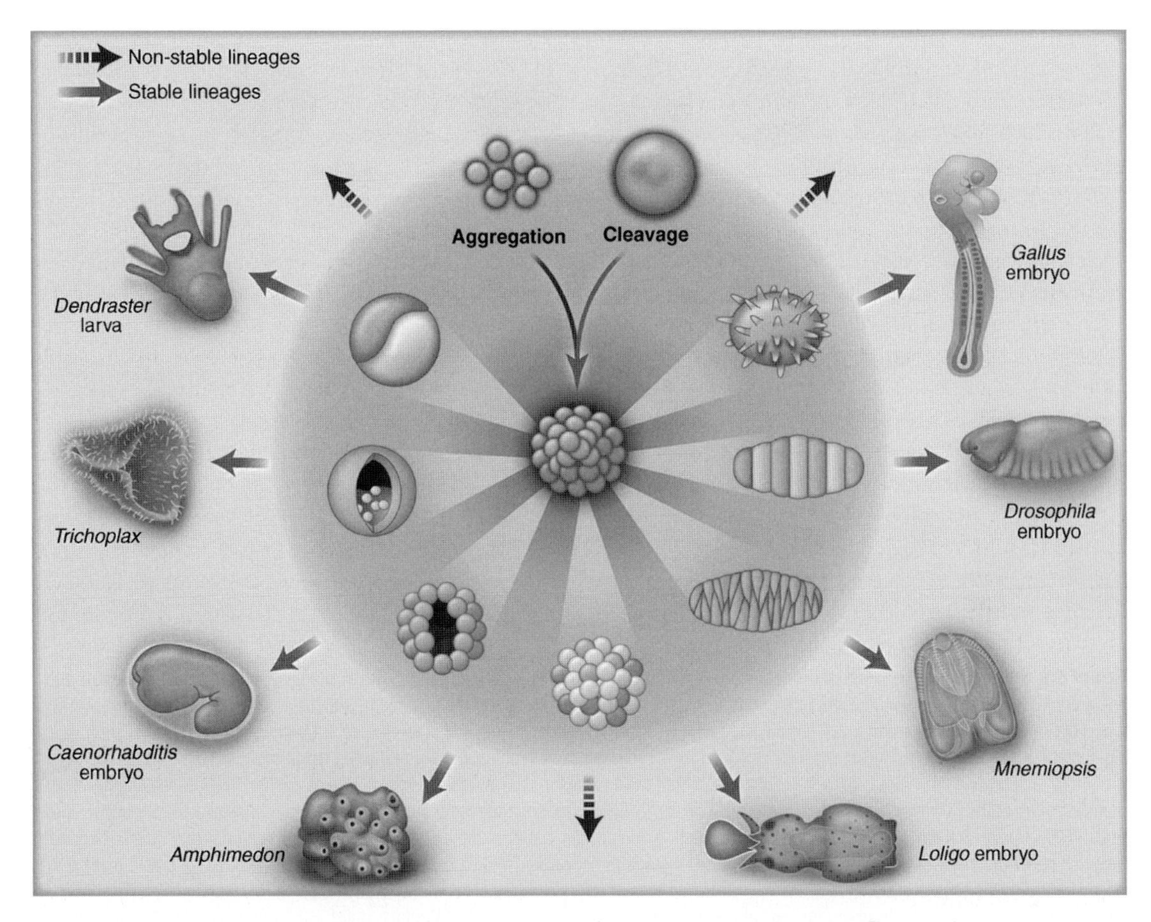

Figure 6.1 A core set of physico-genetic processes, DPMs, underlies morphogenesis of animals. Multicellular entities (center image) were formed by the aggregation of unicellular organisms or the cleavage of enlarged cells, the antecedents of eggs. The inner circle shows morphological motifs generated by some of the key DPMs at the morphogenetic stage (see text). These include (clockwise from top of inner circle) appendages, segments, elongated bodies, and primordia, coexisting alternative cell types, interior cavities, dispersed cells, and multiple layers. Genetically uniform clusters, enabled by the innovation of the egg stage of development, produced stable lineages (unbroken arrows), whereas chimeric clusters did not (broken arrows). A side effect of the egg stage of development was the generation of EPPs, a source of cytoplasmic heterogeneities that led to more reliable implementation of the DPMs at the morphogenetic stage. Present-day organisms containing some or all of the DPM-based motifs are shown in the outer circle. Clockwise from top right: vertebrate (*Gallus*) embryo, arthropod (*Drosophila*) embryo, ctenophore (*Mnemiopsis*), cephalopod (*Loligo*) embryo, demosponge (*Amphimedon*), nematode (*Caenorhabditis*) embryo, placozoan (*Trichoplax*), and echinoderm (*Dendraster*) larva. Figure adapted, with changes, from Newman (2012).

first cleavage is symmetric rather than asymmetric, and the pattern of subsequent cleavages and assignment of cell fates suggests that anterior–posterior polarity is determined in still a different fashion from either *C. elegans* or *B. xylophilus* (Schulze & Schierenberg, 2008). And in the freshwater nematode *Tobrilus diversipapillatus*, no asymmetric cleavages and no distinct cell lineages are generated until the morphogenetic stage (Schierenberg, 2005). Thus, while anterior–posterior polarity is ultimately an essential aspect of nematode anatomy, the final form of the worm appears to be independent of the way that EPPs are deployed (or not) at the egg stage to establish this polarity. This is not to suggest that the effects of the EPPs are a matter of complete indifference: the cytoplasmic patterns in a given nematode species are reproducible and precise, and their experimental disruption in certain cases has

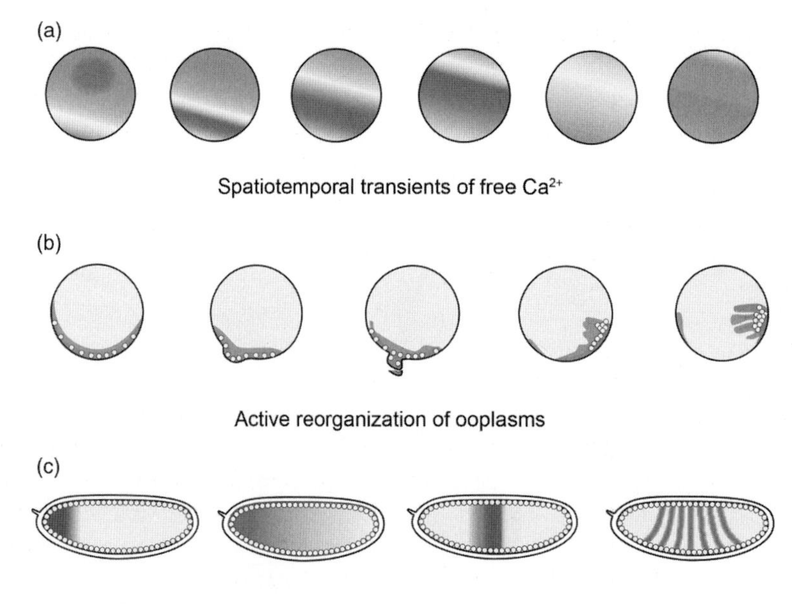

Figure 6.2 Schematic representation of different types of EPP. (a) Patterns of spatiotemporal Ca^{2+} waves in a fertilized ascidian egg. Release of Ca^{2+} from intracellular stores was imaged. Fertilization has occurred at the lower right of the first image, and the whole wave propagation sequence occurs over 72 min. The calcium wave EPP is associated with cytoplasmic reorganization and other molecular changes in several phyla. Drawing based on photographs in Dumollard et al. (2002). (b) Ooplasmic reorganization in an ascidian egg between fertilization and the start of cleavage. The site of fertilization in this case is the upper right quadrant of the left-most panel, and the cytoplasmic flows and contractions leading to the changes are associated with Ca^{2+} waves like those shown in (a). Characterization of the reorganizing ooplasmic components can be found in Sardet et al. (2007), from which this series was redrawn. (c) Generation of molecular patterns in the *Drosophila* egg cytoplasm prior to cellularization of the blastoderm. The first panel shows the distribution of *bicoid* mRNA in the newly laid egg. This maternally deposited pattern is not due to an EPP, but is transformed by an EPP, cytoskeleton-facilitated diffusion into the concentration gradient seen in the second panel. The resulting gradient of Bicoid protein initiates a series of LALI involving the regulation of gap and pair-rule genes and their products (third and fourth panels) in the excitable medium constituted by the layer of unsequestered nuclei (small circles). In this case, the EPP-generated pre-pattern reflects a segmented body plan that had originally evolved utilizing DPMs (see text). Drawing adapted from Forgacs and Newman (2005); see references therein.

been shown to be inimical to successful development (Guedes & Priess, 1997).[5]

EPPs may seem analogous to DPMs in that they are based on material properties and capabilities such as diffusion, viscoelasticity, sedimentation, convection, and chemical excitability, that are common (i.e. generic) to living and (though not typically in combination) some non-living systems (Newman & Comper, 1990). But while DPMs are inherently *multicellular*, EPPs are based on ancient *intracellular* functionalities (such as actinomyosin fluxes and the release of calcium ions from vesicular stores) that came to operate in larger, but still intracellular, spatial domains. Mesoscale physical processes do not characteristically occur on the spatial scale of a single cell (~10 μm), but in a cell enlarged by an order of magnitude or more (as it is in eggs), cell physiological functions that had evolved to operate on the microscale were transformed into novel spatiotemporal organizing effects such as pulsating Ca^{2+} transients and cytosolic flows (reviewed in Newman, 2009, 2011a).

As suggested above, the DPMs, by virtue of the generally progressively inclusive sets of genes they employ and the distinctive morphological motifs they generate, can be considered as defining determinants of the various phyla. Since DPMs function by mobilizing mesoscale physical processes in multicellular clusters, in principle they can act on morphogenetic-stage organisms without the need for any molecular

[5] See Kalinka and Tomancak (2012) for additional evidence for, and implications of, the evolutionary plasticity of early development within other animal phyla.

pre-specification patterns in an egg (Figure 6.1). When the cluster arises from cleavage of an egg, however, there is an opportunity for the egg to be rendered internally heterogeneous by the action of one or more EPPs. In that case, cleavage produces a cluster of cells that differ from one another across the cell mass in a continuous or discontinuous fashion, rendering the resulting embryo reproducibly asymmetric.

Generating such polarities is advantageous for the propagation of a biological type from one generation to the next. Embryos that pass through a polar egg stage present a set of boundary values and initial conditions for the DPMs once they are set into motion at the morphogenetic stage. Although these DPMs will be the same for a given phylotype whether or not the system had been 'prepared' in this fashion, the outcomes of physical processes are much more reliable when their boundary and initial conditions (i.e. the shape of the spatial domain and the activity of the edges in absorbing or reflecting chemical or mechanical variables, and the starting values of chemical or mechanical variables across the spatial domain) are preset rather than random (Sokolnikoff & Redheffer, 1966). Because gene regulatory networks are multi-stable, cell differentiation at the morphogenetic stage can, in principle, self-organize without any pre-existing spatial cues (Huang, 2012; Kaneko, 2006). However, pre-specification of the initial sites of differentiation by ooplasmic factors is a way of directing and standardizing morphological outcomes in the face of the intrinsic parameter sensitivity of the basic physico-genetic organizing processes, and thus stabilizing phylogenetic lineages (Newman, 2011a).

A highly unusual case in animal development, and one that illuminates the relationship between DPMs and EPPs, is the specification of segment locations in long germ-band insects such as *Drosophila* by molecular pre-patterns in the egg and syncytial embryo (Lawrence, 1992; Figure 6.2c). Unlike typical EPPs, these molecular distributions actually presage the anatomy of the developed organism (larva and adult), and their molecular components are, exceptionally, transcription factors. The origin of the long germ-band pre-patterning system in fact appears to have resulted from a heterochronic shift whereby an ancestral, multicellular, DPM-based segmentation mechanism like that seen in short germ-band insects such as the grasshopper *Schistocerca* came to operate before cellularization and was thereby converted to EPP-based patterning (Salazar-Ciudad et al., 2001).

The phylotype of an organism with a given complement of DPMs would be the same regardless of which EPPs were operative at its egg stage of development, or even if the egg remained homogenous and multicellular patterning was only initiated at the morphogenetic stage. While no animal eggs that lack EPPs have been described, some groups, like mammals, appear to lack maternally specified ooplasms (Gardner, 2005), although they exhibit robust post-fertilization cytoplasmic dynamics (Ducibella et al., 2006). When normal patterns of Ca^{2+} oscillations were perturbed in mouse embryos, developmental outcomes such as implantation rate and pre- and post-implantation embryo growth rates were impaired (reviewed in Newman, 2009).

The insertion of an egg stage into the life cycle of protometazoan organisms that may originally have developed from multicellular propagules thus enabled the 'taming' of the generic processes discussed in 'Multicellular systems as loci of 'bio-generic' physical processes'. It did so by causing them to operate on a cell population with stereotypical clade-associated boundary and initial conditions. The emergence of the egg as an evolutionary innovation thus represented a step in the progression from the 'bio-generic' period in the history of development toward the modern 'programmatic' period (Newman, 2011a; Newman et al., 2006).[6]

Evolutionary transformations of generic developmental mechanisms

The previous sections have dealt with the kinds of generic physical determinants of tissue morphogenesis

[6] My contention that development has become increasingly programmatic over the course of evolution has nothing to do with the concept of a 'genetic program' for development. The latter became a mainstay of gene-centric developmental biology and of the modern evolutionary synthesis once the role of DNA in protein synthesis became elucidated (Ernst Mayr, for instance, suggested that 'the machinery for translating the genetic program provides the "general theory of development" which embryologist are looking for most assiduously' (Mayr, 1968)). There are good reasons to reject the notion that the genome constitutes a blueprint or a program for development (Newman, 2002; Nijhout, 1990), but it is uncontroversial that canalizing evolution has led to embryogenesis occurring according to an internally determined sequence of steps.

and pattern formation that originated morphological templates among multicellular forms in the early period of animal evolution. It was seen that certain properties basic to all animal cells—cortical tension, surface and shape polarities, complex, non-linear biochemical dynamics—influenced the ways in which generic physical determinants were realized (see point (i) in 'Introduction'). The previous discussion also described how egg size and shape, and the EPPs—post-fertilization intracellular patterning processes—could refine and standardize production of these generic forms at the morphogenetic stage by specifying boundary and initial conditions for the DPMs, the characteristic physico-genetic determinants of animal development (point (ii)).

None of the described cellular responses or evolutionary changes significantly altered the characteristic morphological outcomes of those generic determinants. The morphological motifs of body plans and organs of the earliest multicellular animals resembled generic forms seen in non-living chemically and mechanically excitable soft materials. Those of present-day animals are not too different, which explains the facility with which they can be mathematically and computationally modeled using basic physical concepts (Forgacs & Newman, 2005).

But the *mechanisms* of development in present-day organisms are very much removed from the mostly generic physical processes that are hypothesized to have produced the original ontogenetic templates. While in some cases the originating generic processes are still discernable, in many other cases evolved connections among disparate basic effects have led to exquisitely controlled mechanisms that have no counterparts in the world of non-living matter (see point (iii) in 'Introduction'). The remaining parts of this section are devoted to a description of two such examples: actively regulated gradient formation and size-scaled segmentation.

Active regulation and scaling of morphogen gradients

Few physical effects of the 'middle scale' are more generic than diffusion, the dispersal of material over a spatial domain simply by 'random walk'. For molecules, the basis of the random movement is Brownian motion, a manifestation of the statistical properties of the thermal energy of the surrounding medium. Diffusion of a molecular species can even out its distribution, or if it emanates from a source, generate a gradient.

If we confine our attention to the diffusion of a molecular species through a cluster of cells, relevant questions are how the molecule's source becomes established (e.g. by environmental induction or self-organization in egg-independent forms, by pre-specification in egg-dependent organisms), and whether the source is localized (e.g. a single cell or small subcluster), or more widespread. The diffusivity of the molecule, which depends in part on its affinity to the extracellular medium in which it is transported, is also of importance. Also relevant is how the diffusible molecule is handled by the cells among which it moves. If there is no degradation or other form of removal, the entire tissue or tissue domain will become uniformly occupied by the molecule, though even transient, non-uniform distributions may be biologically efficacious (see below). If there is a localized source and sink, the resulting gradient will decline in a linear fashion. But if degradation occurs uniformly over the domain, the fall-off will be exponential (Lander, 2007). None of this takes the role of molecular diffusion in cellular systems outside the realm of generic mesoscale physical systems.

For gradients to have a developmental role, different concentrations of the diffusible molecule must elicit different effects from the cells exposed to it. In such cases, the molecule is referred to as a 'morphogen'. If the response to the gradient is binary, the effect is termed 'induction'. Since many non-living chemical systems also jump between two qualitatively different states in response to a continuously varying effector (e.g. the formation of a precipitate when a solution becomes saturated), such switchable systems can still be considered generic. If, instead, the cells respond to different values of the gradient by exhibiting more than two phenotypes, the morphogen is thus said to provide 'positional information'. Although originally conceived of acting via multiple preset, genomic, concentration-dependent thresholds (Wolpert, 1971), such 'gradient-interpretation' phenomena are now seen to be much more complex, involving dynamical feedbacks, self-organization, and combinations of durational and concentration-dependent effects (reviewed in Jaeger et al., 2008; see also Balaskas et al., 2012 and Nahmad & Lander, 2011). However they are realized, morphogen gradient–differentiation response systems are highly non-generic, in that they have no counterpart in non-biological systems. These systems can only have arisen by evolutionary departure from the generic diffusion-based mechanisms, though the morphological templates laid down by the original mechanisms may still persist.

Leaving aside the question of the cellular interpretation of gradients, we can focus on gradient formation itself. Recent work has shown that what appear to be simple concentration gradients in developing embryos can, alternatively, be generated largely generically (i.e. by simple diffusion with a uniform cellular sink) or in a mechanistically intricate fashion, involving active, conditional responses by cells. In the case of FGF8 morphogen gradients in living zebrafish embryos, for example, such gradients are established and maintained by two essential factors: fast, free diffusion of single molecules away from the source through the extracellular medium, and a sink-type function of the receiving cells, regulated by receptor-mediated endocytosis. Using the technique of fluorescence-correlation spectroscopy in the ECM of early zebrafish embryos, Yu et al. (2009) were able to visualize single-molecule dynamics of transport of FGF8, a key morphogen in the gastrulation in this animal (Furthauer et al., 2004). FGF8 was found to have a diffusion coefficient two orders of magnitude greater than that of Decapentaplegic (Dpp), a molecule known to spread by a more complex, receptor-dependent mechanism, in the *Drosophila* larval wing disc (Kruse et al., 2004). The investigators also measured the attenuation of the FGF8 gradient from an injected localized source of fluorescent-tagged protein and found its shape to be consistent with uniform breakdown or uptake by the cells in the morphogen's path. This was confirmed when experimental enhancement of endocytosis decreased the range of the gradient (Yu et al., 2009).

The mechanisms of formation of other developmental gradients, such Dpp in the *Drosophila* wing disc mentioned above, in which the morphogen's distribution is shaped by endocytosis by the target tissue (Kruse et al., 2004) and Wg in the same primordium, where the gradient's robustness depends on its self-enhanced degradation (Eldar et al., 2003), depart significantly from free diffusion. Presumably (though we have no evidence for this at present), these complex, bio-generic (in the sense defined above) and non-generic (dynamical, feedback-dependent) gradient-forming mechanisms would be elaborations, over the course of evolution, of more primitive mechanisms based on diffusion alone.

The described mechanisms of gradient formation, whether generic, quasi-generic, or fully dynamical, do not generally exhibit *scaling*, the preservation of the gradient's shape with changes in the size of the tissue over which it forms (see the discussion of this issue in Jaeger et al., 2008). When bodies or organs change in size by growth during development or genetic alteration of the phenotype, mechanisms such as that described for FGF8 in the zebrafish gastrula would lead to gradients with altered slopes. But the distribution of Dpp in the *Drosophila* larval disc, for example, actually scales such that the gradient slope remains the same over at least a twofold increase in disc size (Wartlick et al., 2011). This is important in the growth control of the disc by Dpp, permitting the morphogen to control cell proliferation in a uniform, position-independent fashion (Wartlick et al., 2011) despite its non-uniform distribution, as well as in Dpp's control of the positions of veins in wings over a range of sizes (Hamaratoglu et al., 2011).

Two recent studies have shown that scaling of the Dpp gradient in the wing disc is brought about by an 'expansion–repression' mechanism. This is a network motif composed of two diffusible, functionally interconnected molecules, a morphogen and an 'expander'. The expander facilitates spread of the morphogen, which in turn regulates the expander's synthesis (Ben-Zvi & Barkai, 2010). In the case of the Dpp system, the 'expander' is the protein Pentagone (Pent). Dpp represses Pent production, limiting its secretion to one end of the morphogenetic field, whereas Pent increases Dpp diffusion across the entire field by downregulating a proteoglycan that binds to it. Pent essentially measures the system size: it continues to accumulate, and thus the Dpp gradient continues to expand, until the Dpp levels are sufficiently high to repress Pent expression, even at the edge of the field (Ben-Zvi et al., 2011; Hamaratoglu et al., 2011).

The expansion–repression mechanism is highly non-generic: it is all but impossible for it to have originated concurrently with the primitive developmental gradients of protometazoan organisms.[7] Nonetheless, this network motif was plausibly acquired in the course of evolution in reinforcement of the shape of a primordial Dpp-like diffusion gradient that lacked the scaling properties of its present-day counterpart.

Segment scaling along the vertebrate axis

A well-established example of the production of a morphological character by generic physical mechanisms acting in concert is the formation of somites, paired blocks of tissue that emerge in a sequential cranial-caudal direction along the vertebrate body axis during embryogenesis. As originally proposed in the 1970s, at the successive points where a periodically changing cell state of the presomitic mesoderm (the 'clock') takes on a critical value in conjunction with the arrival of a slowly moving front of potentially changed cell behaviour traveling along the embryo's length (the 'wavefront'), the cell mass becomes individualized from the adjacent tissue, forming a somite (Cooke & Zeeman, 1976).

Two decades later, experimental evidence was presented for a formally related mechanism for somitogenesis (Palmeirim et al., 1997). This mechanism included an intracellular biochemical clock comprised of the Notch pathway transcriptional switch Hes1 exerting its effects in an oscillatory fashion because of a time-lag between transcription of its mRNA and translation of its protein (i.e. a molecular biological quirk with dynamical side-effects; Lewis, 2003; Monk, 2003), and an inhibitory gradient of the morphogen FGF8, with its source at the embryo's tail tip. The elongation of the embryo led progressively more caudal axial tissue to reside in the subthreshold region of the gradient. A puzzling aspect of the 'wavefront' with no obvious role in the somite-forming mechanism was that the *phase* of the Hes1 oscillation was itself graded along the embryonic axis, leading any given clock time, including the critical value for segment formation (though no actual molecule), to periodically sweep along the embryo's length as the concentrations oscillated in place within the cells (Palmeirim et al., 1997).

Although (apart from embryo elongation) two independent processes, a biochemical periodicity and a gradient, are involved in this patterning mechanism, their coordinated action in producing somites could plausibly be viewed as historically incidental. Biochemical, and hence behavioural, oscillations are physically inevitable features of cell physiology (Reinke & Gatfield, 2006), and their synchronization through weak, non-specific interactions is equally so (Strogatz, 2003). As discussed above, gradients can form by simple diffusion, needing only a localized source of a secreted molecule. While the formation of a series of somites depends on a specific balance between the oscillation period and the gradient shape, unsegmented pre-vertebrates harbouring these two generic processes could have flourished for a long period before this 'somitogenic ratio' was stumbled upon.

While the clock-and-wavefront mechanism could have thus been a determinant in the origination of vertebrate segmentation, it turns out to be oversimplified as a model for somitogenesis in present-day vertebrates. This becomes evident when the scaling behaviour of somitogenesis is considered. It had previously been shown that surgical manipulations of *Xenopus* embryos that reduce embryo size lead to fewer cells being allocated to each somite, which thus become proportionally smaller, though their species-characteristic number and relative position were both conserved (Cooke, 1975, 1978). Lauschke and co-workers recently explored the mechanism of this phenomenon in pre-somitic mouse mesenchyme in tissue culture, where features of the clock-and-wavefront process were maintained (i.e. somites formed at a fixed distance from a localized source of FGF8) and, remarkably, the scaling behaviour was preserved (Lauschke et al., 2013). Employing real-time imaging of Notch activity, these investigators found that absolute segment size is determined by the value of the Hes1 oscillator *phase gradient*, that is, the slope of the phase difference from one point to another along the axis. This slope becomes adjusted in tissue domains of reduced size

[7] See De Robertis (2006) and references therein for an example of a highly complex, evolved set of relationships between simpler gradient-based mechanisms as the basis for a highly regulative developmental phenomenon in amphibian embryos, Spemann's organizer.

such that the full range of phases is unchanged as are the numbers of somites generated.

The mechanism of conservation of the phase gradient is unknown, specifically, how the expression of the clock genes in individual cells becomes recalibrated when the size of the tissue mass is altered. Since the phase gradient is a kinetic relationship between concentration changes of molecules that is not embodied in any one molecule or its concentration, this scaling is different from the Dpp scaling based on molecular gradients described above (Ben-Zvi et al., 2011; Hamaratoglu et al., 2011). And while the scale-independent formation of species-characteristic numbers of somites does not disconfirm the clock-and-wavefront model in its entirety and may indeed have evolved to stabilize its morphological outcomes, it employs non-generic mechanisms that were not plausibly in place in the earliest segmented vertebrate ancestors.

Conclusion

An understanding of development cannot be complete without an account of why organisms have the morphological features they do. A 'laws-of-form' tradition extending back at least to Johann Wolfgang von Goethe, represented by such figures as Étienne Geoffroy Saint-Hilaire, Richard Owen, William Bateson, D'Arcy W. Thompson, and in more recent years, Brian Goodwin, took seriously the connection between morphological evolution and development well before the rise of contemporary evodevo. Nonetheless, the evidence for stereotypical morphological motifs in living tissues was set aside by mainstream developmental biology, which takes existing forms as given, seeking only to describe, in cellular and molecular terms, how they are generated. Comparative developmental biology does better in this regard, since the conservation and variability of body plans and organ forms despite changes in the developmental systems by which they are realized provides insight into the plasticity and constraints of such generative processes. However, the inveterate gene-centrism of such analyses, along with the wide acceptance of the Darwinian modern evolutionary synthesis, which asserts that there are no inherent features or motifs of the morphological phenotype (which is held to emerge incrementally, with functional adaptation being the only creative force), have worked against the realization of a general theory of development (Linde-Medina, 2010; Newman & Linde-Medina, 2013).

Here I have attempted to update the laws-of-form approach by linking the modern physics of materials with contemporary knowledge of genetic determinants, particularly the cell interaction components of the developmental toolkit. But because developmental processes are also historically determined, the law-like behaviours they exhibit are inherently hybrid and complex. In particular, I treat evolutionary time, the biological fifth dimension of development (Newman, 1995), as central to the understanding of form, but not in the uniformitarian, unconstrained manner of the modern evolutionary synthesis: in the framework presented, the relative roles of physics and genetics change over time.

Three stages in the evolution of developmental systems were described. The first was a largely generic physical stage, in which aggregates of unicellular, opisthokont ancestors of the Metazoa mobilized various mesoscale physical processes and effects so as to generate an array of multicellular motifs—multilayering, lumens, folds, segments, appendages, and so forth. The scale and rheological properties of the aggregates render them susceptible to generic organizing principles. But there is also biological specificity involved, even at this early stage, since the genes of the interaction toolkit that are required to mobilize the relevant physical effects are not present in every lineage (Newman, 2011c, 2012). The non-inclusion or loss of some of these genes in various lineages led to the formation of 'proto-phyla' with distinct but overlapping sets of dynamical patterning modules (DPMs). Furthermore, because the subunits of the mouldable materials, cells, had non-generic features (cortical tension, multi-stable compositional states), the action of generic processes would have had additional biological peculiarities. Notwithstanding this multiscale interplay between physical and biological determinants, the resulting forms would have typically resembled those also seen in non-living soft, excitable materials (the hallmark of 'bio-generic' processes).

The second stage discussed took place in multicellular forms generated by division or cleavage of eggs, which were suggested to have been enlarged, modified released cells of the protometazoan 'aggregate organisms' (Newman, 2011a). The *genetic uniformity* of the cellular progeny of the resulting egg or zygote would have promoted the evolutionary stability of these new types of organism (Grosberg & Strathmann, 1998), whereas its *biochemical nonuniformity* would have promoted their developmental reliability. The latter was due in part to the operation of EPPs prior to cleavage, leading to polarities in the clusters of cells of the mid-embryonic 'morphogenetic stage' at which material attributes of the primitive aggregates were recapitulated. In particular, the phylum-characteristic DPMs of these organisms could now be implemented with reproducible (though variable, according to major clade), initial and boundary conditions. The result of this was internalized specification of developmental parameters that in the more primitive forms must have been subject to random effects.

The third and most recent stage of the evolution of development is proposed to have involved the stabilization, integration, and intensification of the processes leading to forms that by now had established themselves as inhabitants of complex ecosystems. By the hypothesis presented here, little that occurred during this period would have disrupted the morphological outcomes that were set in place during the two earlier periods. By the play of trial-and-error within already adapted morphological templates, mechanisms of reinforcement and reliability, such as those represented by the 'expansion–repression' mechanism of gradient scaling, or the phase gradient scaling of segmentation, would have produced forms with ever more dependable developmental programming. The phylum-specific DPMs would still come into play at an embryo's morphogenetic stage but now with the addition of highly non-generic, fail-safe mechanisms.

Box 6.1 lists examples of generic physical processes of mesoscale systems along with examples

Box 6.1

(a) Examples of generic processes of mesoscale physicochemical systems
Liquid phase separation
Molecular diffusion
Viscous flow
Buckling and wrinkling of elastic sheets
Multi-stability in open chemical systems
Chemical oscillation
Reaction–diffusion (Turing-type) instability
Micelle formation
Liquid crystal phase transformations

(b) Examples of bio-generic processes in multicellular aggregates
Differential interfacial tension-regulated tissue boundary and engulfment dynamics
Actively regulated morphogen gradients
Tissue flow via cytoskeletal, cell–cell and ECM rearrangements
Tissue folding via elastic interactions with ECM
Alternative states of differentiation of gene regulatory networks
Gene regulatory and other biochemical oscillations
Cell pattern formation by
LALI and other cell-morphogen reaction-diffusion analogues
Lumen formation via cell surface polarization
Convergent extension via cell intercalation

(c) Examples of non-generic processes in animals and their embryos
Interpretation of morphogen gradients
Scale-invariance of slope of morphogen gradients
Organizer-dependent regulation of morphogenesis and patterning
Scale-invariance of somite number

of the bio-generic versions of these mechanisms embodied in tissues. Also listed are some non-generic developmental processes and mechanisms proposed to have been derived from the bio-generic ones over the course of evolution.

The declining efficacy, implied in this view, of genetic change in altering the forms of bodies and organs once they have been set in place by generic and bio-generic physical processes is closely related to what Rupert Riedl termed 'burden' (Riedl, 1978) and William Wimsatt, 'generative entrenchment' (Wimsatt, 2007). Such phenomena have been claimed to be consistent with the modern synthesis notion that macroevolutionary change is due exclusively to microevolutionary (gradualistic) processes (Schoch, 2010). But when considered in the light of evidence discussed here and elsewhere (e.g. Newman & Bhat, 2009; Newman et al., 2006) that phylum-level diversification occurred early, in morphogenetically uncanalized lineages, burden/generative entrenchment (at least in the version discussed here) can, in fact, be recognized as impediments to gradualistic evolution.

The three-stage scheme provides a framework for conceptualizing the evolution of animal development. It addresses the most primitive origins of multicellular organization: the establishment early on of the characteristic metazoan morphological motifs, with disparate body plans organized by a largely common set of conserved genes with roots in single-celled ancestors. It describes the emergence of eggs as morphological novelties with an optional but important stabilizing role in the generation of form, and indicates why, although the egg is developmentally prior to mid-embryonic stages in the lineages in which it appears, its role in determining phylum-associated morphology is minor compared to events at the multicellular morphogenetic stage, thus explaining the embryonic hourglass (Newman, 2011a).

In addition, the proposed scheme can be seen as the basis of a theory of development itself. This is an ambitious claim, since complex, multiscale systems are not typical 'theoretical objects': entities (like planetary motion, atomic structure, ideal gases) whose behaviour can be characterized by axioms, laws, or mathematical models. To complicate things further, according to the description presented above, historical legacy is a formal determinant of present-day developmental systems, whose forms derive from the generic physical processes that were active at their origination but realized in each generation by partly non-generic processes acquired in the subsequent period. Although many scientific and philosophical theorists reject the notion that historically determined objects can be described by scientific laws (Gould, 1987), other thinkers have grappled with this possibility in terms of interdependence of levels of determination with qualitatively different developmental propensities (Althusser, 1969; Levins & Lewontin, 1985; Marx, 1967 [1867]). The characterization of developmental systems presented in this paper suggests that their multiscale nature and historicity can be productively analysed in relation to one another.

Put another way, while products of evolution, or more generally, of history, are indeed unique, they are also the outcomes of uniform (i.e. unchanging with time) natural processes, which continue to operate within them, albeit 'unevenly'. Thus, while there are no 'historical laws', insofar as we can model and manipulate the determinants of evolved entities (e.g. mesoscale physics, gene regulatory networks in the case of developmental systems), such items are indeed appropriate objects of theory. It is also clear, however, that we cannot expect the same level of predictability and explanatory power from the theoretical analysis of complex, multiscale products of evolution as with theories of physical systems.

With its focus on the morphogenetic stage as an echo of the most primitive multicellular forms, this framework avoids the 'adultocentrism' of most theoretical analyses of development, including the genetic program and positional information concepts. Since phylum-specific morphological motifs can be generated not only from fertilized eggs, but also from cell clusters and asexual propagules, the notion that development is a circumscribed sequence of changes that can be understood only in relation to the formation of a paradigmatic adult from a plan within a single cell, has rightly been criticized as unhelpfully teleological (Minelli, 2003, 2011). The discussion above has shown that the building blocks of animal form are intrinsic material properties of developmental systems, and that their assembly into embryonic, larval, adult, and regenerative forms in present-day organisms is likely to be more an expression of the inherent morphological capabilities of such systems and their evolutionary integration and refinement than of selection for, and genetic encoding in the egg, of anatomical end-points arrived at primarily by gradualistic functional adaptation.

References

Abedin, M., and King, N. (2008). The premetazoan ancestry of cadherins. *Science*, **319**, 946–8.

Althusser, L. (1969). Contradiction and overdetermination. In L. Althusser, *For Marx*. Trans. B. Brewster. Allen Lane, London, pp. 87–128.

Amack, J.D., and Manning, M.L. (2012). Knowing the boundaries: extending the differential adhesion hypothesis in embryonic cell sorting. *Science*, **338**, 212–15.

Baines, A.J. (2009). Evolution of spectrin function in cytoskeletal and membrane networks. *Biochemical Society Transactions*, **37**, 796–803.

Balaskas, N., Ribeiro, A., Panovska, J., et al. (2012). Gene regulatory logic for reading the Sonic Hedgehog signaling gradient in the vertebrate neural tube. *Cell*, **148**, 273–84.

Ball, P. (2009). *Nature's Patterns: A Tapestry in Three Parts*, Oxford University Press, Oxford; New York.

Beloussov, L. (1998). *The Dynamic Architecture of a Developing Organism*. Kluwer Academic Publishers, Dordrecht.

Ben-Zvi, D., and Barkai, N. (2010). Scaling of morphogen gradients by an expansion-repression integral feedback control. *Proceedings of the National Academy of Sciences USA*, **107**, 6924–9.

Ben-Zvi, D., Pyrowolakis, G., Barkai, N., et al. (2011). Expansion-repression mechanism for scaling the Dpp activation gradient in *Drosophila* wing imaginal discs. *Current Biology*, **21**, 1391–6.

Brodland, G.W. (2002). The differential interfacial tension hypothesis (DITH): a comprehensive theory for the self-rearrangement of embryonic cells and tissues. *Journal of Biomechanical Engineering*, **124**, 188–97.

Carroll, S.B., Grenier, J.K., and Weatherbee, S.D. (2004). *From DNA to Diversity: Molecular Genetics and the Evolution of Animal Design*. Blackwell, Malden.

Castets, V., Dulos, E., Boissonade, J., et al. (1990). Experimental evidence of a sustained standing Turing-type nonequilibrium chemical pattern. *Physical Review Letters*, **64**, 2953–6.

Chalmers, D.J. (2006). Strong and weak emergence. In P. Davies and P. Clayton, eds., *The Re-Emergence of Emergence*. Oxford University Press, Oxford.

Conway Morris, S. (2006). Darwin's dilemma: the realities of the Cambrian 'explosion'. *Philosophical Transactions of the Royal Society of London B*, **361**, 1069–83.

Cooke, J. (1975). Control of somite number during morphogenesis of a vertebrate, *Xenopus laevis*. *Nature*, **254**, 196–9.

Cooke, J. (1978). Somite abnormalities caused by short heat shocks to pre-neurula stages of *Xenopus laevis*. *Journal of Embryology and Experimental Morphology*, **45**, 283–94.

Cooke, J., and Zeeman, E.C. (1976). A clock and wavefront model for control of the number of repeated structures during animal morphogenesis. *Journal of Theoretical Biology*, **58**, 455–76.

De Gennes, P.G. (1992). Soft matter. *Science*, **256**, 495–7.

De Robertis E.M. (2006). Spemann's organizer and self-regulation in amphibian embryos. *Nature Reviews Molecular Cell Biology*, **7**, 296–302.

Ducibella, T., Schultz, R.M., and Ozil, J.P. (2006). Role of calcium signals in early development. *Seminars in Cell and Developmental Biology*, **17**, 324–32.

Dumollard, R., Carroll, J., Dupont, G., et al. (2002). Calcium wave pacemakers in eggs. *Journal of Cell Science*, **115**, 3557–64.

Eldar, A., Rosin, D., Shilo, B. Z., et al. (2003). Self-enhanced ligand degradation underlies robustness of morphogen gradients. *Developmental Cell*, **5**, 635–46.

Forgacs, G., and Newman, S.A. (2005). *Biological Physics of the Developing Embryo*. Cambridge University Press, Cambridge.

Foty, R.A., Forgacs, G., Pfleger, C.M., et al. (1994). Liquid properties of embryonic tissues: measurement of interfacial tensions. *Physical Review Letters*, **72**, 2298–301.

Foty, R.A., Pfleger, C.M., Forgacs, G., et al. (1996). Surface tensions of embryonic tissues predict their mutual envelopment behavior. *Development*, **122**, 1611–20.

Freeman, G. (1999). Regional specification during embryogenesis in the inarticulate brachiopod *Discinisca*. *Developmental Biology*, **209**, 321–39.

Furthauer, M., Van Celst, J., Thisse, C., et al. (2004). FGF signalling controls the dorsoventral patterning of the zebrafish embryo. *Development*, **131**, 2853–64.

Gardner, R.L. (2005). The case for prepatterning in the mouse. *Birth Defects Research C Embryo Today*, **75**, 142–50.

Glimm, T., Bhat, R., and Newman, S.A. (2014). Modeling the morphodynamic galectin patterning network of the developing avian limb skeleton. *Journal of Theoretical Biology*. http://dx.doi.org/10.1016/j.jtbi.2013.12.004

Gould, S. J. (1987). *Time's Arrow, Time's Cycle: Myth and Metaphor in the Discovery of Geological Time*. Harvard University Press, Cambridge.

Grosberg, R.K., and Strathmann, R. (1998). One cell, two cell, red cell, blue cell: the persistence of a unicellular stage in multicellular life histories. *Trends in Ecology and Evolution*, **13**, 112–16.

Guantes, R., and Poyatos, J.F. (2008). Multistable decision switches for flexible control of epigenetic differentiation. *PLoS Computational Biology*, **4**, e1000235.

Guedes, S., and Priess, J.R. (1997). The *C. elegans* MEX-1 protein is present in germline blastomeres and is a P granule component. *Development*, **124**, 731–9.

Hamaratoglu, F., De Lachapelle, A.M., Pyrowolakis, G., et al. (2011). Dpp signaling activity requires Pentagone to scale with tissue size in the growing *Drosophila* wing imaginal disc. *PLoS Biology*, **9**, e1001182.

Hammel, J.U., Herzen, J., Beckmann, F., et al. (2009). Sponge budding is a spatiotemporal morphological patterning process: insights from synchrotron radiation-based x-ray microtomography into the asexual reproduction of *Tethya wilhelma*. *Frontiers in Zoology*, **6**, 19.

Hasegawa, K., Futai, K., Miwa, S., et al. (2004). Early embryogenesis of the pinewood nematode *Bursaphelenchus xylophilus*. *Development Growth and Differentiation*, **46**, 153–61.

Hentschel, H.G., Glimm, T., Glazier, J.A., et al. (2004). Dynamical mechanisms for skeletal pattern formation in the vertebrate limb. *Proceedings of the Royal Society of London, Series B Biological Sciences*, **271**, 1713–1722.

Hernández- Hernández, V., Niklas, K.J., Newman, S.A., et al. (2012). Dynamical patterning modules in plant development and evolution. *International Journal of Developmental Biology*, **56**, 661–74

Huang, S. (2012). The molecular and mathematical basis of Waddington's epigenetic landscape: a framework for post-Darwinian biology? *Bioessays*, **34**, 149–57.

Jaeger, J., Irons, D., and Monk, N. (2008). Regulative feedback in pattern formation: towards a general relativistic theory of positional information. *Development*, **135**, 3175–83.

Kalinka, A. T., and Tomancak, P. (2012). The evolution of early animal embryos: conservation or divergence? *Trends in Ecology and Evolution*, **27**, 385–93.

Kaneko, K. (2006). *Life: An Introduction to Complex Systems Biology*. Springer, Berlin-New York.

Keller, R., Shook, D., and Skoglund, P. (2008) The forces that shape embryos: physical aspects of convergent extension by cell intercalation. *Physical Biology*, **5**, 15007.

King, N., Westbrook, M.J., Young, S.L., et al. (2008). The genome of the choanoflagellate *Monosiga brevicollis* and the origin of metazoans. *Nature*, **451**, 783–8.

Krieg, M., Arboleda-Estudillo, Y., Puech, P.H., et al. (2008). Tensile forces govern germ-layer organization in zebrafish. *Nature Cell Biology*, **10**, 429–36.

Kruse, K., Pantazis, P., Bollenbach, T., et al. (2004). Dpp gradient formation by dynamin-dependent endocytosis: receptor trafficking and the diffusion model. *Development*, **131**, 4843–56.

Lander, A.D. (2007). Morpheus unbound: reimagining the morphogen gradient. *Cell*, **128**, 245–56.

Lauschke, V.M., Tsiairis, C.D., François, P., et al. (2013). Scaling of embryonic patterning based on phase-gradient encoding. *Nature*, **493**, 101–5.

Lawrence, P.A. (1992). *The Making of a Fly: The Genetics of Animal Design*, Blackwell Scientific Publications, Oxford-Boston.

Levin, M. (2012). Morphogenetic fields in embryogenesis, regeneration, and cancer: non-local control of complex patterning. *Biosystems*, **109**, 243–61.

Levine, H., and Ben-Jacob, E. (2004). Physical schemata underlying biological pattern formation—examples, issues and strategies. *Physical Biology*, **1**, 14–22.

Levins, R., and Lewontin, R.C. (1985). *The Dialectical Biologist*. Harvard University Press, Cambridge.

Lewis, J. (2003). Autoinhibition with transcriptional delay: a simple mechanism for the zebrafish somitogenesis oscillator. *Current Biology*, **13**, 1398–408.

Linde-Medina M. (2010). Natural selection and self-organization: a deep dichotomy in the study of organic form. *Ludus Vitalis*, **18**, 25–56.

Maître, J.L, Berthoumieux, H., Krens, S.F., et al. (2012). Adhesion functions in cell sorting by mechanically coupling the cortices of adhering cells. *Science*, **338**, 253–6.

Marx, K. (1967; orig. pub. 1867). *Capital: A Critique of Political Economy* (trans. Moore, S. and Aveling, E.). International Publishers, New York.

Mayr, E. (1968). Comments on 'Theories and hypotheses in biology.' *Boston Studies in the Philosophy of Science*, 5, 450–6.

Meinhardt, H., and Gierer, A. (2000). Pattern formation by local self-activation and lateral inhibition. *Bioessays*, **22**, 753–60.

Minelli, A. (2003). *The Development of Animal Form: Ontogeny, Morphology, and Evolution*. Cambridge University Press, Cambridge.

Minelli, A. (2011). Animal development: an open-ended segment of life. *Biological Theory*, **6**, 4–15.

Monk, N.A. (2003). Oscillatory expression of Hes1, p53, and NF-κB driven by transcriptional time delays. *Current Biology*, **13**, 1409–1413.

Müller, G.B., and Newman, S.A. (1999). Generation, integration, autonomy: three steps in the evolution of homology. In G.K. Bock and G. Cardew, eds., *Homology* (Novartis Foundation Symposium 222). Wiley, Chichester, pp. 65–73.

Müller, P., Rogers, K.W., Jordan, B.M., et al. (2012). Differential diffusivity of Nodal and Lefty underlies a reaction-diffusion patterning system. *Science*, **336**, 721–4.

Munro, E., Nance, J., and Priess, J.R. (2004). Cortical flows powered by asymmetrical contraction transport PAR proteins to establish and maintain anterior-posterior polarity in the early *C. elegans* embryo. *Developmental Cell*, **7**, 413–24.

Nahmad, M., and Lander, A.D. (2011). Spatiotemporal mechanisms of morphogen gradient interpretation. *Current Opinion in Genetics and Development*, **21**, 726–31.

Newman, S.A. (1995). Interplay of genetics and physical processes of tissue morphogenesis in development and evolution: The biological fifth dimension. In D. Beysens, G. Forgacs, and F. Gaill, eds., *Interplay of Genetic and Physical Processes in the Development of Biological Form*. World Scientific, Singapore, pp. 3–12.

Newman, S.A. (2002). Developmental mechanisms: putting genes in their place. *Journal of Biosciences*, **27**, 97–104.

Newman, S.A. (2009). E.E. Just's 'independent irritability' revisited: the activated egg as excitable soft matter. *Molecular Reproduction and Development*, **76**, 966–74.

Newman, S.A. (2011a). Animal egg as evolutionary innovation: a solution to the 'embryonic hourglass' puzzle. *Journal of Experimental Zoology Part B Molecular and Developmental Evolution*, **316**, 467–83.

Newman, S.A. (2011b). Complexity in organismal evolution. In C. Hooker, ed., *Philosophy of Complex Systems*. Elsevier, Amsterdam, pp. 335–54.

Newman, S.A. (2011c). The developmental specificity of physical mechanisms. *Ludus Vitalis*, **19**, 343–51.

Newman, S.A. (2012). Physico-genetic determinants in the evolution of development. *Science*, **338**, 217–19.

Newman, S.A., and Bhat, R. (2008). Dynamical patterning modules: physico-genetic determinants of morphological development and evolution. *Physical Biology*, **5**, 15008.

Newman, S.A., and Bhat, R. (2009). Dynamical patterning modules: a 'pattern language' for development and evolution of multicellular form. *International Journal of Developmental Biology*, **53**, 693–705.

Newman, S.A., and Comper, W. D. (1990). 'Generic' physical mechanisms of morphogenesis and pattern formation. *Development*, **110**, 1–18.

Newman, S.A., Forgacs, G., and Müller, G.B. (2006). Before programs: the physical origination of multicellular forms. *International Journal of Developmental Biology*, **50**, 289–99.

Newman, S.A., and Linde-Medina, M. (2013). Physical determinants in the emergence and inheritance of multicellular form. *Biological Theory*, **8**, 274–85.

Newman, S.A., and Müller, G.B. (2000). Epigenetic mechanisms of character origination. *Journal of Experimental Zoology Part B (Molecular and Developmental Evolution)*, **288**, 304–17.

Nijhout, H.F. (1990). Metaphors and the roles of genes in development. *BioEssays*, **12**, 441–6.

Ninomiya, H., David, R., Damm, E.W., et al. (2012). Cadherin-dependent differential cell adhesion in *Xenopus* causes cell sorting in vitro but not in the embryo. *Journal of Cell Science*, **125**, 1877–83.

Palmeirim, I., Henrique, D., Ish-Horowicz, D., et al. (1997). Avian *hairy* gene expression identifies a molecular clock linked to vertebrate segmentation and somitogenesis. *Cell*, **91**, 639–48.

Reinke, H., and Gatfield, D. (2006). Genome-wide oscillation of transcription in yeast. *Trends in Biochemical Sciences*, **31**, 189–91.

Richardson, M. K. (1999). Vertebrate evolution: the developmental origins of adult variation. *Bioessays*, **21**, 604–13.

Riedl, R. (1978). *Order in Living Organisms: A Systems-Analysis of Evolution*. Wiley, New York.

Rodríguez-Fraticelli, A.E., Auzan, M.A., Alonso, M.A., et al. (2012). Cell confinement controls centrosome positioning and lumen initiation during epithelial morphogenesis. *Journal of Cell Biology*, **198**, 1011–23.

Rohrschneider, M.R., and Nance, J. (2009). Polarity and cell fate specification in the control of *Caenorhabditis elegans* gastrulation. *Developmental Dynamics*, **238**, 789–96.

Rokas, A., Kruger, D., and Carroll, S.B. (2005). Animal evolution and the molecular signature of radiations compressed in time. *Science*, **310**, 1933–8.

Salazar-Ciudad, I., Solé, R., and Newman, S.A. (2001). Phenotypic and dynamical transitions in model genetic networks. II. Application to the evolution of segmentation mechanisms. *Evolution & Development*, **3**, 95–103.

Sardet, C., Paix, A., Prodon, F., et al. (2007). From oocyte to 16-cell stage: cytoplasmic and cortical reorganizations that pattern the ascidian embryo. *Developmental Dynamics*, **236**, 1716–31.

Schierenberg, E. (2005). Unusual cleavage and gastrulation in a freshwater nematode: developmental and phylogenetic implications. *Development Genes and Evolution*, **215**, 103–8.

Schoch, R.R. (2010). Riedl's burden and the body plan: selection, constraint, and deep time. *Journal of Experimental Zoology Part B Molecular and Developmental Evolution*, **314**, 1–10.

Schulze, J., and Schierenberg, E. (2008). Cellular pattern formation, establishment of polarity and segregation of colored cytoplasm in embryos of the nematode *Romanomermis culicivorax*. *Developmental Biology*, **315**, 426–36.

Schulze, J., and Schierenberg, E. (2011). Evolution of embryonic development in nematodes. *Evodevo*, **2**, 18.

Shalchian-Tabrizi, K., Minge, M. A., Espelund, M., et al. (2008). Multigene phylogeny of Choanozoa and the origin of animals. *PLoS ONE*, **3**, e2098.

Shen, B., Dong, L., Xiao, S., et al. (2008). The Avalon explosion: evolution of Ediacara morphospace. *Science*, **319**, 81–4.

Sheth, R., Marcon, L., Bastida, M.F., et al. (2012). Hox genes regulate digit patterning by controlling the wavelength of a Turing-type mechanism. *Science*, **338**, 1476–80.

Sick, S., Reinker, S., Timmer, J., et al. (2006). WNT and DKK determine hair follicle spacing through a reaction-diffusion mechanism. *Science*, **314**, 1447–50.

Sokolnikoff, I.S., and Redheffer, R.M. (1966). *Mathematics of Physics and Modern Engineering*, New York, McGraw-Hill.

Steinberg, M.S. (2007). Differential adhesion in morphogenesis: a modern view. *Current Opinion in Genetics and Development*, **17**, 281–6.

Steinberg, M.S., and Takeichi, M. (1994). Experimental specification of cell sorting, tissue spreading, and specific spatial patterning by quantitative differences in cadherin expression. *Proceedings of the National Academy of Sciences USA*, **91**, 206–9.

Stoljar, D. (2009). Physicalism. *The Stanford Encyclopedia of Philosophy*, http://plato.stanford.edu/entries/physicalism/#1

Strogatz, S.H. (2003). *Sync: The Emerging Science of Spontaneous Order*, New York, Theia.

Tsai, M.C., and Ahringer, J. (2007). Microtubules are involved in anterior-posterior axis formation in *C. elegans* embryos. *Journal of Cell Biology*, **179**, 397–402.

Turing, A.M. (1952). The chemical basis of morphogenesis. *Philosophical Transactions of the Royal Society of LondonB*, **237**, 37–72.

Wansleeben, C., and Meijlink, F. (2011). The planar cell polarity pathway in vertebrate development. *Developmental Dynamics*, **240**, 616–26.

Wartlick, O., Mumcu, P., Kicheva, A., et al. (2011). Dynamics of Dpp signaling and proliferation control. *Science*, **331**, 1154–9.

Wimsatt, W.C. (2007). *Re-engineering Philosophy for Limited Beings: Piecewise Approximations to Reality*. Harvard University Press, Cambridge.

Wolpert, L. (1971). Positional information and pattern formation. *Current Topics in Developmental Biology*, **6**, 183–224.

Yu, S.R., Burkhardt, M., Nowak, M., et al. (2009). Fgf8 morphogen gradient forms by a source-sink mechanism with freely diffusing molecules. *Nature*, **461**, 533–6.

Zhu, J., Zhang, Y.T., Alber, M.S., et al. (2010). Bare bones pattern formation: a core regulatory network in varying geometries reproduces major features of vertebrate limb development and evolution. *PLoS One*, **5**, e10892.

Zullini, A. (2012) What is a nematode? *Zootaxa*, **3363**, 63–4.

CHAPTER 7

The landscape metaphor in development

Giuseppe Fusco, Roberto Carrer, and Emanuele Serrelli

Introduction

Sewall Wright's graphical visualization of the *fitness landscape* (Wright, 1932) is reputed to be one of the most famous metaphors in the history of biology (Dietrich & Skipper, 2012). On that hilly landscape, populations of organisms are represented as occupying a specific position (or a set of positions, one for each individual) in a space of descriptors, which depends on their genetic or phenotypic constitution. Different positions of this space are characterized by distinct fitness values, which collectively describe a fitness surface whose shape affects the future evolution of the population itself. Evolutionary change is visualized as the population change in the occupancy of the fitness landscape.

Wright's metaphor has met with criticism from both philosophers and theoretical biologists, who pointed out its ambiguous interpretation, the complications with translating it into rigorous mathematical models, and the inability of its standard three-dimensional visualization to capture the properties of real multidimensional fitness functions (Gavrilets, 2004; Kaplan, 2008; Pigliucci, 2012; Pigliucci & Kaplan, 2006; Provine, 1986). Meanwhile, other scholars have defended the value of the landscape metaphor in spite of technical difficulties with its application, at least as a heuristic (Plutynski, 2008; Ruse, 1996; Skipper, 2004; Skipper & Dietrich, 2012).

Where all agree is that during the last eighty years the fitness landscape metaphor has played a central role in the formal mathematical modelling of adaptation and speciation and in the didactics of the basic principles of evolutionary theory (Svensson & Calsbeek, 2012). Furthermore, the notion of fitness landscapes, aptly modified and expanded, has found numerous applications outside evolutionary biology, as for instance in biochemistry, computer sciences, engineering, and economics (Gavrilets, 2010). More importantly here, it inspired Conrad Hal Waddington (1939, 1940) to adopt a landscape visualization in a different biological discipline which, at a different temporal scale, also deals with time dynamics, i.e. development. This chapter reviews and discusses the use of the landscape metaphor in development, rather than in evolution, analysing its relationship with experimental work and theoretical modelling.

We start by defining a landscape as a function of multiple variables and show how this can be interpreted as a dynamical system. From the perspective of dynamical systems modelling, we move to analyse Waddington's 'epigenetic landscape' and landscape representations in current developmental biology literature. Then we delve into the problem of models and metaphorical representations in science, which stands out as a crux for assessing the use of landscapes in development, and analyse the somewhat parallel stories of Wright's and Waddington's landscapes. We conclude with some ideas on developmental landscapes in the context of visualization in science, with a focus on theoretical work in developmental biology.

Towards a Theory of Development. Edited by Alessandro Minelli and Thomas Pradeu

What is a landscape?

A precise definition of 'landscape' is the necessary starting point for any investigation on this visualization in scientific research.

Let us define a landscape as a mathematical function which associates the values of a set of independent variables (*indV*s) to the value of one numerical dependent variable (*depV*) over an ordinary Euclidean space. In technical terms, a landscape is thus a *function of multiple variables*.

For a set of n *indV*s, such a function describes a *hypersurface* of dimension n (the same number of dimensions of the space of *indV*s) embedded in a multidimensional space of $n + 1$ dimensions. In the special case of just one *indV* ($n = 1$), the function can be represented as a one-dimensional curve in a two-dimensional space (the Cartesian plane; Figure 7.1a). When $n = 2$, the function assumes the form of a two-dimensional surface in an ordinary three-dimensional Cartesian coordinate system (Figure 7.1b).

The latter case provides the analogy with geographic landscapes. The two *indV*s are interpreted as geographic coordinates whose values specify a position in a two-dimensional space (e.g. the surface of an ideal globe), while the value of the *depV* represents the elevation with respect to the mean sea level. This analogy, by extension, gives reason for the use of the label 'landscapes' for these kinds of functions, irrespective of the number of *indV*s and of the actual possibility of visualizing the relationship as a geographic landscape. Note that, although in mathematics a function can also be termed a *map*, it is not true that any function is a landscape, as the homonymy with (topographic) maps, which are representations of geographic landscapes, might suggest. For instance, a function which associates each point of its domain to a set of values (rather than only to one) is a map that is not a landscape, under the above definition.

A key property of landscape functions (henceforth, landscapes), as we have defined them, is that they are *scalar fields*, i.e. functions that assign a real numerical value to every point in a space. In the regions of the domain where the scalar field is differentiable, this is associated with a *vector field*, that is, its *gradient*. The gradient of a scalar field is a function that assigns to every point of its domain a vector which points in the direction of the greatest rate of increase of the scalar field and has a magnitude equal to the rate of that increase (Figure 7.2). In other words, the vectors indicate the direction and degree of inclination of the maximum slope at each point in the landscape.

Landscapes and dynamical systems

Because of the association of scalar fields with vector fields, landscapes have close connections with dynamical systems.

As a mathematical formalization, a *dynamical system* consists of a rule that governs the temporal evolution of a set of *system variables*. At any given time, a dynamical system has a *state*, defined by the values of its system variables, and the *temporal*

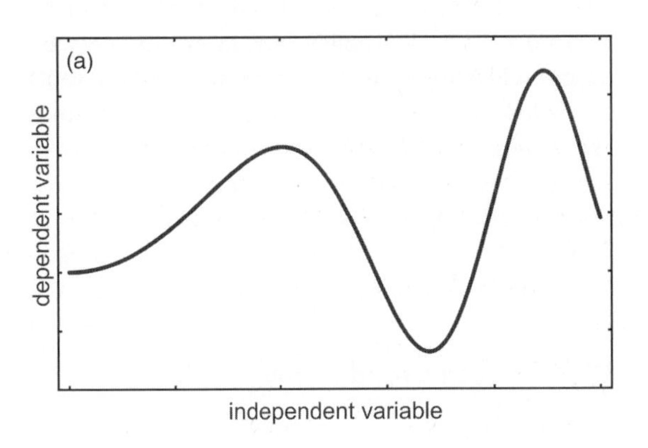

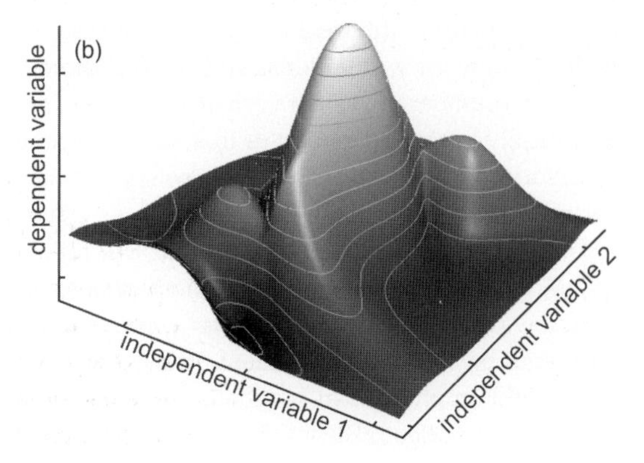

Figure 7.1 Landscape functions in (a) two- and (b) three-dimensional space.

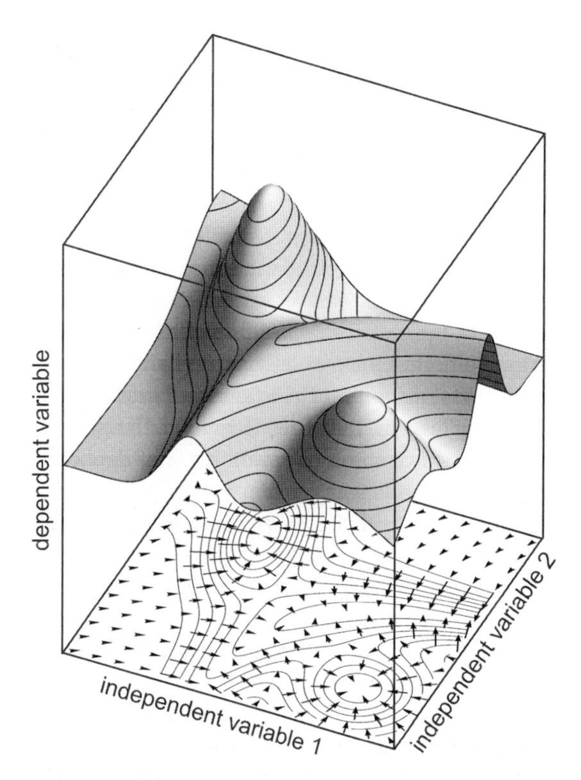

Figure 7.2 A scalar field in two variables (the landscape surface) with the associated vector field that is its gradient. This is depicted as a projection on the bottom plane.

evolution rule makes it possible to derive from that state the past and future states of the system. The state of the system can be represented by a specific point position in an appropriate *phase space* (or *state space*), the axes of which represent the system variables. Dynamical systems customarily take the form of systems of differential equations or systems of finite difference equations, and at low dimensionality (number of system variables ≤ 3), they are graphically portrayed as curves that show the time trajectories of the system in the phase space (*phase portrait*; Figure 7.3).

The key point for the discussion that will follow is that the equations defining a dynamical system describe a vector field which, under certain conditions, can be seen as the gradient of an associated scalar field. In physics, such a scalar field is known as *potential function* (also *scalar potential* or *potential surface*), and a system governed by a potential function has the property that the rate of change of its states is proportional to the derivative of the function (in other words, proportional to its gradient), while the minima and maxima of the function are equilibrium states.

Dynamical systems whose expression is the gradient of a potential function (actually, by convention, with opposite sign) are called *gradient systems* (Hirsch et al., 2004), but there are dynamical systems that do not have a potential function, as it is the case of most real physical systems, which are far from thermodynamic equilibrium. In other words, there are dynamical systems, often called *non-equilibrium* systems, which cannot be faithfully represented as landscapes, while in general any ordinary landscape can potentially be interpreted as a dynamical system. As we will see, the asymmetric relationship between landscapes and dynamical systems has profound implications for an understanding of the use of landscape visualizations for the modelling of developmental processes.

One can also note that the interpretation of a landscape as a dynamical system is generally coupled with a supplementary graphical contrivance. The landscapes which simply depict the relationship between a set of variables and that do not describe the time evolution of any system (e.g. a genotype–phenotype map), which we can call *static*, are depicted as 'uninhabited lands'. On the contrary, the landscapes loaded with a dynamical system interpretation, which we can call *dynamic*, are depicted as 'lands inhabited by entities' whose movements are governed by the shape of the landscape itself. In the classic iconography, not only in biology, these are landscapes populated by rolling balls (e.g. in chemistry) or by swarms of climbing points (e.g. in evolutionary biology) which describe the temporal evolution of the modelled system. In addition, or as an alternative, the surface of the landscape is covered by arrows or stream symbols which show the time evolution of the system from different points in the space. 'Dynamic landscapes' are the kind of landscape put to work in the study of developmental processes.

Waddington's landscapes

Undoubtedly, the most famous application of a landscape visualization in thinking about development is Conrad Hal Waddington's (1940, 1957)

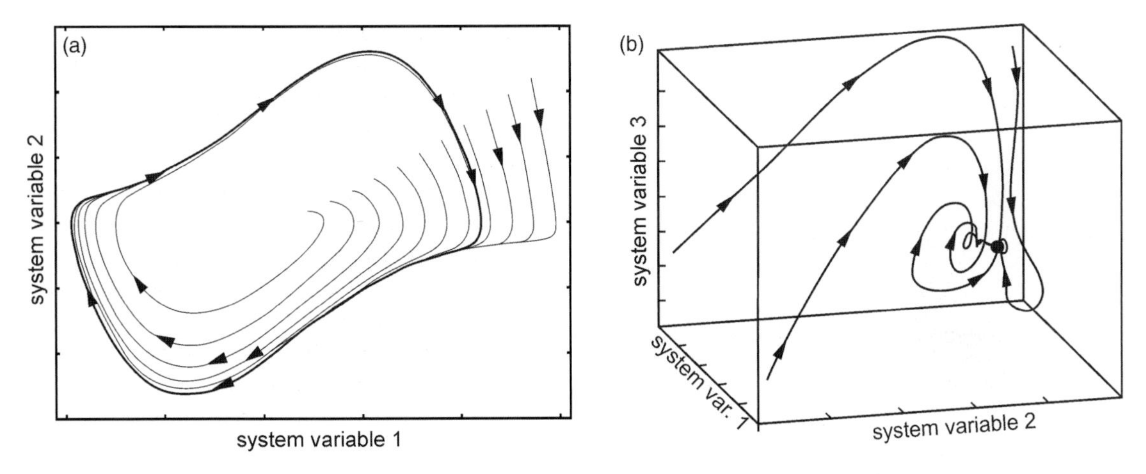

Figure 7.3 Dynamical systems of (a) two and (b) three variables represented in their phase space. In (a), several of the system trajectories lead to the same limit periodic stable attractor, a limit cycle (Van der Pol oscillator, adapted from Wikimedia Commons). In (b), several system trajectories lead to the same stable point attractor.

'epigenetic landscape' (Figure 7.4a), which served as a pivotal concept in his attempt to conceive an explanatory framework encompassing the organizational processes of development ('epigenesis', in contrast to 'preformationism', Maienschein, 2012) and the activity of genes ('genetics'). Today, the term 'epigenetic' tends to be used with a different meaning with respect to that intended by Waddington, i.e. to indicate phenomena of heritable changes in gene expression that are not due to changes in DNA sequence (Eccleston et al., 2007). To avoid ambiguities we will refer to 'Waddington's landscape'. As we will see, this label is further justified by the fact that in contemporary literature, in particular on cell differentiation, landscape visualizations are often accompanied by specific reference to Waddington's work.

Waddington himself, in successive works (1939, 1940, 1956, 1957), gave different interpretations of his landscape visualization (for careful historical reconstructions and philosophical scrutiny see Baedke, 2013; Caianiello, 2009; Fagan, 2012; Gilbert, 1991, 2000; Peterson, 2010; Slack, 2002). Here we summarize the main issues of his concept, which are of particular relevance with reference to its current use in the context of a dynamical system approach to the study of development. Indeed, several authors have seen in Waddington's work a pioneer application of the dynamical systems theory (Franceschelli, 2009, 2011; Kauffman, 1987; Saunders, 1989, 1993; Slack, 2002).

A picture of Waddington's landscape, first described in words (1939), appeared in *Organisers and Genes* (1940) in the form of a painting in the frontispiece of the book. In this book Waddington conceptually examined the embryological knowledge of his time: grafting experiments and other manipulations had demonstrated that the 'organizer', a specific region of the embryo, could deviate contiguous regions of the developing embryo towards forming different tissues and organs. However, in order to be effectively influenced, those parts of the embryo had to be in a specific state of 'competence' which gets progressively lost during development.

Relying on a solid tradition of visualization in embryology (Gilbert, 1991; Griesemer & Wimsatt, 1989), Waddington first envisioned the development of any 'embryo part' as a cascade bifurcation diagram, where, through a sequence of developmental decisions, the part is driven from an undifferentiated state towards one of its alternative possible fates, represented by the tips of the diagram. In this view, the action of the organizer is more a sort of 'evocation' than an 'induction', the emphasis being on the potency and competence of the embryo part that only needs specific triggers at particular times.

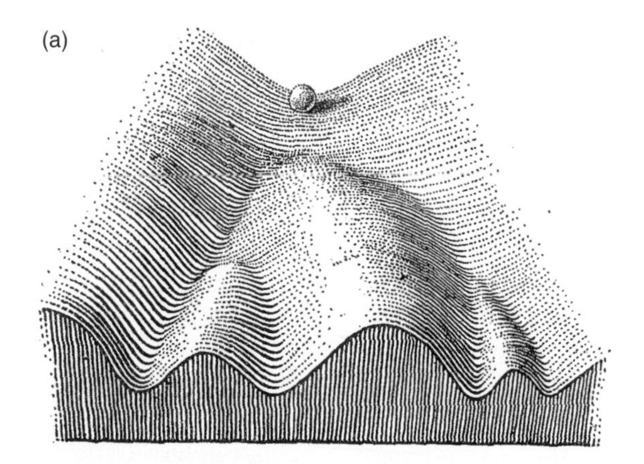

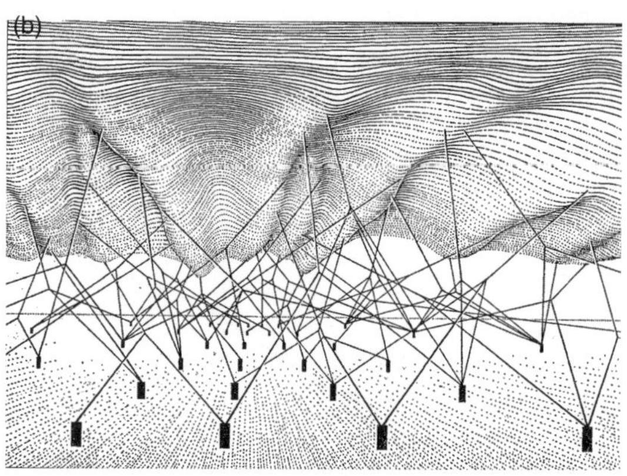

Figure 7.4 Waddington's (1957) depiction of his 'epigenetic landscape' (reprinted with permission). (a) The marble represents a biological system (e.g. a cell) at the verge of taking a developmental path toward one of a set of alterative more differentiated states represented by the three ending depressions at the base of the slope (Waddington, 1957: 29). (b) A vision from behind an epigenetic landscape. The shape of the slope is determined by tension of several interconnected guy-ropes (interacting gene products) that are attached to pegs stuck in the ground (genes) (Waddington, 1957: 36).

Waddington's landscape is clearly a reduction to three dimensions of this first intuition, through the transformation of a branching-track diagram into a system of bifurcating valleys. This transformation allowed Waddington to animate the track graph with notions like equilibrium, disequilibrium, and disturbance. The familiar behaviour of water streaming by gravitation provided Waddington with the means of conjugating several ideas, namely that embryo's parts (i) are in dynamic disequilibrium (like water running downstream) with a progressive loss of potential, (ii) follow a developmental track which, as a whole, is more or less stable ('the normal developmental track is one towards which a developing system tends to return after disturbance' (Waddington, 1940: 93)), and (iii) generally decrease their sensitivity to disturbances, from periods of high sensitivity where regulation is possible ('a valley with gently sloping sides') to periods of strong canalization ('the valley as having vertical sides'). These concepts, with different degree of importance, survived in later developmental biology studies (Gilbert, 2000).

Looking for the 'evocator', i.e. the key causal factor within the organizer, Waddington argued for its chemical nature. He further argued for a chemical explanation of development in general, where concentrations of different chemicals are causally relevant to developmental pathways and decisions. In Waddington's view, chemical compounds were gene products, and what we today would call gene expression profiles, in their turn dependent on other molecules, were thought of as the proximal cause of developmental trajectories. The dependence of gene product concentrations on 'the dosage of the genes' (allele dosage) represented a key passage in his attempt to relate genes and development. By giving a pioneering image of development regulated by chemical interactions involving gene products, Waddington grasped the intuition of gene regulatory networks with a view strikingly similar to that in modern systems biology (Fagan, 2012).

Waddington's landscape was described in more detail in *The Strategy of the Genes* (1957), where its most famous graphical instances are found (Figure 7.4). In the image, which can be considered the 'icon' of the developmental process of differentiation (Figure 7.4a), the graphic represents a tilted moulded surface, down which a marble is going to roll. The rolling marble's path corresponds to the development of some part of an organism from an early undifferentiated state to a mature differentiated state. The landscape topography presents a system of diverging valleys that become shallower and coalesce towards the top, while becoming deeper and fanning out towards the bottom of the slope. The bottom edge sees a series of depressions representing alternative differentiated states of the system. The particular shape chosen for the slope also conveys other ideas about development. The progressive reduction of the number of possible final differentiating states that occurs as the marble rolls downslope represents the progressive restriction of competence and potency of the system that accompanies differentiation. The progressive increase in the elevation of the crests that separate the different developmental pathways represents the process of canalization, the fact that the system becomes increasingly buffered against development disturbances. The different developmental options available at the beginning of the slope could be followed in response to evocating factors that are not represented in the landscape, like some environmental factors, producing a typical representation of the phenomenon of developmental plasticity. Although it is often assumed that the rolling marble represents a developing cell, in fact, and in Waddington's view, it can represent any developing system under the effect of a number of relevant factors, such as gene products or inducing signals.

Waddington's (1957) landscape is a genuine representation of a developing system described in a space of state variables. For a differentiating cell, the height of the surface is proportional to some potential variable ('developmental potential') associated to each combination of the underlying descriptive variables (e.g. concentrations of different substances or gene products) in the cell. The tilt of the surface shows the spontaneous tendency for the system to change its state along one of the available pathways. Although in this graphical representation the surface is in some way external to the marble/cell, in the modelled system the slope is actually determined by the characteristics of the developing cell, while non-represented (*hidden*) variables that play the role of external perturbations or induction may be either inside or outside the developing cell.

Even when unanimously interpreted as the representation of a dynamical system, the metaphor is still open to different understandings. For instance, for Slack (2002), the axes represent concentrations of all the substances, or all the gene products, in the cell, but because of the existence of inducing signals (not represented in the landscape), which can only influence a cell's development while the cell is competent to respond (i.e. uphill), different cells/marbles will roll down different pathways to end up at different states of terminal differentiation. Instead, for Fagan (2012), the horizontal axis projecting outward to the viewer represents time, the horizontal axis parallel to viewer represents the phenotype, and the vertical axis represents the 'order of development'.

In an effort to make the metaphor more explicative, Waddington (1957) provided a view of the 'underside' of the landscape, to show its supposed relationship to the genes (Figure 7.4b). The shape of the surface is determined by the pull of numerous guy-ropes which are in their turn controlled by genes, represented as pegs fixed to the ground. Guy-ropes represent gene products, and their connections represent their reciprocal interactions, which form a network that directly determines the shape of the landscape, i.e. of the dynamic of development.

Summing up, with his landscapes Waddington provided a simple mechanical analogy for the complex biochemical and genetic dynamics that occur in organisms during development (Slack, 2002). The surface embedded in the state space of an organism's molecular components is an effective representation of a dynamical system, potentially able to describe the change in time of a developmental system at any level of biological organization, from the cell (or even from systems within a cell) to the whole organism.

Subsequent modelling in developmental biology, although not directly stemming from Waddington's approach, nonetheless continues to put forward his graphical representations and to refer to his pioneer work.

Landscapes in current developmental biology

Current primary literature in developmental biology makes use of a diversity of graphical visualizations, depending on the specific subject (e.g. cell differentiation, pattern formation, gene expression) and the arena of the argumentation (e.g. experimental report, local dynamic modelization, theoretical generalization). Beyond their predictable occurrence in developmental biology textbooks, if only for historical reasons, Waddington's landscapes can also be found in the current primary literature, in both experimental and theoretical developmental biology papers. They have been brought into play in studies on pattern formation (Lepzelter & Wang, 2008), cell signalling (Sekine et al., 2011), and programmed cell death (Zinovyev et al., 2013); however, they mainly tend to occur in cell differentiation studies, with a further focus on stem cell biology (see below). This is a direct consequence of the landscape model's connection with dynamical systems.

Mathematical modelling in developmental biology, in particular at the level of specific developmental processes, like cell differentiation or pattern formation, is largely implemented through a dynamical system approach. Here developmental biology enters an intimate relationship with *systems biology* (see Jaeger and Sharpe, this volume), an emerging interdisciplinary approach to the study of biological systems which focuses on the complex interactions among different components of the system. The system, in essence, is seen as a network of relations, with gene expression, metabolic networks and cell signalling networks as well-known examples. The formal representation of the system can take several alternative formalizations, from ordinary differential equations to directed graphs, Boolean networks, and Bayesian networks (Fagan, 2012; Klipp et al., 2009). All these different formalisms have a specific scope and range of application. For instance, differential equations define continuous, deterministic models, while other kind of formalisms make it possible to cope with discrete deterministic models or with stochastic models. The different scope of these formalizations is not of particular relevance for the argument we are developing here, and we can limit ourselves to simply note that the use of differential equations is the most common formalization for dynamical systems in general and biology dynamical systems in particular.

In many papers the mathematical modelling of the dynamical system, typically as a system of equations, is accompanied by either or both of two kinds of graphical representations: wiring diagrams and phase portraits. Wiring diagrams are conventional pictorial representations of networks. Network nodes represent interacting entities (e.g. molecules or genes) while the connecting edges represent the interactions among the different nodes, generally with a simple symbolism to discriminate different kinds of relations (e.g. activation vs repression). Quantitative aspects of the interactions and spatial organization of the interacting entities are generally not represented. Wiring diagrams provide a means to grasp the topology of the interactions, that is, the interdependence of system variables which are detailed in the equations, while avoiding the possibility of getting lost in the complex expression of the latter. However, while the wiring diagram represents the system 'machine' behind a given developmental process, it does not make it possible to see how the system changes in time. Thus wiring diagrams are generally accompanied with a different representation of the same system, i.e. the phase portrait. Phase portraits are geometric representation of the trajectories of a dynamical system in its phase space (Figure 7.3). Each curve represents the time evolution of the system starting from a different set of initial conditions (a different point in the phase space). The trajectories reveal the existence of system *attractors*, i.e. the sets of states towards which the systems tend to move over time. An attractor can be a point, a finite set of points, a curve, or even a complicated set with a fractal structure known as a strange attractor. The zone of the phase space where the system is driven towards the attractor is called its *basin of attraction*. Within the limits of the three-dimensional illustration (not easy to overcome on a paper sheet), phase portraits can represent a diverse bestiary of systems behaviours, which is wider than the set representable with a landscape. First, in a three-dimensional phase portrait, three system variables can be represented, rather than two as in a three-dimensional landscape plot, in which the vertical axis is needed to represent the associate potential value of variable value combinations. Second, there are dynamical behaviours of the systems that cannot be faithfully accounted for by a landscape, such as limit cycles and chaotic behaviours (see 'Landscapes and dynamical systems').

Nonetheless, frequently phase portraits are followed by a landscape representation, or by a hybrid visualization which mixes phase portrait and landscape together. To understand their precise role in the current literature, we will start from a few illustrative examples of recent studies on cell differentiation. Actually, the use of Waddington's landscapes is not exclusive to such area of study (see Baedke, 2013).

However, since the aim of this contribution is not a meta-analysis of the current use of the landscape visualization but rather a close examination of the rationale behind its involvement in 'development thinking', limiting the selected examples to cell differentiation does not affect our general argument.

In a review on stem cell dynamics, Enver et al. (2009) are explicit in that they make a 'metaphorical' use of the landscape representation (Figure 7.5a). Using the traditional imagery of a marble rolling down a slope, they give the vertical dimension a thermodynamic interpretation as the free energy of the system in different states. However, in their opinion the valleys of Waddington's landscapes 'missed the scope for relatively stable, if transitory, intermediate cell types observed during the differentiation' (Enver et al., 2009: 388). The valleys (actually barely perceivable) of their landscape are thus punctuated by shallow depressions, representing as many intermediate temporary stable states. Mathematically, all the depressions are seen as attractors, i.e. equilibrium states towards which the dynamical system tends to move. However, as is apparent from the trajectories of the system represented in the graph, although certain paths are more likely than others, the authors specify that for a cell 'the possibility must also exist of moving in the reverse or alternative directions' or that a cell can 'move from one attractor to another by different routes' (Enver et al., 2009: 389).

In a methodological study on mathematical modelling of stem cell differentiation, Wang et al. (2010) do not consider the landscape representation as an accessory metaphor, but they have to go beyond the original landscape concept in order to make it significant in rigorous modelling. While they recognize that the idea of a potential function is particularly useful for equilibrium systems, where the potential is knowable a priori, non-equilibrium systems (in practice, most biological systems) cannot, in general, be modelled as the gradient of a potential. The intuition of some potential, although widely used metaphorically, does not make it possible to move to a precise mathematical formalization. However, at the same time, standard dynamical system analysis cannot account for the fact that developmental processes exhibit, at certain scales at least, a consistent directionality in time. In their words, the '"arrow of time" in the collective change of gene expression across multiple stable gene expression patterns (attractors) is not explained by the regulated activation, the suppression of individual genes which are bidirectional molecular processes, or by the standard dynamical models of the underlying gene circuit which only account for local stability of attractors'. Thus, to capture the global dynamics of this non-equilibrium system and gain insight in the time asymmetry of state transitions, they compute a 'quasi-potential landscape' of the stochastic dynamics of gene circuits that govern cell-fate commitment. This is a function which combines a gradient potential with another force ('curl flux') that stem from the non-integrability of the system. In a fol-

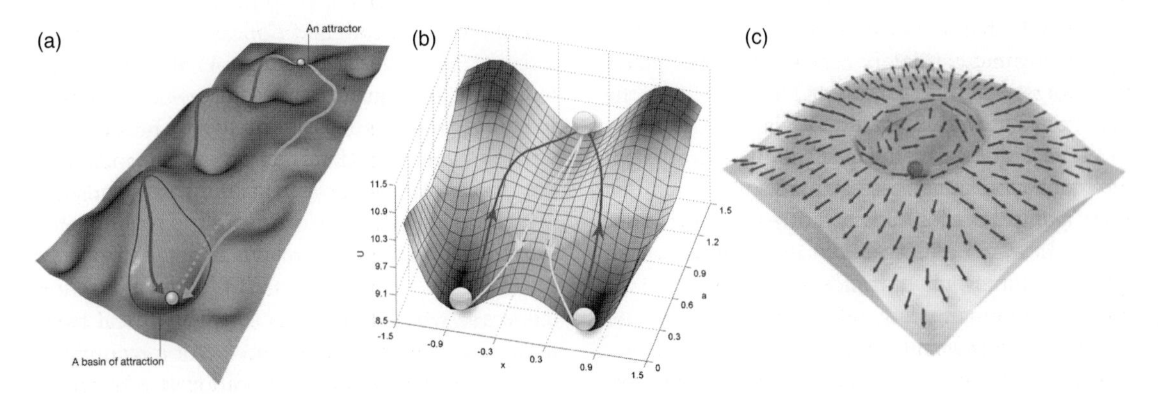

Figure 7.5 A few examples of contemporary developmental landscapes (reprinted with permission): (a) from Enver et al. (2009); (b) from Wang et al. (2011); (c) from Furusawa and Kaneko (2012). In all cases, the behaviour of the developing systems does not faithfully follow what would be expected under a strict gravitational analogy (see text).

lowing paper from the same research group (Wang et al., 2011), underlining the fact that due to the *qualitative* nature of Waddington's landscape it is not very clear how to connect it to the results of experimental work, the authors suggest a formalization which aims at 'quantifying' Waddington's landscape. Focusing on cell differentiation, they use the analytical tool based on the combination of a gradient with a curl force devised in their previous works (Wang et al., 2008, 2010) to construct a 'quantified Waddington landscape' (Figure 7.5b). However, despite superficial similarity, a number of differences between the new landscape and the original Waddington's landscape must be pointed out: (i) in the quantified landscape the temporal stabilization of the uncommitted stem cell state is permitted; (ii) cell-fate decision does not necessarily happen at the hilltop, as developmental paths can start bifurcating even when the system is in a basin of attraction; (iii) 'developmental paths clearly do not follow gradient paths that the gravity driven metaphor of Waddingon would predict', as the curl force makes the developmental path deviate from the steepest descending gradient path; and (iv) while in Waddington's landscape the possible reverse path is supposed to be the same as the forward path, in this quantified landscapes 'the developmental paths are clearly distinct from the retrodifferentiation paths'.

In a dynamical systems perspective on stem cell biology, Furusawa and Kaneko (2012) argue that some core property of 'stemness', like differentiation from a stable state, 'cannot easily be described by Waddington's landscape'. Through a dynamical systems approach, they describe a model in which fluctuating and oscillatory gene expression, 'the essence of stemness', are accounted for. To explain the difference between the traditional view of cell differentiation and the one suggested in their work, they make use of a landscape visualization (Figure 7.5c). On these landscapes, however, the streams of vectors indicating the direction of the system in each state are not always in conformity with the curvature of the landscape, as one would expect if these were to be intended as its gradient field.

As noted by Fagan (2012) and exemplified by the case studies above, landscape visualization does not represent a predictive tool, but rather a visual aid for the derivations of the model. In other words, model construction of cellular systems does not start from the definition of a potential function (the landscape), as is the case under certain condition for fitness landscapes (but see Rice, 2004).

Summing up, landscape visualizations are still present in the current literature, although they come in different 'flavours' and with a great disparity in their interpretative load. This varies from allegoric pictures, like in the editorial commentary (Iovino & Cavalli, 2011) of an experimental work of Thomson et al. (2011) in the same journal issue, where landscapes are not even mentioned, to their full rehabilitation as models, conditional on their mathematical quantification, in the works of Wang et al. (2008, 2010, 2011).

Landscapes, between models and metaphors

Contemporary modelling research in developmental biology, in particular on cell differentiation, makes frequent reference to Waddington's landscape. On the one hand, historical continuity is emphasized. On the other hand, this is supplemented by a statement of progress with respect to Waddington's ideas, and many authors identify such progress with the move from a 'mere metaphor' to a more sound conception. For Ferrell (2012), 'Waddington's landscape [. . .] is more than just a metaphor', but at the same time he argues that the classical shape of its surface, with diverging valleys, is correctly illustrative of some developmental processes but not others, like cell-fate induction, for which he proposes an alternative general shape. Furusawa and Kaneko (2012) claim that today to 'characterize the attractors of stem and differentiated cells quantitatively' we need not just further experiments but also 'theoretical formulations that go beyond Waddington's epigenetic landscape'. For Wang et al. (2011), 'The Waddington landscape is no longer a metaphor. It is physical and quantifiable by the underlying probability landscape.' And Huang (2012) emphasizes that the 'quasi-potential landscape with attractors' is 'a mathematical entity that has a molecular basis and is not a mere metaphor'. It seems that the qualification of a landscape as either a metaphor or a mathematical model is a crucial passage for any evaluation of the use of landscapes in developmental biology.

In everyday discourse, as well as in some philosophy of science studies, the most natural counterpart of metaphors are models: there is an intuitive difference between representations that are 'only metaphors' and others that are 'models in their own right'. However, there is little consensus on any diagnostic feature to distinguish between the two. Mathematization has been proposed as a distinctive feature of models (Lewontin, 1963), but scientific models come in many kinds, including visual and material objects (Downes, 1992), and even organisms (Ankeny & Leonelli, 2011). A narrow selection of represented aspects of the world and the lack of testability were proposed as distinctive marks of metaphors (Kirchner, 1990), but extreme simplification can be found in mathematical models too (Lewontin, 1963), and the application of such models to the world can always be seen as a metaphorical re-description (Hesse, 1966). A famous analysis by Levins (1966) emphasized that model building requires making certain trade-offs between realism, precision, and generality. Levins' scheme can be used to accommodate several kinds of representation as models without the need of setting metaphors apart (Calcott, 2008 vs Pigliucci & Kaplan, 2006 on adaptive landscapes).

In the case of developmental landscapes, we think that the most suitable framework to set the model/metaphor contrast is one based on a criterion technically known as *autonomy*. Morgan and Morrison (1999) emphasize the autonomy of models with respect to both theory and observational data. Models are seen as 'mediators', and their autonomy consists in the possibility of performing intensive research on the model itself, elevated to a 'stable target of explanation' (Keller, 2002: 115). For instance, mathematical analysis of models built through a dynamical systems approach (see 'Landscapes in current developmental biology') has brought about discoveries about dynamical systems themselves, quite independently from the original questions to which they were applied. The produced knowledge can be brought back, at different times, to the 'representational target' that had inspired the model but also to a changing 'representational scope' that can be much wider than the original target (Ankeny & Leonelli, 2011), as exemplified by the fact that the same mathematical model can apply to completely unrelated real systems. A model M can thus be defined as a representation of the system S that can be worked upon with significant autonomy from S and can be usefully employed for answering questions on S and, with every probability, also on a wider set of systems S_i. The set S_i, the representational scope of model M, can change as science progresses. In contrast, the distinctive mark of a metaphor is not that it conveys a very limited set of aspects of the represented system (although, of course, it does), but rather that it cannot be investigated, deepened, and modified independently to accrue knowledge. It can be further amended and/or complemented only as the empirical or theoretical understanding of S progresses. The constitutional subordination of metaphors with respect to the system they represent is not in contrast with the positive heuristic function emphasized by some theorists. Indeed, summarizing, stimulating, and guiding research upon target systems, as opposed to upon themselves, is all metaphors can do. Constitutional subordination further implies that internal consistency is neither a requirement nor an assumption of metaphors. Models, unlike metaphors, need some kind of internal consistency that is at the basis of their constitutional autonomy.

The degree of autonomy criterion is a suitable framework for addressing the problem of landscapes in development. However, it produces a continuum of representations, from the most rigorous of formal models to the most allegoric of metaphors, rather than providing a clear boundary between the two categories (see Kaplan, 2008 for an opposite view). Both models and metaphors are representations of real systems, both emphasize some feature of the real world while deliberately neglecting others, both are good for specific purposes only, and both are potentially misleading in that accepting a specific representation necessarily influences and defines the questions that are considered to be important (Gavrilets, 2004, 2010).

Further insights on the model/metaphor relationship can be gained by looking into the parallel histories of the two major landscape metaphors in biology. At the beginning of the chapter, we mentioned fitness landscapes in evolutionary biology, their success, and the debates that surrounded them in recent years. Historical reconstructions of Waddington's work (Caianiello, 2009; Gilbert, 1991) demonstrated that Waddington's landscape was inspired by Wright's work (1932) through Joseph Needham (1936). Such connection was motivated by Waddington's search for relating development to genetics (Gilbert, 1991).

Beyond historical links between landscapes in development and evolution, we highlight here a curious form of antisymmetry between the history of the two ideas since their respective introduction in scientific literature. Evolutionary landscapes were introduced by a mathematician, Wright, as a pictorial representation of a mathematical model. Then, generations of students of evolution, working in totally different fields, from genetics to palaeontology (Dobzhansky, 1937; Simpson, 1944), adopted the figure as a basis for their non-mathematical theorizations. Some commentators describe the employment therein as a useful heuristic (Ruse, 1996; Skipper, 2004; Skipper & Dietrich, 2012). In contrast, developmental landscapes were introduced by an experimental embryologist as a depiction of his empirical observations to serve as a conceptual tool. The figure was then taken up by biologists and biophysicists with more mathematical skills as a complement to their formal models and simulations about specific developmental processes. Both Wright and Waddington were partly responsible for the later uses of their respective intuitions. Wright (1988) approved the usage in contexts as different from the native context as palaeontology and was charged with levity for that (Pigliucci, 2008). Waddington (1957) himself moved to systems theory and presented his landscape in a completely different context with respect to his previous work (1940). At that time, powerful mathematical and theoretical tools were already available (although without the computational power we have today), but the empirical knowledge, e.g. on genes and their role in development, was by far insufficient for model construction in development.

Interestingly, the two landscape metaphors are opposites in terms of what concerns their native context. Wright's model was a logico-mathematical system of Mendelian populations, with huge dimensionality that was reduced to a three-dimensional representation (Serrelli, 2011). The landscape metaphor here is thus a partial representation of a complex mathematical model, which, however, cannot capture the properties of real multidimensional landscapes, as shown by more recent theoretical work (Gavrilets, 2004, 2010). Waddington, instead, originally had no mathematical models to select from, and the messages entrusted to the metaphor were directly derived from empirical observations of competence, evocation, equilibrium, disequilibrium, disturbance, discrete end-states, and the like in developing tissues of embryos of different species.

In the studies examined in this chapter, developmental landscapes, despite being potentially misleading, nonetheless correctly orientate the attention of the reader towards specific messages chosen by the researcher. Landscape visualizations are metaphorical in that they are not autonomous objects of research and inference, and they show a limited selection of features of the developing systems, although at the same time they are the best visual approximation for the equations that describe them.

Landscapes and 'vision'

Our perusal of recent literature, although far from being exhaustive, has shown that Waddington's landscape is 'alive and well in contemporary developmental biology' (Gilbert, 2000: 734). However, contemporary developmental landscapes differ profoundly from the original representations. It is a semantic question to ask if these 'new landscapes' are still 'Waddington's landscapes', but asking how much of a landscape there is in them is not. On these strange landscapes, or 'quasi-landscapes', one is not authorized to see a marble rolling down the slope. Strange things can happen on such surfaces: marbles emerging from a pit, taking a path different from the one with maximum slope, going uphill, moving from one place just to be back some time later. In addition, the landscape surface is only a part of the graphical representation of a dynamics. This needs to be complemented with vector arrows or stream signs that can even be at odds with the shape of the landscape on which they sit. On this landscape one has to relax the natural spontaneous gravitational interpretation of the represented dynamics, but at the same time it is exactly the fact that everybody has a personal experience of the gravitational force that makes this representation so eloquent.

This is a little-appreciated, misleading trait of landscape representations in general, not only in visualizations of development. The landscape representation deceptively suggests that the phenomenon/system could be described in terms of a potential function that is maximized/minimized at equilibrium; however, this function does not always exist (Rice, 2004). Another, mostly neglected and potentially misleading trait is that, even if a landscape faithfully represents the directions of change of the system from any state, the kinematics of a rolling marble does not necessarily apply. For instance, while a marble in a gravitational setting would increase its speed rolling down a constant slope, the marble of Waddington's landscapes would roll down the same trajectory at a constant speed. At a local minimum of the landscape, a gravitational marble would stop accelerating, whereas Waddington's marble would stop moving (Ferrell, 2012).

Nevertheless, landscape graphical representations are able to convey information not immediately apparent in a phase portrait. For instance, they can effectively express the directionality of the process at large scale, the degree of stability of local equilibria, and the fact that certain trajectories are more probable than others. Some dynamical systems can still be represented as landscapes, for instance stipulating that a 'stable state' is not necessarily exactly a point, but can also be a periodic or aperiodic oscillation within a comparatively small region of their phase space.

In general, a landscape cannot be a faithful representation of a real system. This is especially true of open systems (i.e. systems open to matter and/or energy flow), such as a developing organism or a part of it. But this is not new, as Waddington himself described with words dynamics and interactions which could have never been portrayed on his landscape. Indeed, to stress his view on the role of genes in development, Waddington (1957) produced a second graph (Figure 7.4b). However, whereas in the most famous graph (Figure 7.4a) the figure can still be ideally translated into a formal model, providing specific identity to the three axes, in this second graph this possibility is lost. The ropes traverse the space inserting at the base of the space (why there?) and under the surface. Actually, the ropes (genes products, cell metabolites) should already be represented by the axis of the space, while their insertion points have no physical value; and they are supposed to exert forces that are not represented in the space. Thus Waddington's attempt to convey more explicit reference to the generative forces (or relevant factors) for the particular shape of the landscape results in a less accurate representation.

One can search the reasons behind the persistency of the landscape metaphor in the scientific literature. It appears that the idea of a landscape is floated every time there is the need to talk about a map, an association between quantities, irrespective of the mathematics one can develop on it or the actual possibility of giving precise physical meaning to the axes of the space. Reasons for that can possibly be found in the architecture of our cognition system, or in our difficulty in giving up the tradition of a cherished metaphor. We refrain from pushing these speculations further, and suggest instead a parallel with a historic controversy, partially still ongoing within mathematics. This is about the role of geometric vision in mathematics.

The French school known by the collective name of Nicolas Bourbaki was the leading group of mathematicians who aimed at a complete algebraization of geometry and analysis through the construction of extremely abstract theories, in which the possibility of 'visualizing' mathematical objects is considered to be unnecessary. In contrast, other schools of thought remained more closely linked to an intuitive and geometric vision of mathematics. For example, Vladimir Arnol'd (1998), the author of fundamental works on differential equations and dynamical systems, argued that geometry algebraization, inflated axiomatization and abstraction as an end in itself, leads mathematics to nowhere. In his opinion, the geometric and physical vision of mathematics has a constructive role in the process of mathematical discovery. This is not only a question of method, as there are fundamental mathematical theorems, as for instance the classification theorem for surfaces, which are an achievement of both mathematics and physics together. This discussion was central for most of the last century; however, today the need for a geometric vision of problems, before, during, and after their algebraic differential development, goes beyond all these contrasting arguments. The centenarian development of differential geometry and its centrality as a discipline are no longer in question (Alekseevskij et al., 1991); on the contrary, they are both reputed determinant players in the construction of a 'new geometry' that has yet to come (Yau & Nadis, 2010).

Conclusion

It seems that the landscape metaphor will continue to stay with us, at least for a while. However, it is more difficult to envision its future role in exploring more inclusive levels of abstraction and theorization, as for instance in the search for a comprehensive theory of development. In fact, cell differentiation, in the study of which landscapes have been chiefly employed, cannot be assumed to be a general model for the dynamics of other developmental processes, like cell proliferation, movement, and death, production and consumption of extracellular material, morphogenesis, pattern formation, and growth.

As we have seen, Waddington originally introduced the landscape metaphor in developmental biology as a tool for unifying under the same explanatory framework different developmental processes and their control. Can this be considered an attempt to search for a general theory of development? The answer, not surprisingly, depends on the meaning one attaches to the word 'theory'. Unfortunately, philosophy of science offers more problems than solutions about the nature of theories, in particular about any meaningful distinction between 'theories' and 'models'. Despite a recent rise of interest in defining what theories are (Griesemer, 2013; Morrison, 2007), after several decades during which the philosophy of biology has been mainly focusing on models and modelling in theorizing (Downes, 1992), a reasonable consensus has certainly not been achieved yet. Pragmatically, avoiding venturing into the question of what a scientific theory is or should be, which is far beyond the scope of the present chapter, most would probably agree that Waddington's general model of development fits into what we could call a 'minimal concept of theory', that is, a rational abstract generalization of a set of natural phenomena. However, his conceptualization would not pass the test for a more demanding 'paradigm concept of theory', which requires a theoretical edifice with strong explanatory power for a very large set of natural phenomena as well as high predictive performances to the level of very detailed observations and measures. In common understanding, a general model of a natural phenomenon does not as such qualify as a theory of that phenomenon.

As a matter of fact, independently from the epistemological evaluation on what Waddington was historically aiming at, landscapes have proven to be effective visualization tools for investigating only specific developmental processes, without allowing straightforward formalization into mathematical models. Overall, landscapes seem to be too limited a form of abstraction to stand as a pivotal metaphor in the search for a comprehensive theory of development. However, only future research will be able to assess whether the landscape metaphor can effectively extend its scope to other developmental processes or have any role in a conceptualization of development as a whole.

Acknowledgments

We thank B. Calcott, D. Chinellato, F. Gross, and J. Jaeger for their insightful comments on a previous version of this chapter. ES is grateful to the Centre for the Foundations of Science, University of Sydney, where, thanks to a visiting research fellowship, he carried out a significant part of the studies for this contribution.

References

Alekseevskij, D.V., Shvartsman, O.V., and Solodovnikov, A.S. (1991). *Geometry I. Basic Ideas and Concepts of Differential Geometry*. Springer, Berlin.

Ankeny, R., and Leonelli, S. (2011). What's so special about model organisms? *Studies in History and Philosophy of Science Part A*, **42**, 313–23.

Arnol'd, V.I. (1998). On teaching mathematics. *Russian Mathematical Surveys*, **53**, 229–36.

Baedke, J. (2013). The epigenetic landscape in the course of time: Conrad Hal Waddington's methodological impact on the life sciences. *Studies in History and Philosophy of Biological and Biomedical Sciences*, 44, 756–773.

Caianiello, S. (2009). Adaptive versus epigenetic landscape. A visual chapter in the history of evolution and development. In S. Brauckmann, C. Brandt, D. Thieffry, and G.B. Müller, eds., *Graphing Genes, Cells, and Embryos: Cultures of Seeing 3D and Beyond*. Preprint Series. Max Planck Institute for the History of Science, Berlin, pp. 65–78.

Calcott, B. (2008). Assessing the fitness landscape revolution. *Biology and Philosophy*, **23**, 639–57.

Dietrich, M.R., and Skipper Jr, R.A. (2012). A shifting terrain: a brief history of the adaptive landscape. In E. Svensson and R. Calsbeek, eds., *The Adaptive Landscape in Evolutionary Biology*. Oxford University Press, Oxford, pp. 3–15.

Dobzhansky, T. (1937). *Genetics and the Origin of Species*. Columbia University Press, New York.

Downes, S.M. (1992). The importance of models in theorizing: a deflationary semantic view. In D.M. Hull, K. Okruhlik and M. Forbes, eds., *PSA 1992: Proceedings of the 1992 Biennial Meeting of the Philosophy of Science Association*, Vol. 1. Philosophy of Science Association, East Lansing, pp. 142–153.

Eccleston, A., DeWitt, N., Gunter, C., Marte, B., and Nath, D. (2007). Introduction epigenetics. *Nature*, **447**, 395–440.

Enver, T., Pera, M., Peterson, C., and Andrews, P.W. (2009). Stem cell states, fates, and the rules of attraction. *Cell Stem Cell*, **4**, 387–97.

Fagan, M.B. (2012). Waddington redux: models and explanation in stem cell and systems biology. *Biology and Philosophy*, **27**, 179–213.

Ferrell Jr, J.E. (2012). Bistability, bifurcations, and Waddingtons epigenetic landscape. *Current Biology*, **22**, R458–66.

Franceschelli, S. (2009). Dynamics of the unseen. Surfaces and their environments as dynamical landscapes. In C. Soddu, ed., *GA2009, XII Generative Art International Conference*. Domus Argenia Publisher, Milano, pp. 236–47.

Franceschelli, S. (2011). Morphogenesis, structural stability, and epigenetic landscape. In P. Bourgine and A. Lesne, eds., *Morphogenesis. The Origin of Patterns and Shapes*. Springer, Berlin, pp. 283–294.

Furusawa, C., and Kaneko, K. (2012). A dynamical-systems view of stem cell biology. *Science*, **338**, 215–17.

Gavrilets, S. (2004). *Fitness Landscapes and the Origin of Species*. Princeton University Press, Princeton.

Gavrilets, S. (2010). High-dimensional fitness landscapes and the origins of biodiversity. In M. Pigliucci and G.B. Müller, eds., *Toward an Extended Evolutionary Synthesis*. MIT Press, Cambridge, pp. 45–79.

Gilbert, S.F. (1991). Epigenetic landscaping: Waddington's use of cell fate bifurcation diagrams. *Biology and Philosophy*, **6**, 135–54.

Gilbert, S.F. (2000). Diachronic biology meets evo-devo: C. H. Waddington's approach to evolutionary developmental biology. *American Zoologist*, **40**, 729–37.

Griesemer, J.R. 2013. Formalization and the meaning of 'theory' in the inexact biological sciences. *Biological Theory*, **7**, 298–310.

Griesemer, J.R., and Wimsatt, W.C. (1989). Picturing Weismannism: a case study of conceptual evolution. In M. Ruse, ed., *What the Philosophy of Biology Is*. Kluwer, Dordrecht, pp. 75–137.

Hesse, M. (1966). *Models and Analogies in Science*, University of Notre Dame Press, Notre Dame.

Hirsch, M.W., Smale, S., and Devaney, R.L. (2004). *Differential Equations, Dynamical Systems, and An Introduction to Chaos*, 2nd ed. Elsevier Academic Press, San Diego.

Huang, S. (2012). The molecular and mathematical basis of Waddington's epigenetic landscape: a framework for post-Darwinian biology? *Bioessays*, **34**, 149–57.

Iovino, N., and Cavalli, G. (2011). Rolling ES cells down the Waddington landscape with Oct4 and Sox2. *Cell*, **145**, 815–17.

Kaplan, J. (2008). The end of the adaptive landscape metaphor? *Biology and Philosophy*, **23**, 625–38.

Kauffman, S.A. (1987). Developmental logic and its evolution. *BioEssays*, **6**, 82–7.

Keller, E.F. (2002). *Making Sense of Life: Explaining Biological Development with Models, Metaphors, and Machines*. Harvard University Press, Cambridge.

Kirchner, J.W. (1990). Gaia metaphor unfalsifiable, *Nature*, **345**, 470.

Klipp, E., Liebermeister, W., Wierling, C., Kowald, A., Lehrach, H., and Herwig, R. (2009). *Systems Biology: A Textbook*. Wiley, Weinheim.

Lepzelter, D., and Wang. J. (2008). Exact probabilistic solution of spatial-dependent stochastics and associated spatial potential landscape for the bicoid protein. *Physical Review E*, **77**, 041917.

Levins, R. (1966). The strategy of model building in population biology. *American Scientist*, **54**, 421–431.

Lewontin, R.C. (1963). Models, mathematics and metaphors. *Synthese*, **15**, 222–44.

Maienschein, J. (2012). Epigenesis and preformationism. In E.N. Zalta, ed., *The Stanford Encyclopedia of Philosophy (Spring 2012 Edition)*, http://plato.stanford.edu/archives/spr2012/entries/epigenesis.

Morgan, M.S., and Morrison, M., eds. (1999). *Models as Mediators. Perspectives on Natural and Social Science*. Cambridge University Press, Cambridge.

Morrison, M. (2007). Where have all the theories gone? *Philosophy of Science*, **74**, 195–228.

Needham, J. (1936). *Order and Life*. The MIT Press, Cambridge.

Peterson, E.L. (2010). *Finding mind, form, organism, and person in a reductionist age: the challenge of Gregory Bateson and C.H. Waddington to biological and anthropological orthodoxy, 1924–1980*. Ph. D. Thesis, University of Notre Dame, http://etd.nd.edu/ETD-db/theses/available/etd-04132010-142514/

Pigliucci, M. (2008). Sewall Wright's adaptive landscapes: 1932 vs. 1988. *Biology and Philosophy*, **23**, 591–603.

Pigliucci, M. (2012). Landscapes, surfaces, and morphospaces: what are they good for? In E. Svensson and R. Calsbeek, eds., *The Adaptive Landscape in Evolutionary Biology*. Oxford University Press, Oxford, pp. 26–38.

Pigliucci, M., and Kaplan, J. (2006). *Making Sense of Evolution. The Concepual Foundations of Evolutionary Biology*. Chicago University Press, Chicago.

Plutynski, A. (2008). The rise and fall of the adaptive landscape? *Biology and Philosophy*, **23**, 605–23.

Provine, W.B. (1986). *Sewall Wright and Evolutionary Biology*. University of Chicago Press, Chicago.

Rice, S.H. (2004). *Evolutionary Theory: Mathematical and Conceptual Foundations*. Sinauer Associates, Sunderland.

Ruse, M. (1996). Are pictures really necessary? The case of Sewall Wright's 'adaptive landscapes'. In B.S. Baigrie, ed., *Picturing Knowledge: Historical and Philosophical Problems concerning the use of Art in Science*. University of Toronto Press, Toronto, pp. 303–337.

Saunders, P.T. (1989). Art and the new biology: biological forms and patterns. *Leonardo*, **22**, 33–8.

Saunders, P.T. (1993). The organism as a dynamical system. In F. Varala and W. Stein, eds., *SFI Studies in the Science of Complexity. Lecture Notes*, Vol. 3. Addison Wesley, Reading, pp. 41–63.

Sekine, R., Yamamura, M., Ayukawa, S., Ishimatsu, K., Akama, S., Takinoue, M., Hagiya, M., and Kiga, D. (2011). Tunable synthetic phenotypic diversification on Waddington's landscape through autonomous signaling. *Proceedings of the National Academy of Sciences USA*, **108**, 17969–73.

Serrelli, E. (2011). *Adaptive Landscapes: A Case Study of Metaphors, Models, and Synthesis in Evolutionary Biology*. PhD thesis in Education and Communication Sciences, University of Milano Bicocca, http://hdl.handle.net/10281/19338.

Simpson, G.G. (1944). *Tempo and Mode in Evolution*. Columbia University Press, New York.

Skipper Jr, R.A. (2004). The heuristic role of Sewall Wright's 1932 adaptive landscape diagram. *Philosophy of Science*, **71**, 1176–88.

Skipper Jr, R.A., and Dietrich, M.R. (2012). Sewall Wright's adaptive landscape: philosophical reflections on heuristic value. In E. Svensson and R. Calsbeek, eds., *The Adaptive Landscape in Evolutionary Biology*. Oxford University Press, Oxford, pp. 16–25.

Slack, J.M.W. (2002). Conrad Hal Waddington: the last Renaissance biologist? *Nature Reviews Genetics*, **3**, 889–95.

Svensson, E., and Calsbeek, R. (2012). *The Adaptive Landscape in Evolutionary Biology*. Oxford University Press, Oxford.

Thomson, M., Liu S.J., Zou L.-N., Smith Z., Meissner, A., and Ramanathan, S. (2011). Pluripotency factors in embryonic stem cells regulate differentiation into germ layers. *Cell*, **145**, 875–89.

Waddington, C.H. (1939). *An Introduction to Modern Genetics*. MacMillan, New York.

Waddington, C.H. (1940). *Organisers and Genes*. Cambridge University Press, Cambridge.

Waddington, C.H. (1956). *Principles of Embryology*. MacMillan, New York.

Waddington, C.H. (1957). *The Strategy of the Genes*. Allen & Unwin, London.

Wang, J., Xu, L., and Wang, E. (2008). Potential landscape and flux framework of nonequilibrium networks: robustness, dissipation, and coherence of biochemical oscillations. *Proceedings of the National Academy of Sciences USA*, **105**, 12271–6.

Wang, J., Xu, L., Wang, E. K., and Huang, S. (2010). The potential landscape of genetic circuits imposes the arrow of time in stem cell differentiation. *Biophysical Journal*, **99**, 29–39.

Wang, J., Zhang, K., Xu, L., and Wang, E.K. (2011). Quantifying the Waddington landscape and biological paths for development and differentiation. *Proceedings of the National Academy of Sciences USA*, **108**, 8257–62.

Wright, S. (1932). The roles of mutation, inbreeding, crossbreeding and selection in evolution. *Proceedings of the Sixth Annual Congress of Genetics*, **1**, 356–66. Reprinted in W.B. Provine (1986) *Sewall Wright: Evolution: Selected Papers*. University of Chicago Press, Chicago, pp. 161–77.

Wright, S. (1988). Surfaces of selective value revisited. *The American Naturalist*, **131**, 115–23.

Yau, S.-T., and Nadis, S. (2010) *The Shape of Inner Space*. Basic Books, New York.

Zinovyev, A., Calzone, L., Fourquet, S., and Barillot E. (2013). How cell decides between life and death: mathematical modeling of epigenetic landscapes of cellular fates. In V. Capasso, M. Gromov, A. Harel-Bellan, N. Morozova, and L.L. Pritchard, eds., *Pattern Formation in Morphogenesis. Problems and Mathematical Issues*. Springer, Berlin, pp. 191–205.

CHAPTER 8

Formalizing theories of development: a fugue on the orderliness of change

Scott F. Gilbert and Jonathan Bard

A pluralism of developmental perspectives

This essay must be tempered throughout with humility. In the past 50 years, developmental biology has recapitulated in rapid order the Industrial Revolution's succession of creator from person (organism), to apparatus (cell, molecule), to algorithm (program). In 50 more years, we will be very lucky if our essays are considered 'prescient,' because it is doubtful that they will be considered 'science' by those standards (see Flüsser, 1987).

We also should not be confined by the 'progressive' flow of such displacement from organism to program. Indeed, the notion that these different levels of agency succeed and displace one another is a Modernist notion that should be avoided. Creation, as Paul Weiss (1967, 1977) noted in his early systems theories of development, is found at all levels—the molecular, the cellular, the tissue, the organismal, and the ecological[1]—through the integration and recombination of lower-level entities into higher-level orders, and through the selection of viable possibilities by the upper-level agents. In studying this re-ordering, it is important to remember that while lower-level orders are the components of higher-level orders, the higher-level orders provide the context/niche for selecting possible lower-level structures (see Auletta, 2011; El-Hani & Emmeche, 2000; Ellis, 2012; Longo et al., 2012; Soto et al., 2009).

We should also respect a plurality of explanatory perspectives (Pirsig, 1974; Winther, 2011). In his analysis of part-whole explanations, Winther (2011) catalogues three major modes of developmental explanation: (i) structuralist (top-down) explanation, in which emergent organization is what needs to be explained and mathematical-logical formalisms carry the weight of explanation; (ii) mechanistic (bottom-up) explanation, in which parts and their causal interactions can explain developmental phenomena; and (iii) historical explanation, where development (both of parts and wholes) is placed in a larger, evolutionary, narrative. This paper sees such perspectives as being 'in resonance' with one another. The metaphor is that of electrons in a benzene ring. No perspective, alone, provides a complete account of developmental phenomena.

Relations and downward causation

This chapter takes a systems approach to development in that it tries to step back from the normal minutiae of developmental phenomena and asks how should one start to unpack the complexity of development in a way that captures both the parts and the whole; and a first step is to look at the relationship between them. Higher-order structure provides the 'interpretation' of the lower-level parts and processes. Using a linguistic analogy, the statement

[1] Interestingly, the molecular level, which is 'lower' than the biological levels, may have become a biological level after its appropriating a particular context within the cell. Newman (2012; this book) hypothesizes that the morphological properties of cell division, migration, and ordering were originally physical ('generic') properties of semi-solid deformable materials that later became taken over and canalized by the genome ('genetic'). At this later stage, the molecular level could become part of the biological levels of organization.

Towards a Theory of Development. Edited by Alessandro Minelli and Thomas Pradeu

'The party leaders were split on the platform' demonstrates that words not only define the sentence but that the sentence also defines the meaning of each word. Similarly, the supposedly true headline 'Prostitutes appeal to the Pope' (Russell-Rose, 2011) shows that context determines the meaning of the sentence. Bone morphogenic protein 4 (BMP4) can be a signal for growth, differentiation, or apoptosis within the same organism. What it does depend on is the historical context of the cells receiving it. (This and the other cited developmental examples are discussed in Gilbert, 2010). As Leo Rosten (2003) remarked, the sentence 'I should buy two tickets for her concert?' has seven different meanings depending on which word is emphasized!

Development is all about the interpretation of relationships. The fertilized egg inherits DNA; it does not inherit 'genes'. Genes and gene products are constructed anew in each cell in the developing embryo by the relationships between DNA, transcription factors, and RNA-splicing factors. Only certain regions of the DNA are constructed into genes, and different regions of the genome can be genes in different cell types. Note that the 'gene' is a higher-order structure than DNA, and that the interpretation of 'what is a gene' is done by the cell, an even higher-order structure (Stotz et al., 2006). As John Stamatoyannopoulos (2012: 1603), one of the leaders of the ENCODE project, recently summarized, 'Although the gene has been conventionally viewed as the fundamental unit of genomic organization, on the basis of ENCODE data it is now compellingly argued that this unit is not the gene but rather the transcript . . . On this view, genes represent a higher-order framework . . . creating a polyfunctional entity that assumes different forms under different cell states'.

Oyama (1985) has famously called this 'the ontogeny of information'. The organism does not inherit a 'program' as much as it inherits DNA and a cytoplasmic interpretation device (Gilbert, 1991; Nijhout, 1999). The same programmed music score can be interpreted in numerous ways by different orchestras. Every performance is different, even from the same score and the same orchestra. Indeed, it must be. Compare, for instance the recording of Pachelbel's *Canon* played by the English Chamber Orchestra under the baton of Johannes Somary with the same piece played by Musica Antiqua Köln, directed by Richard Goebel. Moreover, a concert A of 440 Hz is heard very differently when played by a cello or a trumpet. Even the interpretation of concert A differs geographically: concert A is 440 Hz in the United States and Britain, while it is usually 442 Hz in continental Europe. The interpretation of the score differs even in the pitch of the notes.

So there must be interaction between score and instrument (and orchestra, more largely), and there must be interaction between DNA and transcription factors. That the performance of a phenotype depends on its wider context has been long known by embryologists (see Gilbert & Sarkar, 2000) and is manifest in four major categories:

(i) *Plasticity*. Temperature-dependent pigmentation in butterflies, nutrition-dependent caste determination in hymenopterans, and site-specific sex determination in certain invertebrates were all known to early embryologists (see Hertwig, 1894). More recently, it has been seen that almost all, if not all, organisms have some developmental plasticity, and the inherited DNA determines a repertoire of phenotypes, not a specific phenotype. The environment can instruct which of the possible phenotypes to form. Species have evolved such that their genomes are responsive to environmental agents (see Gilbert & Epel, 2009). It is worth noting that the model systems often used in developmental biology have been specifically selected for their canalization (i.e. a lack of environmental agency) so that the genetics of development can be elucidated (Bolker, 2012; Gilbert, 2009).
(ii) *Organicism*. The parts of the organism determine the development of the whole and the whole developing organism reciprocally determines the properties of its parts. Lenoir (1982) has argued that the founders of modern embryology, Döllinger, Pander, von Baer, and Rathke, subscribed to the organicism set forth in Kant's *Critique of Judgment* (quoted in Lenoir, 1982: 25). Said Kant: 'The first principle required for the notion of an object conceived as a natural purpose is that the parts, with respect to both form and being, are only possible through their relationship to the whole . . . Secondly, it is required

that the parts bind themselves mutually into the unity of a whole in such a way that they are mutually cause and effect of one another.' Oskar Hertwig (1892), one of the leaders of embryology, proposed organicism as the true middle ground between reductionism and vitalism. He wrote that the parts of the organism develop in relation to each other, that is, the development of the part is dependent on the development of the whole. This was reiterated by Hans Spemann, who wrote, 'We are standing and walking with parts of our body which could have been used for thinking had they developed in another part of the embryo' (Spemann, 1943: 158–159; transl. by Horder & Weindling, 1986: 219). This emerging order was also thought to be critical to any philosophy of development by Paul Weiss, who said, 'Wherever we study such emergent order, we recognize it to be of tripartite origin, involving (1) elements with an inner order, (2) their orderly interactions, and (3) an environment fit to sustain their ordered group behavior' (Weiss, 1955: 296).

(iii) *Phenotypic heterogeneity*. The same mutation can produce a different phenotype in different individuals (Nijhout & Paulsen, 1997; Wolf, 1997, 2002). Phenotypic heterogeneity comes about because genes are not autonomous agents. Rather, genes interact with other genes and gene products, becoming integrated into complex pathways and networks. Thus, in addition to developmental plasticity dependent upon an environment external to the cell, genes can function differently depending on other genetic parameters. Bellus et al. (1996) found that the effects of the same mutant FGFR3 gene on limb development differed from person to person, with the phenotypes ranging from relatively mild anomalies to potentially lethal malformations. Similarly, the effects of particular mutant genes causing holoprosencephaly differ in the different family members having the same mutant gene (Dubourg et al., 2004; Marini et al., 2003). The severity of a mutant gene's effect often depends on the *other* genes, whose products have become part of the environment of the gene, as well as on environmental factors, and it will take a systems approach to find out how.

(iv) *Co-development*. All this regulation occurs through normal physico-chemical interactions. No higher-level process occurs in any other way. However, selection into viable networks and functional circuits occurs at a higher level, permitting only a subset of possible networks to evolve. As Leibniz (1697), one of the philosophers who most influenced Darwin, realized, while all permutations may be possible, very few will be compossible. By this he meant that not all possibilities could be actualized, because not all parts can function together to make coherent wholes. Ecosystems are examples of compossible systems: a squirrel and a whale are both possible, but not compossible in the same habitat. The fourth example of such higher-level phenomena, then, is the 'holobiont' created by the interactions of the 'host' with its symbionts. The host and symbiont are united anatomically, physiologically, immunologically, developmentally, and even evolutionarily (Gilbert et al., 2012; Nyholm and McFall-Ngai, this volume; Pradeu, 2011). Metabolic pathways initiated in the microbial symbiont get completed in the host, and vice versa; developmental pathways initiated in the host become completed by the symbiont, while the symbiont's metabolism is altered by signals from the host. Indeed, the gut and immune system of mammals is often completed by chemical signals originating from bacterial cells. We are literally 'becoming together' with the outside environment. Microbes are part of our post-embryonic developmental patterning, and the microbiome is our eleventh organ system.

This tells us that downward causation can be brought about in several ways. First, entities at higher levels place constraints on which lower-level interactions are viable and maintainable. Second, the parts must be compossible to form a greater whole. The bacteria that constitute our gut microbiota are not selected for their species; rather, they are selected for their functions (Faust et al., 2012; HMPC, 2012). Third, the higher-level entities also interpret the lower-level agents: a signal for apoptosis in one cell is a signal for proliferation in another. Fibroblast growth factors promote growth in some cases and prevent growth in others. The transcription activated by paracrine factors depends upon the receiving cell's developmental history.

And there is a fourth mechanism: the higher-level structures give the physical location in which the lower-level modules function. For literal 'top-down' causation, one can't beat the dorsal-ventral patterning of vertebrates or *Drosophila*. But for these processes to occur—for the top to become distinguished from the bottom—one needs the placement of mRNAs in particular places within the cell. The *gurken* mRNA has to be placed dorsally in the *Drosophila* oocyte; the *Vg1* message has to be placed ventrally in the amphibian oocyte. Chordin must be made by the dorsal cells of the vertebrate embryo; while the homologous protein must be made in the ventral cells of the fly embryo. And both arise by the interactions of numerous tissues (see El-Hani, manuscript submitted). Indeed, the ventral cells of *Drosophila* arise only because the Dorsal protein, a transcription factor, is placed into the ventral cells' nuclei by interactions between the oocyte and the ventral follicle cells. If the Dorsal protein enters all cells, the entire embryo is ventralized. And this is regulated by the positioning of the *gurken* mRNA in the future dorsal region of the oocyte. Thus, the higher-level cell structure can regulate the places of transcription factor-gene regulation and so generate patterning.

Relations and upward causation

This phenomenon of 'downward' causation meets with and interacts with the phenomena of 'upward' causation. Two principles must be recalled in every discussion of upward causation in embryology. First, there is Haraway's principle (2008: 25–26) that 'relationships are the smallest possible pattern for analysis'. Information is not about essence; it is about relations. This is germane to the above discussion and will be continued in the discussion below. Second, development acts almost exclusively through stereocomplementarity (Gilbert & Greenberg, 1984). Stereocomplementarity is the interaction between shapes, and it is one of the great unifying principles of biology. It is literally 'fitness', that is, things that fit together: enzymes/substrates; antibodies/antigens; DNA/transcription factors; paracrine factors/receptors; sperm/egg; the interlocking components of signal transduction pathways; the interlocking components of ribosomes. Keys must fit only into certain locks. It is all 'copulation', literally, the binding together. Thus, information in development is about the interaction of complementary shapes. Our 'information' is in-form-ation. That is, information takes shape; it is not abstract, although the rules for interactions may become so. Rather, even though we may represent information flow with arrows to indicate causation, direction, and temporality, we are really discussing the interactions of shaped objects.

There are ways other than stereocomplementarity through which nature transfers information. Mechanical transduction is used occasionally in development, especially in the production of the circulatory system and skeletal elements (Culver & Dickinson, 2010; Gilbert & Epel, 2009; Tang et al., 2004). Frequency, which is used in echolocation and insect mating systems, is rarely used in development, the major examples being the predator-induced hatching of red-eyed tree frog larvae and the settlement of coral larvae (Vermeij et al., 2010; Warkentin et al., 2006). Stereocomplementarity is the major way that information is embodied in developmental processes. And stereocomplementarity implies reciprocal relation. The stereocomplementary molecules mediating gamete recognition are the fastest diverging proteins known (Palumbi, 2009). And for each change on the protein of one gamete, there has to be a corresponding change on the other.

So one might ask: what is the stereocomplementary relation that defines 'reality' for the embryo? What type of interaction determines whether an entity is a real (i.e. functional or morphological) unit for development? Let us consider that the primary unit of reality for the embryo is the enhancer–transcription factor relation. If a gene for a marker protein (such as β-galactosidase or green fluorescent protein) is ligated to a promoter, enhancer traps can determine what the embryo considers 'real'. This reality might not correspond to adult 'reality'. Surely, enhancers will activate genes in the retina and the gut tube. But one enhancer will activate genes in the *medial* rib, while a different enhancer will activate genes in the *lateral* portion of the rib (Guenther et al., 2008). Apparently, 'lateral rib' is an anatomical construction unit recognized by the embryo. Similarly, enhancer traps of the *Drosophila* embryo shows that the embryo comprises numerous compartments that are not apparent in the adult, but which are building units of the embryo (Buszczak et al., 2007).

But this is just the primary relationship and is in the nucleus. In order to understand or model cells and, indeed, embryos, one has to relate what happens in the nucleus to what happens in the cell cytoplasm and cell membrane.

Scoring development: we all live in recursive subroutines, recursive subroutines, recursive subroutines

When one starts to think about the principles that guide embryogenesis, one might be given the impression that all the decades of developmental biology research have shown is that the development of a particular simple tissue depends partly on its parent tissue (lineage) and partly on its neighbours (signalling). While its development can be understood with hindsight, there are still no formal predictors or rules as to what might happen. Development still lacks the sorts of underlying principles that make physics tractable and laws that gives quantitative predictions; one reason is that there are no elementary particles, and another is that any laws have proved elusive. Worse, it lacks a natural notation for writing the score, and that is one area that is touched on here.

The best that we have been able to do is to borrow the language of physics and try to constrain as much of development as possible into differential equations that describe change and predict performance, a tradition that started with the classic paper of Turing (1952) on molecular pattern formation. Today, there is a considerable amount of research in this area (see Barkai & Perrimon, 2011). Nevertheless, while there have been impressive successes in modelling a few phenomena such as signalling pathways (e.g. Witt et al., 2011), *Drosophila* segmentation (Ingolia, 2004) and somitogenesis (Goldbeter & Pourquié, 2008), general principles that help understand development have yet to emerge from this approach other than to affirm the integral importance of upwards and downwards causation.

The main reason for this is that development is very complicated and that what seems a simple development change is actually underpinned by the coordinated activity of hundred of proteins. What one soon notes is that, whatever the embryo, change is based on a relatively small set of protein networks whose outputs are the processes that drive patterning, signalling, proliferation, differentiation, and morphogenesis, and these themes are used over and over again (Bard, 2013). This simplicity stands in strong contrast to the complexity of the full developmental score with its swarms of genes, molecules, and tissues. Table 8.1 gives some idea of the numbers of these components for the mouse and human. The figures for protein-coding genes and proteins are well known; the number of developmental networks and output actions comes mainly from Gilbert (2010) and the numbers of tissues from Bard (2012). Figure 8.1 indicates how these events are integrated.

The number of processes is surprisingly small, and they fall into two groups (Table 8.1 and Table 8.2). First, there are the gene regulatory networks (Levine & Davidson, 2008), which comprise ~10 signal-activated networks and an unclear number of patterning and timing networks. These control the second group, which we can think of as process networks that actually lead to phenotypic change: here there are 5–10 pathways associated with proliferation, ~3 apoptosis networks, 5–10 morphogenetic networks, and a hierarchy of differentiation pathways (Bard, 2012). The number of high-level differentiation pathways is less clear because major cell types have subtypes, but one pointer here comes from the options available to neural crest cells. These include mesenchymal cells (bone, muscle, cartilage, fibroblasts), epithelia of various sorts, neurons and neuron-support cells, and melanocytes but not the other major lineage of blood cells and their many subtypes. There are perhaps ~10 main

Table 8.1 Levels and numbers.

Protein-coding genes	35 000
Proteins	~70 000
Developmental networks	~60
Output processes	~60
Simple tissues	~10 000

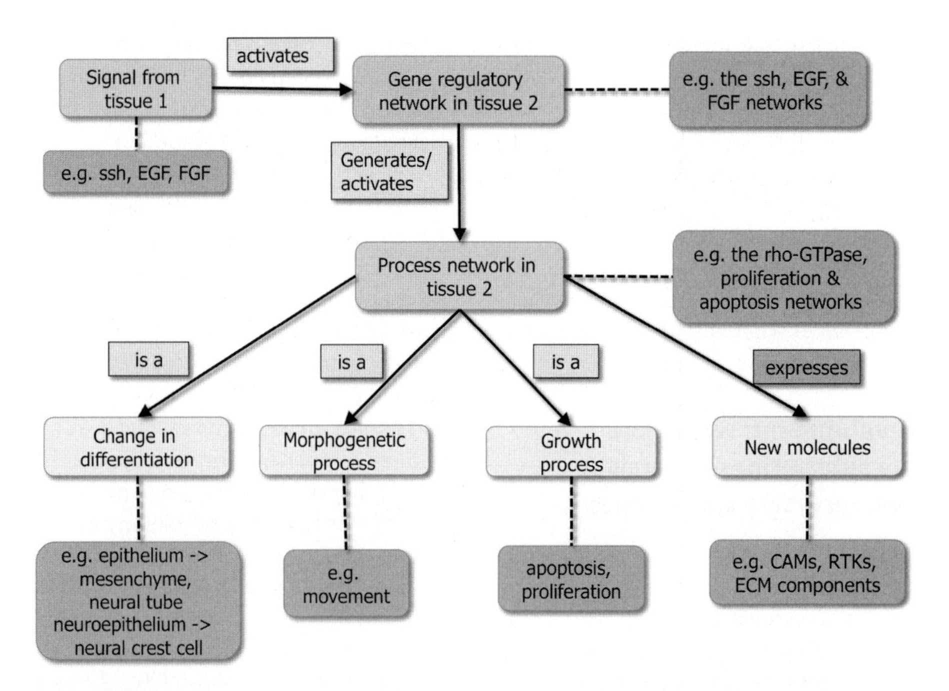

Figure 8.1 Graph showing the effects of signalling pathways. Examples are in darker grey boxes. The 'is a' link represents a typing or classification. Examples are in grey boxes. (From Bard, 2011b. Reproduced with permission from John Wiley & Sons.)

Table 8.2 Some major networks whose output are the processes that drive development.

Gene regulatory networks	Process networks	
Signalling	***Differentiation to***	***Morphogenesis***
ERK/MAPK	haematopoiesis lineage	boundary formation
FGF, JAK/STAT	erythroid lineage	(Eph-ephrin)
Notch–delta	lymphocyte lineage	epithelium
Shh, SMAD	myeloid lineage	branching
TGFβ, VEGF, Wnt	epithelium	folding
	mesenchyme	migration
Patterning	chondrocyte	rearrangement
Hox patterning	fibroblast	mesenchyme
RTK patterning	muscle	adhesion
Notch oscillator system	osteoblast	migration
signalling gradients (e.g. Shh)	neuron	
etc.	neuron-support cell	***Apoptosis***
	pigment-producing cell	caspase, fas
Timing		cellular apoptosis
Nothing is known of these		
		Proliferation
		cyclin+downstream events

cell differentiation routes. Taking a broad brush to the topic, there are ~50 major processes that underpin development (perhaps 60 if we allow for the possibility that a few more will be discovered), with some having several outputs (Figure 8.2).

These leitmotifs are used over and over again in each complex, multicellular animal as it develops. The fine details are not of course the same in each: evolutionary change means that the exact details of the networks and their outputs vary from organism to

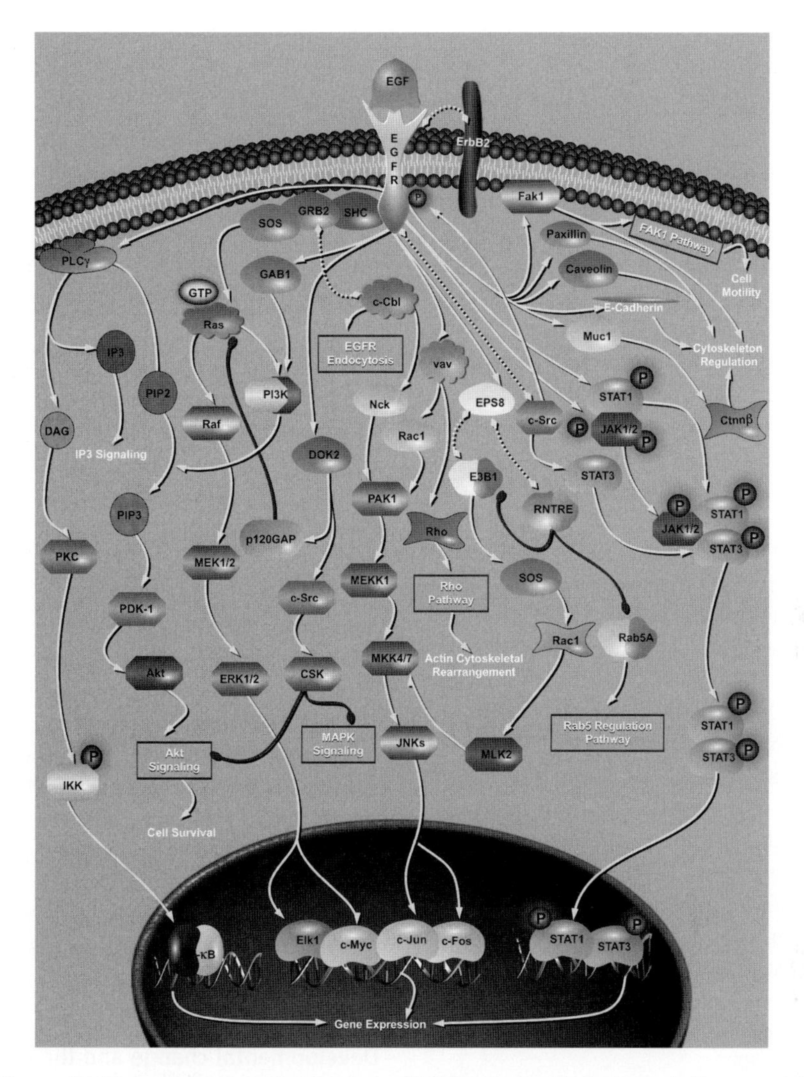

Figure 8.2 The EGF network (>60 proteins) activates the cell cycle. (modified from www.sabiosciences.com/pathwaycentral.php)

organism and from tissue to tissue. It should also be pointed out that the repertoire of developmental networks (Table 8.2) excludes the many more networks that 'run' the biochemical, physiological, and neurological systems. Nevertheless, if there is any underlying simplicity to be found in developmental biology, it centres around a basic set of molecular networks[2] whose outputs are the processes that drive embryogenesis (Figure 8.1). Not that these networks are simple: they contain ~10–50 interacting proteins (see http://www.sabiosciences.com/pathwaycentral.php and Figure 8.2). While elucidating the components and the organization of these networks has been a triumph of the last decade of research in molecular genetics, we still don't know how they work qualitatively, let alone quantitatively.

[2] What are called networks here are more commonly called pathways. The former is the preferable term because these assemblages of proteins often include alternate routes and end-points.

Processes are the subroutines of development

It is easiest to see how frequently the same processes are used by looking at the emerging anatomy of a vertebrate, in which similar structures are produced across the embryo. In the mouse, for example, there are ~200 long bones, >50 vertebrae, and many examples of muscles, ligaments, neuronal nuclei and ganglia, and bifurcating tubes. Similar structures are produced over and over again with minor, locale-specific features that do no more than tinker with the numerical parameters of the process. The central difference between a femur and a phalange, for example, is only one of scale: the former is ~40 times the length of the latter. The development of classes of standard modules is ubiquitous across complex organisms and reflects the regular and frequent use of the processes that build these modules, albeit that their activities are modulated by local molecular constraints.

Modular development has an interesting implication within a systems context. There has been some discussion in the literature as to whether part at least of the genome should be viewed, metaphorically at least, as a database of genomic information available to developing cells (e.g. Noble, 2010). Indeed, it is hard not to visualize the networks that generate dynamic processes as being the output of genomic subroutines that are used in many different contexts. This metaphor can be taken a little further: as program subroutines have outputs that depend on their input parameters, so the output of process subroutines depend on the details of the cell types in which they are expressed. Each organism's development arises from an evolutionarily canalized set of compossible subroutines (Huang et al., 2009; Kauffman, 1987).

A formal language for development

The language of differential equations is sadly of limited applicability for development: we just don't know enough about the participants, their interactions, or their rate constants to be able to use the alphabet of mathematics to describe what is going on. The events of development do, however, give us some clues as to how to start describing things with some formality. Development involves events at levels from the genome through gene expression, signalling, networks, and processes, to tissues. In this list, it is clear that processes stand out as different: while genes, protein, networks, cells, and tissues reflect states, processes reflect activities: they drive state changes. While the former are nouns, the latter are verbs!

There turns out to be an area of mathematics known as graph theory that captures this difference. A mathematical graph is nothing like a data graph because it doesn't deal with numerical data. It turns out that many complex stories can be decomposed into a series of small facts of the general form

<state 1> <relationship 1> <state 2>[3]

Each is, for obvious reasons, known as a triplet and a given state can be involved in two or more triplet. For a given story, the set of linked triplets comprise the mathematical graph. For development, these triplet relationships are mainly of the form

<noun a> <verb x> <noun b>

where the nouns may be anything from a tissue to a cell to a network to a molecule, and verbs reflect processes (differentiates into, migrates, apoptoses, etc.). In practice, each triplet can be seen as a simple fact, with the relationship often being the *activity* or *process* that drives the change (e.g. *<SHH><activates><the shh signalling pathway>*, where shh stands for the signalling protein sonic hedgehog). The other core relationship is <is_a> and this is used as a classification tag (e.g. *<ectoderm> <is_a> <epithelium>*). Here, it is worth noting that Figure 8.1 is actually a formal graph.

Developmental change and the notation of graphs

The use of graphical notation to describe developmental change turns out to be useful in several ways (Bard, 2011b, 2013). First, the representation is visual; second, the format is web accessible and can be linked to other resources; third, the format lends itself to being updated as new information is discovered; fourth, making the graph highlights gaps in knowledge and so suggests experiments; fifth, it shows the centrality of the relation as the fundamental unit of development. Nevertheless, the format does have limitations. It is not easy, for example

[3] In graph theory, the standard terminology is *<node> <edge> <node>*.

to represent the internal structure of chemical reactions and biochemical pathways: they either need the insertion of dummy intermediates or they require a richer formulation than triplets (see www.sbgn.org/). A more difficult problem is including the full complexity of a developmental event: representing the networks underpinning the processes other than by their names would be unwieldy. In practice, this could only be done by listing the triplets and handling them computationally.

It turns out that many developmental phenomena can be represented as a graph where the nodes are biological entities scaling from proteins upwards and the edges are relationships (Bard, 2011b); and there are several advantages in doing this:

(i) They unpack the complexity of development by reducing it to a set of simple but integrated facts, albeit that the set may be quite large.
(ii) Extra triplets can be added as new parts of the story are discovered.
(iii) Where nodes have ontology IDs, links to associated data can be included.
(iv) IDs from PubMed can be used as citations of facts

There are, however, further advantages in representing the graphs as diagrams that show the general organization. For developmental biology, these include:

(v) They emphasize that control of development is widely distributed.
(vi) Gaps in the diagram highlight areas where further work is needed.
(vii) Colour can be used to reduce the complexity of the narrative.

Together, these advantages mean that the mathematical graph can be seen as a terse, updatable review of a developmental event. Further, because databases are continually being upgraded with new data, the ID links ensure that the associated data is also up to date. This is not to suggest that the graphical notation should be seen as a step towards a more general theory but rather that the formalism articulates the sort of clarity that makes theorization one step easier.

The information required to make a graph of how change takes place in an embryo comes from experimentation, and not only involves signals and the activation of processes but also a clear understanding of what these processes do. Some of this information is not yet available, and the resulting graphs will skate over some details (e.g. a network can be represented by a single node rather than by a intricate sub-graph whose details may not be germane to the problem being considered). Things are more complicated where morphogenesis takes place, as the final structure will not only depend on signals and networks but on such as physical activity by cells that is constrained by the geometry of the tissues (Figure 8.3). In a sense though, making the triplets and so producing the formal representation of the developmental event is relatively straightforward.

Integrating all this information for a real example in a single clear diagram is often hard to do, partly because so much is going on and partly because it can be difficult to maintain the sense of the dynamics. One trick that is helpful here is to embed molecular nodes within the blocks for the tissues. Another is to use different colours for different aspects of the diagram. Note that a classic graph, the London Underground map, where the relationship is <*connects with*>, uses colours to distinguish paths through the network.

An example clarifies this. All developing tissues, once they reach a critical size, need a blood supply, and this is achieved by the local mesenchyme secreting the signal protein vascular endothelial growth factor (VEGF). Research on mouse embryos has shown that this signal diffuses into the local environment, providing a concentration gradient that decreases with distance. Receptors on nearby blood vessels bind this signal, signal transduction activates the proliferation pathway locally, Notch–Delta activation ensures that the new cells form a single capillary, and this extends up the concentration gradient towards and into the original tissue (for reviews, see Chung & Ferrara, 2011; Suchting et al., 2007). There are some 30 small facts associated with this event, and each can be described as a triplet (with further triplets linking these facts with the original publications, as stored in PubMed, Gene Ontology and GXD, the mouse gene-expression database).

The key elements of the data on angiogenesis are shown in a graphical representation in Figure 8.4.

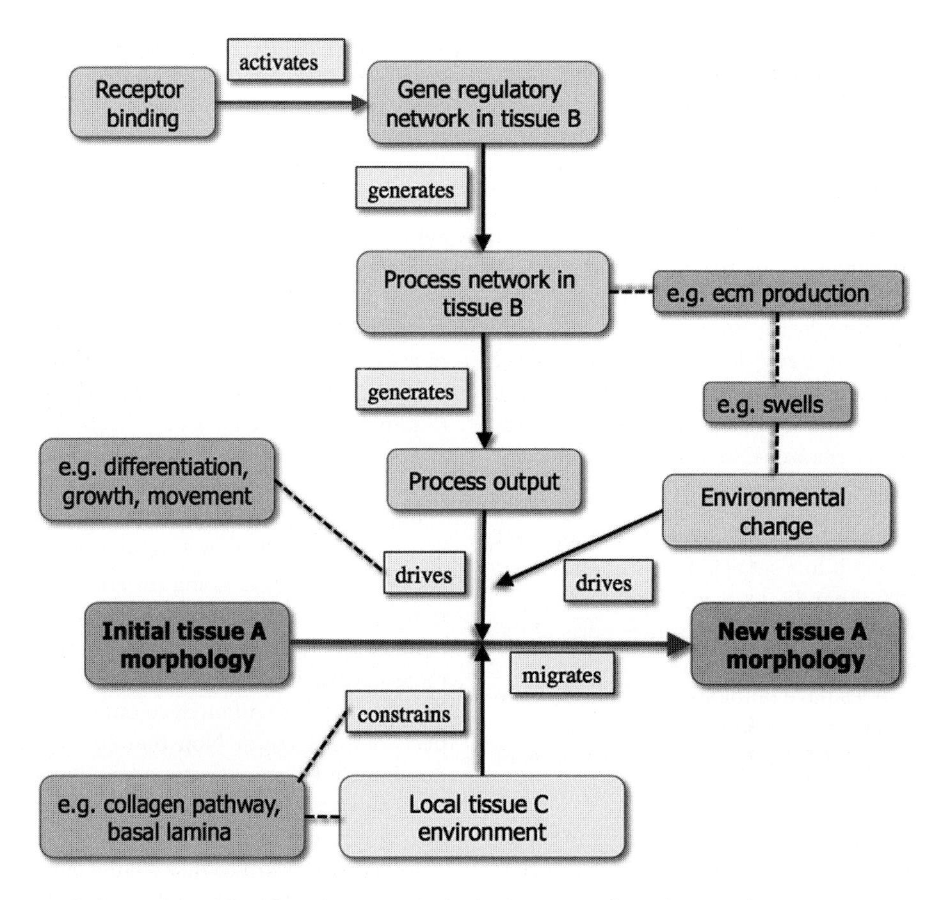

Figure 8.3 A graph showing the modelling of morphogenesis either by the downstream effects of gene activity (upper example is the effect of extracellular matrix production) or through existing boundary effects in the environment (lower example is a collagen track used for contact guidance). (From Bard, 2011b. Reproduced with permission from John Wiley & Sons.)

The use of shading and the embedding of molecular nodes within tissue ones enable the key features to be easily grasped. While links to gene-expression data and PubMed citations could be added, they would make the graph unwieldy, but could be included, with some trouble but little difficulty in a formal listing of the triplets. One advantage of the pictorial representation is that it becomes easy to see gaps in the story. Obvious questions that are yet to be answered are: how is blood circulation established, what is the range of the VEGF gradient, and how is Notch activated?

It should be emphasized that this graph is produced to demonstrate that the complexities of development can be represented in a compact visual format rather than a computational entity. While mathematicians will correctly assert that using a single triplet is inadequate to describe complex chemical interactions, the diagram does however provide an intuitive and clear understanding or what is going on. Equally important, this representation shifts the focus from the signal that activates angiogenesis to the actual process of angiogenesis. The different shades represent tissue states, processes, and networks, and it is worth pointing out that, in the graphical context, all nodes and all levels have equal status—there is no preferred level, a well-known property of systems biology analyses (Noble, 2010).

Discussion and conclusion

As was said in the opening section, this chapter takes a systems approach to development in that

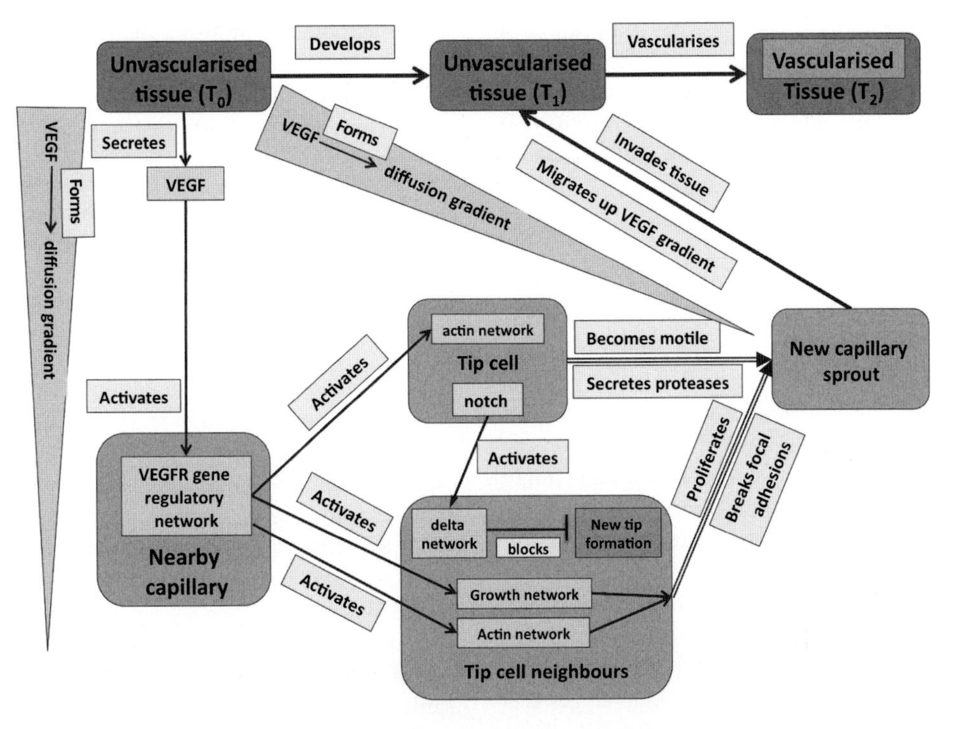

Figure 8.4 A graph describing some of the core events underpinning angiogenesis. Tissues are in darker boxes while molecular events are in lighter boxes, and processes are in pale boxes. (From Bard, 2011b. Reproduced with permission from John Wiley & Sons.)

it tries to step back from the normal minutiae of developmental phenomena. Perhaps the key point that it makes is that development always involves all levels from the genome to the environment, with causation working in both directions mainly to activate ~60 networks whose outputs are the processes that drive development (see Saetzler et al., 2011).

In the wider context, a theory of development cannot be a subset of a theory of genetics because much of development is not run by the genome. Genomic activity is neither cell- nor tissue-autonomous but acts as a resource to be activated by signals from other tissues. Any theory of the development of a tissue involves the prior history of that tissue, knowledge of the tissue's environment, and a description the geometry of that tissue's environment. The music is written in several parts. In a wider context, Waddington (1975) was less than impressed with a simple genomic description of development, noting that these were three perspectives on 'diachronic biology'. Gilbert et al. (1996) used the notion that development is the first derivative of gene expression, anatomy, and physiology, and that evolution is the first derivative of development. In this view, genetics is the means by which the same processes of development become inherited from one generation to the next, and evolution is seen as changes in the developmental processes, thereby giving new anatomical or physiological properties. The purpose of this discussion is to look at some of the implications of this approach.

Development as integrated processes

At first sight, this emphasis on processes might seem to contradict one of the few general principles of systems biology, that there is no preferred level of control. This focus on processes is not, however, a matter of control or levels, but accepts the reality that it is only processes that generate actions and that these can affect anything from a gene up to a tissue. Actions are verbs and everything else is a noun! Processes can thus affect change at any level, and in development the core processes are those that cause anatomical changes. An implication of processes being actions is that it becomes possible to describe developmental events as mathematical graphs where the nodes are biological entities (at any level from molecule to tissue) while the edges are the processes. One might almost go as far as to say that such graphs are the natural language for formally describing development.

Perhaps the greatest limitation of the use of graphs here is that it is very hard to incorporate into them the effects of mutation. The limitation is important in two contexts: first, mutation is a key tool in exploring the function of molecular networks, and the most that the graphical representation can do is to indicate where such experimentation might be helpful; second, mutation is the driver of evolutionary change, and an ability to represent this sort of change graphically would be useful. Nevertheless, focussing on processes has interesting evolutionary implications, mainly because evolutionary change is essentially mutation-induced developmental change that has been selected (changes that are lost are normally viewed as congenital abnormalities), as has been apparent for almost a century (Goldschmidt, 1927).

Simple inspection shows that the key mutations that have driven anatomical change over the last few hundred million years are those that affect the dynamics of developmental processes. In vertebrates, these are for minor patterning and growth: there is good reason to suggest that, once vertebrates reached land, most further change was essentially quantitative. While mutations affect the function of individual proteins, the actual downstream phenotype depends on the role of that protein within the networks in which it plays some role. It is certainly sensible to suggest that mutations affect the dynamics of the networks and so modulate the output process (Bard, 2010). This focus on processes has a further advantage: mutations that affect the output of processes are integral to the network, so that those particular mutations are easily inherited (Bard, 2011a).

Development as performance

If there is any analogy for development, per se, it is performance. Performance is a mixture of score, interpretation, and improvisation. The notion of development as musical performance was mentioned earlier in this essay, and this conceit can be traced back at least as far as Karl Ernst von Baer (1864: 281):

> For that reason, I believe I can compare the various life-processes to musical thoughts or themes and call them creative ideas, which construct their own bodies themselves. What we call in music harmony and melody is here type (the combination of parts) and rhythm (the sequence of forms).

Comparing development to a symphony or a rhapsody is not an uncommon trope (see for example, Keim, 2012; Marino, 2004; Qiu, 2006). Schelling (1802; Schelling & Schott, 1989) famously remarked that 'architecture is frozen music', and to those of us for whom anatomy is architecture (as in the word *Bauplan*), the music of development is not frozen at all.

In music theory, a chord is a 'simultaneity', a series of different notes, each of which is played at the same time as the other pitches of its group. Thus, a chord progression is called a 'succession of simultaneities'. Chord progressions are the homologies of music. They are the underlying unity amidst the apparent diversity. The I-VI-IV-V progression (e.g. C-A^{m}-$D^{m/}$-G^{7}) originated in Western music in the 40s. It is the underlying progression of *Heart and Soul, The Way You Look Tonight*, and hundreds of others. The I-IV-I-V (C-F-C-G) theme is also characteristic of Western music, although it is a much earlier clade. It was very common in Elizabethan English music, and it is still extant, where this progression forms the basis for *Goodnight, Ladies* and *The Lion Sleeps Tonight*. There are only so many chords that work together. It's not what's possible. It's what's compossible.

Evolution occurs by changing development. Improvisation—playing something novel with other musicians—is not complete freedom. Rather, it is a mutual understanding of the chord progressions (Gorow, 2002). A good improviser has to know the cord changes, even if he or she decides to experiment with them. Each improvisation has to work within the musical context provided by the other performances. This is the mutually constructed niche that 'enables' the particular improvisation (Longo et al., 2012). So not everything is possible.

But within the rules and within the context, there is an infinite number of possibilities. Each animal has most of the same notes. But it is where you play the notes (in combination with what other notes), how long you play them, and how loud you play them, that matter. Homologies are the chord progressions of evolution. Each species is its own song. Each individual is a performance of that song, with its own idiosyncratic improvisation on the score. Graphs may provide the notation by which we can visualize the score. And we must remember that each score is not merely for a song, but for a choreographed performance of interacting shapes, a dance. Like dance, development is brought about by the interacting of pliable surfaces. The 'idea' of the dance is not the dance, the score, or the fleshy agents constructing it. The dance is its performance. Scoring such choreography has been very difficult and continues to be an active endeavor (Benesh & Benesh, 1983; Neagle & Ng, 2003).

Thus, development is an ongoing performative act. It involves a score (DNA), an orchestra for interpretation (to choose what DNA is a gene, what the function of BMP4 is in any particular cell, etc.), and improvisation (regulating gene expression such that most knockout mice have minimally altered phenotypes; altering anatomy by changing gene expression patterns). Like an ensemble group, no conductor is needed—just some ion transport as sperm meets egg is enough to start the show. The relationships between cell surfaces generate morphogenetic fields, tissues, and organs. The body builds itself as it develops, each whole becoming a part of something larger that it generates, and each whole defining the context of its parts. Development is a creative choreography of molecules, cells, tissues, organisms, and ecosystems. As each organism is a new developmental performance, we are left with Yeats' (1929) question, 'How can we know the dancer from the dance?'

Acknowledgements

SFG is funded by the Academy of Finland and a faculty research grant from Swarthmore College. He thanks S. Friedler and G. Levinson for their discussions, and A. Soto for her provocative critiques of the draft.

References

Auletta, G. (2011). *Cognitive Biology: Dealing with Information from Bacteria to Minds*. Oxford University Press, New York.

Bard, J. (2010). A systems view of evolutionary genetics. *Bioessays*, **32**, 559–63.

Bard, J. (2011a). The next evolutionary synthesis: from Lamarck and Darwin to genomic variation and systems biology. *Cell Communication and Signaling*, **9**, 30.

Bard, J. (2011b). A systems biology formulation of developmental anatomy. *Journal of Anatomy*, **218**, 591–9.

Bard, J. (2012). A new ontology (structured hierarchy) of human developmental anatomy for the first 7 weeks (Carnegie stages 1–20). *Journal of Anatomy*, **221**, 406–16.

Bard, J. (2013). Driving developmental and evolutionary change: a systems biology view. *Progress in Biophysics and Molecular Biology*, **111**, 83–91.

Barkai, N., and Perrimon, N. (2011). The era of systems developmental biology. *Current Opinion in Genetics and Development*, **21**, 681–811.

Bellus, G.A., Gaudenz, K., Zackai, E.H., et al. (1996). Identical mutations in three different fibroblast growth factor receptor genes in autosomal dominant craniosynostosis syndromes. *Nature Genetics*, **14**, 174–6.

Benesh, R., and Benesh, J. (1983). *Reading Dance: The Birth of Choreology*. McGraw-Hill, New York.

Bolker, J. (2012). There's more to life than rats and flies. *Nature*, **491**, 31–3.

Buszczak, M., Paterno, S., Lighthouse, D., et al. (2007). The Carnegie protein trap library: a versatile tool for *Drosophila* developmental studies. *Genetics*, **175**, 1505–31.

Chung, A.S., and Ferrara, N. (2011). Developmental and pathological angiogenesis. *Annual Review of Cell and Developmental Biology*, **27**, 563–84.

Culver, J.C., and Dickinson, M.E. (2010). The effects of hemodynamic force on embryonic development. *Microcirculation*, **17**, 164–78.

Davidson, E.H. and, Levine, M.S. (2008). Properties of developmental gene regulatory networks. *Proceedings of the National Academy of Sciences USA*, **105**, 20063–6.

Dubourg, C., Lazaro, L., Pasquier, L., et al. (2004). Molecular screening of SHH, ZIC2, SIX3, and TGIF genes in patients with features of holoprosencephaly spectrum: mutation review and genotype-phenotype correlations. *Human Mutation*, **24**, 43–51.

El-Hani, C.N. (2014). Downward determination as a propensity-changing non-causal relation. In press.

El-Hani, C.N., and Emmeche, C. (2000). On some theoretical grounds for an organism-centered biology: property emergence, supervenience, and downward causation. *Theory in Biosciences*, **119**, 234–75.

Ellis, G.F.R. (2012). Top-down causation and emergence: some comments on mechanisms. *Journal of the Royal Society Interface Focus*, **2**, 126–40.

Faust, K., Sathirapongsasuti, J.F., Izard, J., et al. (2012). Microbial co-occurrence relationships in the human microbiome. *PLoS Computational Biology*, **8**:e1002606.

Flüsser, V. (1987). *Vampyroteuthis infernalis*. Atropos Press, New York.

Gilbert, S.F. (1991). Commentary: cytoplasmic action in development. *Quarterly Review of Biology*, **66**, 309–16.

Gilbert, S.F. (2009). The adequacy of model systems for evo-devo. In A. Barberousse, M. Morange and T. Pradeu, eds., *Mapping the Future of Biology: Evolving Concepts and Theories*. Springer, Dordrecht, pp. 57–68.

Gilbert, S.F. (2010). *Developmental Biology*, 9th ed. Sinauer Press, Sunderland.

Gilbert, S.F., and Epel, D. (2009). *Ecological Developmental Biology*. Sinauer Associates, Sunderland.

Gilbert, S.F., and Greenberg, J.P. (1984). Intellectual traditions in the life sciences II. Stereocomplementarity. *Perspectives in Biology and Medicine*, **28**, 18–34.

Gilbert, S.F., Opitz, J.M., and Raff, R.A. (1996). Resynthesizing evolutionary and developmental biology. *Developmental Biology*, **73**, 357–72.

Gilbert, S.F, Sapp, J., and Tauber, A.I. (2012). A symbiotic view of life: we have never been individuals. *Quarterly Review of Biology*, **87**, 325–41.

Gilbert, S.F., and Sarkar, S. (2000). Embracing complexity: organicism for the 21st century. *Developmental Dynamics*, **219**, 1–9.

Goldbeter, A., and Pourquié, O. (2008). Modeling the segmentation clock as a network of coupled oscillations in the Notch, Wnt and FGF signaling pathways. *Journal of Theoretical Biology*, **252**, 574–85.

Goldschmidt, R. (1927). *Physiologische Theorie der Vererbung*. Springer, Berlin.

Gorow, R. (2002). *Hearing and Writing Music: Professional Training for Today's Musician*, 2nd ed. September Publishing, Gardena.

Guenther, C., Pantalena-Filho, L., and Kingsley, D.M. (2008). Shaping skeletal growth by modular regulatory elements in the *Bmp5* gene. *PLoS Genetics*, **4**:e1000308.

Haraway, D.J. (2008). *When Species Meet*. University of Minnesota Press, Minneapolis.

Hertwig, O. (1892). Urmund und Spina bifida. *Archiv für mikroskopische Anatomie*, **39**, 353–503.

Hertwig, O. (1894). *The Biological Problem of To-day: Preformation or Epigenesis?* Macmillan, New York.

HMPC (Human Microbiome Project Consortium) (2012). Structure, function and diversity of the healthy human microbiome. *Nature*, **486**, 207–14.

Horder, T.J., and Weindling, P.J. (1986). Hans Spemann and the organiser. In T.J. Horder, J.A. Witkowski and C.C. Wylie, eds., *A History of Embryology*. Cambridge University Press, Cambridge, pp. 183–242.

Huang, S., Ernberg, I., and Kauffman S.A. (2009). Cancer attractors: a systems view of tumors from a gene network dynamics and developmental perspective. *Seminars in Cell and Developmental Biology*, **20**, 869–76.

Ingolia, N.T. (2004). Topology and robustness in the *Drosophila* segment polarity network. *PLoS Biology*, **2**: e123.

Kauffman, S.A. (1987). Developmental logic and its evolution. *BioEssays*, **6**, 82–7.

Keim, B. (2012). Cellular symphony of embryo development. http://www.wired.com/wiredscience/2012/06/embryo-development-video/

Leibniz, G.W. (1697). On the ultimate origination of things. Quoted in B.C. Look, *Leibniz's Modal Metaphysics*. Stanford Encyclopedia of Philosophy. http://plato.stanford.edu/entries/leibniz-modal/

Lenoir, T. (1982). *The Strategy of Life: Teleology and Mechanics in Nineteenth Century German Biology*. D. Reidel, Dordrecht.

Longo, G., Montévil, M., and Kauffman, S. (2012). No entailing laws, but enablement in the evolution of the biosphere. *ACM Proceedings of the genetic and evolutionary computation conference, GECCO'12, Philadelphia, 7–11 July 2012*, Arxiv 1201.2069.

Marini, M., Cusano, R., De Biasio, P., et al. (2003). Previously undescribed nonsense mutation in SHH caused autosomal dominant holoprosencephaly with wide intrafamilial variability. *American Journal of Medical Genetics*, **117A**, 112–15.

Marino, M. (2004). Biography of Philip A. Beachy. *Proceedings of the National Academy of Sciences USA*, **101**, 17897–9.

Neagle, R.J., and Ng, K.C. (2003). *Machine-Representation and Visualisation of a Dance Notation*. Proceedings of Electronic Imaging and the Visual Arts—London July 2003.

Newman, S.A. (2012). Physico-genetic determinants in the evolution of development. *Science* **338**, 217–9.

Nijhout, H.F. (1999). Control mechanisms of polyphenic development in insects. *Bioscience*, **49**, 181–92.

Nijhout, H.F., and Paulsen, S.M. (1997). Developmental models and polygenic characters. *American Naturalist*, **149**, 394–405.

Noble, D. (2010). Biophysics and systems biology. *Philosophical Transactions of the Royal Society of London A*, **368**, 1125–39.

Oyama, S. (1985). *The Ontogeny of Information: Developmental Systems and Evolution*. Duke University Press, Durham.

Palumbi, S.R. (2009). Speciation and the evolution of gamete recognition genes: pattern and process. *Heredity*, **102**, 66–76.

Pirsig, R.M. (1974). *Zen and the Art of Motorcycle Maintenance: An Inquiry into Values*. William Morrow, New York.

Pradeu, T. (2011). A mixed self: the role of symbiosis in development. *Biological Theory*, **6**, 80–8.

Qiu, J. (2006). Epigenetics: the unfinished symphony. *Nature*, **441**, 143–5.

Rosten, L. (2003). *The New Joys of Yiddish*. Random House, NY. Excerpt at http://everything2.com/title/YiddishNew York.

Russell-Rose, T. (2011). Prostitutes appeal to Pope: text analytics applied to search. Accessed at http://isquared.wordpress.com/2011/07/28/prostitutes-appeal-to-pope-text-analytics-applied-to-search/

Saetzler. K., Sonnenschein, C., and Soto, AM. (2011). Systems biology beyond networks: generating order from disorder through self-organization. *Seminars in Cancer Biology*, **21**, 165–74.

Schelling, F.W.J. (1802). Philosophie der Kunst;quoted in Bright, M. (1984). *Cities Built to Music*. Ohio State University Press, p. 82.

Schelling, F.W.J. von, and Stott, D. W (1989). *The Philosophy of Art/Friedrich Wilhelm Joseph Schelling; edited, translated and introduced by Douglas W. Stott; foreword by David Simpson*. University of Minnesota Press, Minneapolis.

Soto, A.M., Rubin, B.S., and Sonnenschein, C. (2009). Interpreting endocrine disruption from an integrative biology perspective. *Molecular and Cellular Endocrinology*, **304**, 3–7.

Spemann, H. (1943). *Forschung und Leben*. Engelhorn, Stuttgart.

Stamatoyannopoulos, J.A. (2012). What does our genome encode? *Genome Research*, **22**, 1602–11.

Stotz, K.C., Bostanci, A., and Griffiths, P.E. (2006). Tracking the shift to 'postgenomics'. *Community Genetics*, **9**, 190–6.

Suchting, S., Freitas, C., le Noble, F., et al. (2007). The Notch ligand Delta-like 4 negatively regulates endothelial tip cell formation and vessel branching. *Proceedings of the National Academy of Sciences USA*, **104**, 3225–30.

Tang, G.H., Rabie, A.B., and Hagg, U. (2004). Indian hedgehog: a mechanotransduction mediator in condylar cartilage. *Journal of Dental Research*, **83**, 434–8.

Turing, A.M. (1952). The chemical theory of morphogenesis. *Philosophical Transactions of the Royal Society of London B*, **237**, 37–72.

Vermeij, M.J., Marhaver, K.L., Huijbers, C.M., et al. (2010). Coral larvae move toward reef sounds. *PLoS One*, **5**:e10660.

von Baer, K.E. (1864). Welche Auffassung der Natur ist die richtige? und wie ist diese Auffassung auf die Entomologie anzuwenden? In K.E. von Baer, *Reden gehalten in wissenschaftlichen Versammlungen und kleinere Aufsätze vermischten Inhalts. Erster Theil. Reden*. H. Schmitzdorff, St. Petersburg, pp. 237–84.

Waddington, C.H. (1975). *Evolution of an Evolutionist*. Edinburgh University Press, Edinburgh.

Warkentin, K.M., Caldwell, M.S., and McDaniel, J.G. (2006). Temporal pattern cues in vibrational risk assessment by embryos of the red-eyed treefrog. *Agalychnis callidryas. Journal of Experimental Biology*, **209**, 1376–84.

Weiss, P. (1955). Beauty and the beast: life and the rule of order. *Scientific Monthly*, **81**, 286–99.

Weiss, P. (1967). One plus one does not equal two. In G.C. Quarton, T. Melnechuk, and F.O. Schmitt, eds., *The Neurosciences: A Study Program*. Rockefeller University Press, New York, pp. 801–21.

Weiss, P. (1977). The system of nature and the nature of systems: empirical holism and practical reductionism harmonized. In K.E. Schaefer, H. Hensel and R. Brady, eds., *Toward a Man-Centered Medical Science*. Futura, Mt. Kisko, pp. 17–63.

Winther, R.G. (2011). Part-whole science. *Synthese*, **178**, 397–427.

Witt, J., Barisic, S., Sawodny, O., et al. (2011). Modeling time delay in the NFκB signaling pathway following low dose IL-1 stimulation. *EURASIP Journal on Bioinformatics and Systems Biology*, **3**, 1–6.

Wolf, U. (1997). Identical mutations and phenotypic variation. *Human Genetics*, **100**, 305–21.

Wolf, U. (2002). Genotype and phenotype: genetic and epigenetic aspects. In A. Grunwald, M. Gutmann, and E.M. Neumann-Held, eds., *On Human Nature: Anthropological, Biological, and Philosophical Foundations*. Springer, New York, pp. 111–20.

Yeats, W.B. (1929). Among school children. Accessed at http://poetry.about.com/od/poems/l/blyeatsamongchildren.htm

CHAPTER 9

General theories of evolution and inheritance, but not development?

Wallace Arthur

Introduction: five assumptions

Let's start by taking a close look at this book's title, and contrast its commendable brevity and 'singularity' with five important assumptions that are implicit in it. *Towards a Theory of Development* implies the following: first, that a single theory of development is at least possible; second, that we do not yet have one; third, that it is desirable to seek one; fourth, that we have a consensus on what is meant by 'a theory' in the biological sciences; and finally that we likewise have a consensus on the meaning of 'development', so that we can draw a clear line between developmental and non-developmental biological processes.

It is no surprise that not all of the contributors agree on all of these points. Views vary both on an individual basis and between different 'camps'. By the latter, I mean the disciplinary (and subdisciplinary) allegiances of the various authors. Some are biologists, and others are philosophers of biology. Among the biologists, some are very definitely developmental biologists while others are harder to categorize, though many of these are students of evo-devo. Within each of these disciplines there are further subdivisions: 'camps within camps', if you will.

I would describe myself as a student of evo-devo, and specifically one whose academic training was originally in the area of evolutionary ecology and population genetics. This training will have influenced my views; and it will also influence the structure of this chapter, because I will start on what is for me well-trodden ground—evolution—and move from that starting point to development. This will enable me to begin with the deceptively simple idea of a 'theory of evolution'—a phrase that is often bandied about without careful consideration of what it means, albeit this is truer of the popular science literature than it is of the specialist literature on evolutionary biology. From this beginning, I intend to take a course that will deal with the 'five assumptions' listed in the opening paragraph, though not necessarily in the order in which they were given there, progress towards a focus on the causal structure of development, and then consider the possible links between such structure and the possibility of a 'theory of development' that is implicit in this book's title.

The theory of evolution

It would be odd for a book in the area of evolutionary biology to have a title *Towards a Theory of Evolution*. The reason for this is that most biologists, and probably most laypeople too, think, or in some cases are convinced, that we already have a theory of evolution and thus that 'towards' would be an inappropriate word. But do we really have such a theory, and if so, what is it?

It is necessary here to deal with the question of what 'a theory' means in biology and indeed in science more generally. There are of course multiple usages of 'theory', both scientific and otherwise. However, this is not the place for an exhaustive listing and comparison of these. Rather, I wish to concentrate on one particular usage: where 'a theory' is very different from a typical hypothesis, both

Towards a Theory of Development. Edited by Alessandro Minelli and Thomas Pradeu

in its breadth and in the degree to which it is accepted as being correct.

Since 'a typical hypothesis' is too vague for my purposes, let's use a specific hypothesis to contrast with 'a theory'. In recent publications resulting from work done by members of my own lab and their collaborators, the following hypothesis has been advanced: during speciation events involving geophilomorph centipedes, natural selection modifies developmental reaction norms—specifically, the one relating to the effect of temperature on trunk segment number (Vedel et al., 2008, 2009, 2010). Note that this is a narrow hypothesis in that it refers only to one character in one smallish group of closely related species. Note also that, at the time of writing, 'we' (either the group who proposed the hypothesis or the evolutionary biology community in general) have no idea whether or not it is correct.

Now, contrast these features—restricted breadth and uncertainty of correctness—with their counterparts in 'the theory of evolution'. For this purpose, we can use a particular version of the theory: 'natural selection is the main cause of evolutionary modification'. This, as many readers will recognize, is a variant of an often-used quotation from Darwin's (1859) *Origin of Species.* Note the breadth—the theory is intended to apply to all species, both extant and extinct, of all kinds, including animals, plants, and the other kingdoms, and at all times since the origin of life some three and a half or four billion years ago to the present. Note also that most biologists would regard the theory as being correct.

However, there is a difficulty with 'main' in the version of the theory given above (and in Darwin's version). It is certainly advisable to include this word, because in biology theories are rarely if ever *universally* applicable as they sometimes are in the physical sciences. Typically, there are 'exceptions that prove the rule'. Darwin himself was aware that some variations might be neither beneficial nor detrimental ('neither useful nor injurious' was Darwin's phrase) and so would be what we now call 'selectively neutral'. He saw these as a small minority of cases, so they did not challenge natural selection's place as the 'main' agent of change. But according to the neutral theory of molecular evolution that was articulated in the second half of the 20th century (Kimura, 1968, 1983) the majority of variations at the molecular level are selectively neutral (or 'nearly neutral'). The fate of such variants is determined not by natural selection but rather by genetic drift. After much debate on this theory (a good account of the early stages of the debate can be found in Lewontin, 1974), a sort of stalemate has been reached whereby it is acknowledged by almost all evolutionists that genetic drift has a greater evolutionary role than was appreciated before Kimura's work. However, there is still no clear consensus on the percentage of changes that it is responsible for at, say, the level of DNA base sequences. So, whether selection is the 'main' agent of evolutionary change at the molecular level is debatable.

Neutrality is not the only challenge to how 'main' an agent of change natural selection is. In Darwin's day, the Lamarckian process of inheritance of acquired characters was a more obvious challenge. And, in the last few decades, the possibility that the developmental variation acted upon by natural selection is in some way biased or structured rather than random, and that it thus may be an agent influencing the direction of evolutionary change, has been put forward by several authors, most famously Gould and Lewontin (1979) (see also Arthur, 2004 and Gould, 1989). These and other challenges notwithstanding, most biologists would probably still accept the Darwinian theory that selection is the main cause of evolutionary change.

A final point that needs to be made about Darwin's theory is that it is mechanistic. Natural selection is a *mechanism* through which populations, and ultimately species, may change in multiple characters at both molecular and morphological levels. And indeed it is a *testable* mechanism. This provides an important contrast with early debates on a theory of development, as we will shortly see.

Theories of development: a historical approach

In the 1600s and 1700s there was a debate about the nature of development between the groups of thinkers referred to as the preformationists and the epigeneticists. The former thought that a miniature adult was present in the sperm (spermists) or egg (ovists) and that its development consisted solely of growth. The latter thought that there was no such miniature adult present in either of the types of gamete. Instead, there was thought to be information of some sort that could be interpreted as a set of instructions about how to build an adult of the species concerned. The prevailing current view is that the epigeneticists were right, the preformationists wrong, though the argument is more complex than might be concluded from the simplified version given above. For two very different, but equally thorough, accounts of this issue, see Pinto-Correia (1997) and Robert (2004).

At first sight, this pair of opposing theories might be considered as having a relationship that was roughly equivalent to the opposition between Lamarckian and Darwinian theories. But this apparent equivalence does not stand up to close scrutiny; the reason why this is the case has everything to do with mechanisms and our ability to bring evidence to bear on whether or not they exist. Lamarck proposed a mechanism that has been looked for and found to be lacking, despite intermittent claims to the contrary. Darwin proposed a mechanism that has been looked for and found, both in nature and in the laboratory. Selection experiments are commonplace in population genetics and have allowed not just a confirmation of the existence of selection but a quantification of its magnitude and its effects.

In contrast, the preformationist and epigeneticist theories of development, when first proposed, gave no clear mechanisms that could be searched for and either found or not found. Preformationism virtually precludes a mechanism anyhow, except for one relating to growth. And although now, centuries later, we understand quite a lot about how the information present in a fertilized egg is used to initiate the developmental process, nothing of this (e.g. cytoplasmic localizations of transcription factors) was known to the original epigeneticists. So although, historically, there was a battle between two opposing theories of development, the lack of clearly articulated mechanisms by either of the two camps meant that the 'battle' was more philosophical than scientific.

Towards a theory of development in the 21st century

Might it be possible to have a single, mechanistic, theory of development, comparable to what can be considered to be Darwin's single, mechanistic theory of evolution? This is the question to which I now turn. As a preamble to the main discussion in this section, though, we should return to one of those 'five assumptions' that I identified at the outset: to be specific, the assumption that there is a consensus on the meaning of 'development'.

Rather than attempting to define development, and thus to delineate those biological processes that are developmental ones from those that are not, I will adopt the simpler approach of saying what 'development' definitely includes. It includes all cases in which a fertilized egg undergoes a series of changes, including growth via cell proliferation, and ends up producing a multicellular adult organism that is capable of producing gametes (which can then participate in further fertilizations). 'Development' may of course include many other processes that differ from the above, such as vegetative reproduction (Minelli, this volume) and regeneration (Vervoort, 2011). Although I will not discuss these further explicitly, it seems likely that much of what follows applies to them as well as to egg-to-adult development.

So, back to the possibility of a single mechanistic theory of development. An important issue at this point is the level of detail to consider. Too detailed an approach will lead to a failure to generalize. For example, the statement that 'all developmental processes are driven by homeobox genes and their protein products' would be easily disproved—despite the very great importance of this group of genes and proteins in the development of all animals (see Duboule, 1994 and various online updates for a classification of the many families of homeobox genes). On the other hand, if I were to take a very broad (or 'undetailed') approach and to state that 'all developmental processes involve the transfer of positional information' (in the sense of Wolpert, 1969), this would be much harder to disprove. But, even if this statement turned out to be right, it is not telling us anything about the nature of the positional information and so, arguably, does not represent a truly mechanistic theory. These examples illustrate the need for a 'middle way' in terms of the level of detail that a mechanistic theory of development should adopt.

Development is often divided into two main components: within cell and between cell. For a cell to become differentiated into a particular type (e.g. muscle, nerve), many internal changes need to take place. For a population of cells to develop into the correct three-dimensional shape, such as the spindle shape of a biceps muscle or the flatter shape of the muscles spanning our foreheads, the changes that are needed are not within the cells concerned, which are of the same type in both cases, but rather within the overall population of cells to which they contribute. Processes at this level are referred to as pattern formation and/or morphogenesis. To make some progress on the quest for a mechanistic theory, I am now going to sacrifice generality to the extent of focussing on cell differentiation and putting aside, for the moment, the between-cell aspect of development.

It can be argued that we already have a general theory of cell differentiation. If we go back about 50 years, there were two main possible ways in which undifferentiated cells could be considered to turn into differentiated ones: selective gene loss (permanent and irreversible) or variable gene activity (temporary and at least potentially reversible). The set of experiments that determined which of these alternative mechanistic theories was correct was conducted by the British biologist John Gurdon and his colleagues (see Gurdon et al., 1958).

These experiments are well known, so a brief summary is all that is needed here. Working with the frog *Xenopus*, Gurdon and co-workers enucleated an egg and injected into it the nucleus of a differentiated cell from later in development. Of course, they did this many times, not just once. Although some of the hybrid cells they produced in this way died, others not only survived but started to develop; and some of these went on to produce normal tadpoles and even adult frogs. This result demonstrates conclusively that the complement of genes needed for development had not been reduced during differentiation. Rather, some genes had just been switched off; and they were able to be switched on again by agents in the egg cytoplasm.

So, there has for some time been proof of the 'variable gene activity theory' of cell differentiation, as it was dubbed by Davidson (1986). Or has there? There is a large jump from *Xenopus* to animals in general, let alone multicellular organisms in other kingdoms. Although later experiments on other species have produced similar results, most famously 'Dolly' the sheep (Wilmut et al., 1997), there is still a large leap of faith to be made from a select few model systems to the animal kingdom and beyond. But it is at least possible that cell differentiation works by the reversible process of gene switching rather than the irreversible one of gene loss *in general*.

The term 'variable gene activity theory' might be taken to imply that it is *genes* that are the key to understanding the process of cell differentiation. But we should be careful not to adopt such an unbalanced view. Whatever the identity of the agents in the egg cytoplasm that switch inactivated genes back on again (many of them are proteins), they, instead of the genes that they act upon, could be regarded as the key players. However, such a view is equally unbalanced. Cell differentiation and other developmental processes are the result of *an interplay* between genes and other molecules, an interplay that is in many cases also influenced by environmental factors.

The causal structure of development

It is now time to broaden out again and consider development in its totality rather than focussing specifically on cell differentiation. One way of approaching the possibility of a theory of development that is broader in its scope than the 'variable gene activity theory' is to think of development in terms of *causal links*, as follows.

The idea of cause and effect, or, to put it another way, the causal link between two entities, is fundamental to science. In the context of development, cause and effect operate in many ways and at different levels.

At the molecular level, an example of a causal link is the relationship between the binding of a transcription factor protein (cause) and the switching on of the target gene (effect). However, this description does not reflect the complexity of reality. Usually, several transcription factors and co-factors are involved. The actual process of transcription involves RNA polymerase and other molecules. And the product of transcription (pre-mRNA) is just an intermediate in a longer chain of downstream events. These include intron splicing, transfer out of the nucleus to cytoplasmic ribosomes, translation into protein, post-translational modification of the protein, and movement of the protein, perhaps to other cells. Then there are the consequences of the protein's function, such as the multiple developmental events that ensue from the secretion of the protein Sonic Hedgehog from cells in the notochord of a vertebrate embryo. These 'developmental events' include differentiation of many cells into particular cell types in particular regions. So we have now, almost without noticing, moved from the level of molecules within cells to the more macroscopic level of 'regions', which is really another label for cell populations or blocks of tissue.

At a higher level still are the actions of various hormones that influence growth rates throughout the body. It is probably these, or factors with which they interact, that get modified evolutionarily to produce what D'Arcy Thompson called 'transformations' (Thompson, 1917).

Regardless of the level of operation or the exact mode of action, causal links can be represented in abstract form by arrows and other simple general symbols. I will use such an abstract approach here, rather than going into molecular details, which are at least partly specific to each different developmental context. This approach should enable discussion of the causal structure of development in general terms.

The simplest abstract representation of cause and effect is:

$$C \rightarrow E$$

This leaves the nature of both entities, and of the causal influence that connects them, unspecified. All it contains is the asymmetry of the relationship, denoted by the direction of the arrow.

One of the givens about development is that it contains a very large number of causal links. This is because something that is from one point of view an effect (such as E above) is, from another point of view, a cause. For example, if C is a transcription factor and E the expression of a target gene, and if the target gene makes another type of transcription factor that switches on a third gene, then we have a coupled pair of causal links, as follows:

$$C \rightarrow E/C \rightarrow E$$

This kind of notation will become too cumbersome for systems with many causal links, so we can simply give each entity a letter and let the arrows indicate causality:

$$a \rightarrow b \rightarrow c \rightarrow d \rightarrow e \rightarrow f \rightarrow g \rightarrow h \rightarrow i \rightarrow j$$

The next problem is that the way in which causal links are connected up in development is far more complex than the linear pattern shown above. There are examples of hierarchical patterns of connection (diverging arrows), combinatorial ones (converging arrows) and negative feedback (as when a gene product accumulates and acts to switch off the gene that made it; $a \leftrightarrow b$).

The overall development of even a simple animal doubtless involves multiple instances of all three of these patterns. Together, they are almost certainly more common than linear causal sequences (as in a to j above). Thus, while the use of a series of letters was an advance on our starting point, where many entities would have to be doubly labelled (C/E) because they were both cause and effect, its alphabetic nature is inappropriate for non-linear (i.e. most) patterns of connection of causal links. In complex patterns of connection (let's call them networks), we can simply use a standard symbol (such as O) for an object/entity (gene, transcription factor, morphogen, hormone, and so on), just as we use a standard symbol ($\rightarrow$) for the link between one entity and another. Some simple networks are illustrated, using this system of representation, in Figure 9.1; a more complex one is shown in Figure 9.2.

When I first became interested in evo-devo, before that name was coined and (just) before the homeobox was discovered, I used this form of representation and argued that development was primarily a hierarchical pattern of causal links (the 'morphogenetic tree'), with other patterns (e.g. negative feedback ones) being less important (Arthur, 1984). Subsequently, I refined both my representation and my argument, so that non-hierarchical effects were made explicit (Arthur, 1997). Meanwhile, other authors (e.g. Raff, 1996) were stressing the importance of modularity, both in development itself and in its evolution. It is now clear that most if not all developmental systems are modular to a degree; that is, they are characterized by spatially restricted modules, inside each of which there is a high density of causal links, whereas the density of links between the modules is lower.

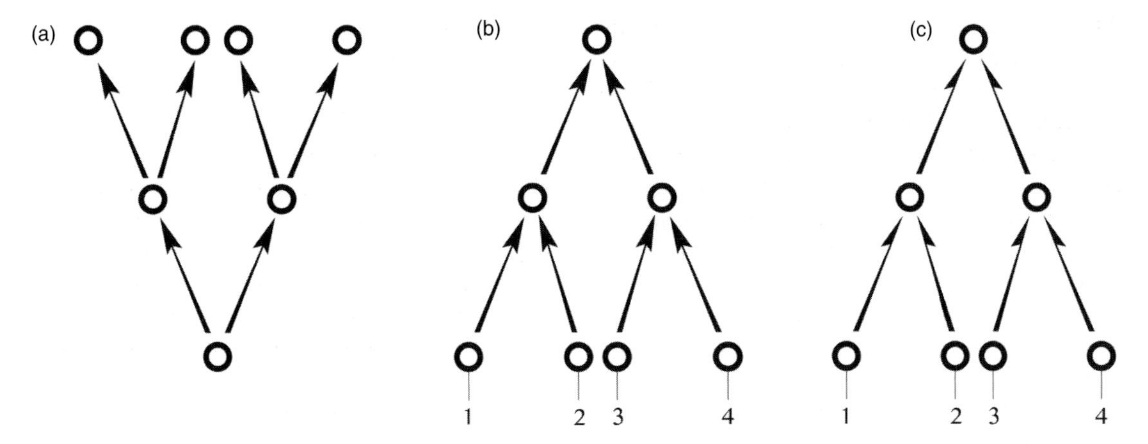

Figure 9.1 Three abstract causal networks in development. (a) Hierarchy—a divergent pattern of causal links. (b, c) Two different convergent patterns. (b) Redundant control: here the situation is 'either-or'. (c) Combinatorial control: here the situation is 'both-and': both half-arrows are required to produce the effect. The difference between (b) and (c) is that while (c) requires all four causes in order to produce the end-effect, in (b) any of the causes 1–4 will suffice. The advantages of these two convergent patterns are effectively opposites of each other. Combinatorial control ensures that some developmental event does not happen unless a series of prior events have all happened; redundant control ensures that the later developmental event will go ahead as long as any of the earlier ones have done so. One of these types of causality may be more appropriate in some developmental contexts and the other, in different contexts.

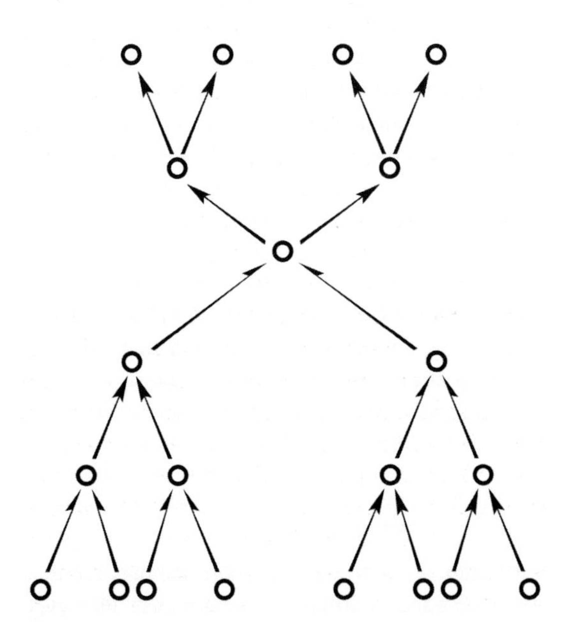

Figure 9.2 A complex causal network involving hierarchical, combinatorial, and redundant elements. Note that the hierarchy (top) will not be activated unless both the earlier combinatorial/redundant 'pyramids' have taken effect. The particular casual pattern shown is, of course, just one of very many possibilities if there are 21 interacting developmental entities, as here.

The American geneticist Eric Davidson and his colleagues have made a detailed examination of the causal structure of development at the level of what they call gene regulatory networks or GRNs (Davidson, 2010; Peter & Davidson, 2009; 2011; see also Morange, this volume). One of their conclusions (Davidson, 2010: 911) is that 'GRNs differ in their degree of hierarchy, and also in the types of modular sub-circuit of which they are composed'. This conclusion could probably be expanded and extended to developmental processes more generally: they vary in the degree of importance of hierarchical, modular, and other aspects of their causal structure.

Now the central problem comes into focus. It is probably true that all developmental processes include hierarchical, modular, combinatorial, and feedback elements. But is there any rule that specifies the relative importance of these? Is there any limit to which, for example, a developmental system can vary along the dimension of multiply hierarchical to non-hierarchical? At present we do not know the answers to such questions; indeed, it is not even clear how they would best be approached experimentally.

In the absence of obvious experimental approaches, one possibility would be to go down a different road: that of mathematical modelling (see

Jaeger and Sharpe, this volume). In the same way that von Dassow et al. (2000) used a modelling approach to examine the robustness of a particular developmental system—the segment polarity module in *Drosophila*—it should be possible to examine the extent to which developmental systems more generally continue to produce the same outcome despite variations imposed by the modeller on their causal structure. We could start, for example, with a strongly hierarchical system and then, bit by bit, reduce the number of hierarchical linkage-patterns and replace them with other types of pattern. This whole field—mathematical modelling of development—has much potential and has been explored widely, though it does not yet feature as much as it should in most developmental biology texts. Important studies, in addition to those mentioned above, include the early general work of Meinhardt (1982) and the more recent work on the development (and evolution) of mammalian teeth by Jukka Jernvall and colleagues (see, for example, Kangas et al., 2004; Salazar-Ciudad & Jernvall, 2002, 2010).

Despite much work in this field, no single unifying theory has yet emerged. And this fact directs our attention back to the 'five assumptions' with which this chapter started.

Back to the 'five assumptions'

We have dealt with three of these so far, as follows. There may be no consensus on exactly how to define development or the best way to use 'theory' in biology. However, by taking a pragmatic approach, these difficulties can be set to one side. It is clear that as of now we do not have a single theory of development. But this leaves two of the five assumptions that I have not yet discussed after first mentioning them. These are: that a single theory of development is possible; and that we should be actively searching for it rather than being content to unravel the details of specific developmental processes and systems.

It is hard to separate these two assumptions. If an overarching theory of development is not possible, then clearly we should not be wasting time looking for it. But equally, it will never become clear whether such a theory exists unless we *do* look for it. Then again, after what amount of looking, without finding the elusive general theory of development, should we stop and conclude that it does not exist? A parallel with the search for the Higgs boson springs to mind. So, where do we go from here? The following section offers an idiosyncratic approach.

Looking for pairs of opposites

It sometimes seems that a general theory only makes sense in the light of a contrast with an opposing theory. Whether this is true is debatable, but nevertheless looking for pairs of opposing theories can be helpful in interpreting the history of the formulation of theories. It may also be helpful in searching for new general theories, as I will explain below.

In looking at theories of evolution, we already saw the conflict between Lamarckian and Darwinian theories in the 19th century; we also noted the conflict between Kimura's neutral theory of molecular evolution (late 20th century) and its selectionist (i.e. neo-Darwinian) counterpart. A requirement for any kind of evolution, of course, is inheritance; and how inheritance works is also an issue to which general theories can be brought to bear. Indeed, Mendelian theory (Mendel, 1866) is one of the few cases in the biological sciences where a theory of very general applicability is often stated in the form of laws (in this case two of them)—because quantifiable predictions can be tested by experiments that give quantitative results.

Here is a comment by one of the founders of population genetics, Ronald A. Fisher (1930), on the interesting question of why no one came up with the Mendelian theory of inheritance on an a priori basis:

> It is a remarkable fact that had any thinker in the middle of the nineteenth century undertaken, as a piece of abstract and theoretical analysis, the task of constructing a particulate theory of inheritance, he would have been led, on the basis of a few very simple assumptions, to produce a system identical with the modern scheme of Mendelian or factorial inheritance.

But Fisher is missing a vital point. We now contrast 'blending' and 'particulate' inheritance. However, prior to Mendel, this dichotomy was not apparent. Pre-Mendelian views on inheritance are only seen as 'blending' in retrospect. A contrast between two types of theory is only really obvious when we have seen both and articulated the difference between them.

Let's now return from inheritance to development. We saw that there was an early conflict of opinion between the preformationists and the epigeneticists. We also noted the later conflict between the 'selective gene loss theory' and the 'variable gene activity theory' of one particular aspect of development—cell differentiation. But now it is time to ask whether there are conflicting theories of development that are neither ancient nor partial. In other words, our key question becomes: are there major conflicting theories of development as a whole in the early 21st century?

The answer would at first sight seem to be 'no'. Modern developmental biology texts (e.g. Gilbert, 2010; Slack, 2008) are not structured in the way that they would be if there was a major debate going on between alternative general theories. But perhaps if we look harder and try to envisage conflicting theories we can find them.

Here is an example. As noted above, many developmental processes involve the passing on of 'positional information' in the sense of Wolpert (1969). The well-known gradient in the concentration of the Bicoid protein in *Drosophila* eggs is one instance of this: here, the information provided is to do with position along the anterior–posterior axis.

Although the details of this and many other examples of positional information are now well documented, the question of 'what is the opposite of positional information?' is rarely posed. Indeed, it hardly seems to make sense, because the only answer that springs to mind is the trivial one, 'lack of positional information'. But here is a different possible answer: historical information. Suppose cells do not 'know' their position in space but do 'know' their position in time—in the sense that they know how many mitoses they have come through since some fixed reference point in the past, which could be the zygote or alternatively the last stem cell before one of their ancestral cells became differentiated.

It is clear that historical information exists: telomere reduction through ageing is an example (and an example that applies much more widely than in the mammalian context in which most research has been carried out: see Monaghan, 2010). Would it be possible for a developmental system to work on the basis of historical information alone? The answer to this question is almost certainly 'no'. Even in highly determinate developmental systems, such as that of the nematode model system *Caenorhabditis elegans*, cell–cell interactions that specify positional information seem to be involved—for example in the induction of the vulva (Sommer & Sternberg, 1994). However, the fact that we have been able to pose this question at all reveals a dichotomy—historical–positional—that resembles the dichotomy between blending and particulate inheritance. Perhaps we should say that the transfer of positional information is the main but not exclusive means by which development is driven, thus giving a theory of development akin to Darwin's theory of natural selection as the main driver of evolution? Recall, though, that 'positional information' is not really a mechanism; it is more of a phenomenon.

Other dichotomies relating to the general way in which development works can be envisaged. For example, the 'genetic imperialist' view that development is 'controlled' by genes can be contrasted with the view that developmental genes may be affected by environmental factors—a view that is encapsulated in the 'developmental reaction norm' approach (Schlichting & Pigliucci, 1998). But here we encounter two problems: first, the very existence of this dichotomy is rejected by 'developmental systems theory' (Oyama et al., 2001); and second, even if we accept the genetic–environmental dichotomy, there is much variation among different taxa in the extent to which the environment influences development. For example, temperature-dependent sex determination is common in turtles but rare in mammals. Thus a general theory saying that development either is or is not affected by the environment seems doomed to failure.

The above two dichotomies were not difficult to come up with. But are there others that might be found by looking harder? In particular, is there one that might cause a 'eureka moment' in which we suddenly recognize either that our whole current approach to development is correct and is a 'type X' theory as opposed to a 'type Y' one; or alternatively, that our current approach is wrong because it is of 'type X', whereas actually 'type Y' is where the truth lies?

If any of the contributors to this volume could envisage the kind of dichotomy I am alluding to here between alternative types of theory of development, it would be a major advance—either in our understanding of development (if there was a theory-shift) or at least in what might be called 'our understanding of our understanding' of development if there was simply a confirmation of the current type of theory. The fact that we cannot envisage such a dichotomy is thus hardly surprising. Few books can claim to have precipitated what Kuhn (1970) famously called a paradigm shift—though Darwin's (1859) *Origin of Species* is an exception. But if the current volume can stimulate a search that might lead to such a shift, that would be a major achievement.

Alternative theories of causal patterns?

It may be helpful to return to the abstract view of development, in which it is seen as a network of causal links. As noted earlier, the entire such network, in any multicellular organism, must be complex and irregular. It has no single pattern but rather a combination of patterns: hierarchical, combinatorial, modular, negative feedback, and others. However, even if we consider very simple systems of causality, the number of possible patterns is large: see Figure 9.3 for the possible causal networks that exist when there are only two developmental agents that can interact with each other. Depending on what we allow in terms of (i) only a single type of link, (ii) activating and repressing links, (iii) feedback links, and (d) quantification of the strength of the link, the number of overall possible causal networks ranges from three to infinity. Extending beyond two interacting entities, Figure 9.4 shows a possible causal series in which the generic symbol for a 'developmental agent' (O) has been replaced with G and P for genes and proteins. This brings the picture closer to reality in one respect but not in another: as we noted earlier, a linear series or chain of causal links is too simple.

There is no doubt that activating, repressing, and feedback links all exist and are common in real developmental systems. Equally, there is no doubt that such systems are composed of an enormous number of interacting agents, rather than a mere two. Since the number of causal links in the development of, say, a human, is vast, so—to a much greater degree—is the number of possible networks involving these links. And although the simplified systems shown in Figure 9.3 lack the opportunity for higher-level patterns such as hierarchy (see Figure 9.1) and modularity, real developmental systems incorporate these at many levels.

Does this approach lead to the identification of any meaningful 'alternative theories'? So far, it seems not. The idea that developmental systems might be either (i) exclusively hierarchical in their causality (divergent arrows through time) or

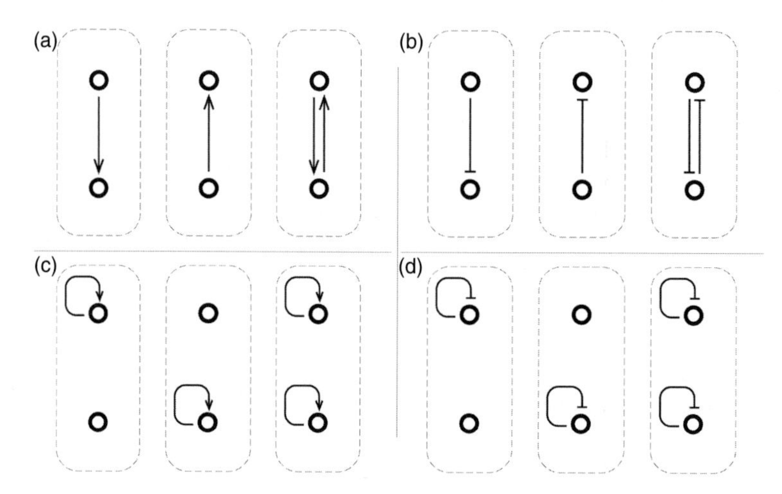

Figure 9.3 Possible causal patterns involving two developmental agents when causal links are: (a) activating; (b) repressing; (c) self-activating (positive feedback: 'runaway' result); and (d) self-repressing (negative feedback: auto-regulatory result). Each of the four cases gives three possibilities. Combining them gives a huge number of possibilities. If variable strength of causal link is included (e.g. by variable thickness of arrow) then the number of possibilities becomes effectively infinite.

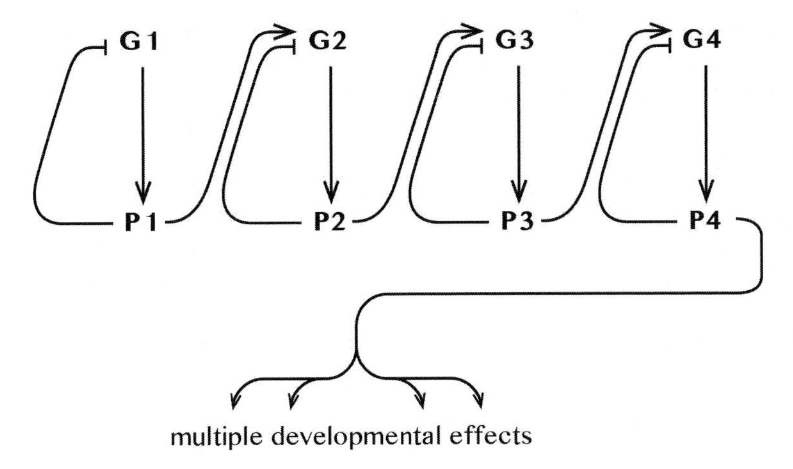

Figure 9.4 A simple, linear, causal series in which the generic symbol (O) has been replaced with G for gene and P for protein. Although only the final gene in the series is shown as having pleiotropic effects (all mediated through its single protein product P4), pleiotropy of developmental genes is probably the rule rather than the exception.

(ii) exclusively combinatorial (convergent arrows) is naïve. We now know enough examples of these patterns to be confident that real developmental systems always involve both. Thus if alternative theories are to be found in the area of patterns of causality, the way in which they are 'alternative' (or 'conflicting') must be more subtle. I believe that searching for such subtleties in the future may be fruitful, and that such a search, combined with a mathematical modelling approach in which causal interactions are quantified rather than being simply labelled as activating or repressing, might yet produce a theory of development. But if asked to guess at the outcome of such endeavours on the basis of work conducted to date, I would tend to support, albeit reluctantly, the view of Gilbert and Bard (this volume) that no single overarching theory of development will be found.

Conclusion

Science is driven by curiosity. In particular, it is driven by curiosity about how things work—'things' including development, evolution, and inheritance. Moreover, science is driven by the quest for simple, general explanations; not complicated, case-by-case ones. Simple, general explanations (alias theories) have a certain beauty or elegance—in this respect both the biological and the physical sciences share a common stance with mathematics. But if there is a conflict between beauty and truth, the former must give way to the latter: a general theory that is simple, beautiful, and wrong must be discarded. When this happens, we have what T.H. Huxley (1873) famously called 'the great tragedy of Science – the slaying of a beautiful hypothesis by an ugly fact'. In relation to a possible single general theory of development, this might be rephrased as follows: 'the preclusion of a beautiful theory by too many heterogeneous facts'. I hope this pessimistic view turns out to be wrong, but at present I see no evidence to suggest that it will.

Acknowledgments

I thank Scott Gilbert, Alessandro Minelli, and Thomas Pradeu for their helpful comments on the manuscript.

References

Arthur, W. (1984). *Mechanisms of Morphological Evolution: A Combined Genetic, Developmental and Ecological Approach*. Wiley, Chichester.

Arthur, W. (1997). *The Origin of Animal Body Plans: A Study in Evolutionary Developmental Biology*. Cambridge University Press, Cambridge.

Arthur, W. (2004). *Biased Embryos and Evolution*. Cambridge University Press, Cambridge.

Darwin, C. (1859). *On the Origin of Species by Means of Natural Selection, or the Preservation of Favoured Races in the Struggle for Life.* J. Murray, London.

Davidson, E.H. (1986). *Gene Activity in Early Development.* Academic Press, Orlando.

Davidson, E.H. (2010). Emerging properties of animal gene regulatory networks. *Nature*, **468**, 911–20.

Duboule, D. (1994). *Guidebook to the Homeobox Genes.* Oxford University Press, Oxford.

Fisher, R.A. (1930). *The Genetical Theory of Natural Selection.* Clarendon Press, Oxford.

Gilbert, S. (2010). *Developmental Biology*, 9th ed. Sinauer, Sunderland.

Gould, S.J. (1989). A developmental constraint in *Cerion*, with comments on the definition and interpretation of constraint in evolution. *Evolution*, **43**, 516–39.

Gould, S.J., and Lewontin, R.C. (1979). The spandrels of San Marco and the Panglossian paradigm: a critique of the adaptationist programme. *Proceedings of the Royal Society of London Series B*, **205**, 581–98.

Gurdon, J. B., Elsdale, T.R., and Fischberg, M. (1958). Sexually mature individuals of *Xenopus laevis* from the transplantation of single somatic nuclei. *Nature*, **182**, 64–5.

Huxley, T.H. (1873). *Biogenesis and Abiogenesis: Critiques and Addresses.* Macmillan, London.

Kangas, A.T., Evans, A.R., Thesleff, I., et al. (2004). Non-independence of mammalian dental characters. *Nature*, **432**, 211–14.

Kimura, M. (1968). Evolutionary rate at the molecular level. *Nature*, **217**, 624–6.

Kimura, M. (1983). *The Neutral Theory of Molecular Evolution.* Cambridge University Press, Cambridge.

Kuhn, T.S. (1970). *The Structure of Scientific Revolutions*, 2nd ed. University of Chicago Press, Chicago.

Lewontin, R.C. (1974). *The Genetic Basis of Evolutionary Change.* Columbia University Press, New York.

Meinhardt, H. (1982). *Models of Biological Pattern Formation.* Academic Press, London.

Mendel, G. (1866). Versuche über Pflanzenhybriden. *Verhandlungen des naturforschenden Vereines in Brünn*, **4**, 3–47.

Monaghan, P. (2010). Telomeres and life histories: the long and the short of it. *Annals of the New York Academy of Sciences*, **1206**, 130–42.

Oyama, S., Griffiths, P.E., and Gray, R.D., eds. (2001). *Cycles of Contingency: Developmental Systems and Evolution.* MIT Press, Cambridge.

Peter, I.S., and Davidson, E.H. (2009). Modularity and design principles in the sea urchin embryo regulatory network. *FEBS Letters*, **583**, 3948–58.

Peter, I.S. and Davidson, E.H. (2011). Evolution of gene regulatory networks controlling body plan development. *Cell*, **144**, 970–85.

Pinto-Correia, C. (1997). *The Ovary of Eve: Eggs and Sperm and Preformation.* University of Chicago Press, Chicago.

Raff, R.A. (1996). *The Shape of Life: Genes, Development and the Evolution of Animal Form.* Chicago University Press, Chicago.

Robert, J.S. (2004). *Embryology, Epigenesis and Evolution: Taking Development Seriously.* Cambridge University Press, Cambridge.

Salazar-Ciudad, I., and Jernvall, J. (2002). A gene network model accounting for development and evolution of mammalian teeth. *Proceedings of the National Academy of Sciences USA*, **99**, 8116–20.

Salazar-Ciudad, I., and Jernvall, J. (2010). A computational model of teeth and the developmental origins of morphological variation. *Nature*, **464**, 583–6.

Schlichting, C.D., and Pigliucci, M. (1998). *Phenotypic Plasticity: A Reaction Norm Perspective.* Sinauer, Sunderland.

Slack, J.M.W. (2008). *Essential Developmental Biology*, 2nd ed. Blackwell, Oxford.

Sommer, R.J., and Sternberg, P.W. (1994). Changes of induction and competence during the evolution of vulva development in nematodes. *Science*, **265**, 114–8.

Thompson, D.A.W. (1917). *On Growth and Form.* Cambridge University Press, Cambridge.

Vedel, V., Chipman, A.D., Akam, M., et al. (2008). Temperature-dependent plasticity of segment number in an arthropod species: the centipede *Strigamia maritima*. *Evolution & Development*, **10**, 487–92.

Vedel, V., Brena, C., and Arthur, W. (2009). Demonstration of a heritable component of the variation in segment number in the centipede *Strigamia maritima*. *Evolution & Development*, **11**, 434–40.

Vedel, V., Apostolou, Z., Arthur, W., et al. (2010). An early temperature-sensitive period for the plasticity of segment number in the centipede *Strigamia maritima*. *Evolution & Development*, **12**, 347–52.

Vervoort, M. (2011). Regeneration and development in animals. *Biological Theory*, **6**, 25–35.

Von Dassow, G., Meir, E., Munro, E.M., et al. (2000). The segment polarity network is a robust developmental module. *Nature*, **406**, 188–92.

Wilmut, I., Schnieke, A.E., McWhir, J., et al. (1997). Viable offspring derived from fetal and adult mammalian cells. *Nature*, **385**, 810–3.

Wolpert, L. (1969). Positional information and the spatial pattern of cellular differentiation. *Journal of Theoretical Biology*, **25**, 1–47.

CHAPTER 10

Cell differentiation is a stochastic process subjected to natural selection

Jean-Jacques Kupiec

Introduction

Although, during antiquity, a few materialists postulated the existence of atoms moving randomly, scientific theory was generally determinist until the nineteenth century. It was only from the 1850s that probabilistic theories applied to natural phenomena began to be widely accepted: first, with Darwin's theory of natural selection postulating 'indefinite and fluctuating variations' of organisms, soon recognized as random mutations by geneticists, and second, with the rise of statistical physics that followed Boltzman's work.

In biology, two exceptions to this determinist hegemony can be mentioned concerning ontogenesis: in antiquity, Empedocles put forth a sort of primitive Darwinian mechanism resorting to a mixture of chance and selection to explain the generation of animals and, in the late 19th century, Wilhelm Roux published *Der Kampf der Theile im Organismus* [*The struggle of the Parts in the Organism*] (1881) in which he suggested a phenomenon of Darwinian competition between the components of organisms to explain functional adaptation. The ideas of these two authors, however, neither became predominant nor exerted a significant influence. After abandoning his theory, Roux became a founder of experimental embryology and the proponent of another theory of development based on preformationism and embryo mosaïcism. According to this determinist theory, embryonic cells were supposed to carry determinants of adult body parts split between daughter cells after each cell division (Roux, 1895). In the same determinist vein, August Weismann postulated the existence of a highly organized microscopic material structure in the germinal cells that he called 'germinative plasma'. This structure, which foreshadowed DNA, was thought to control development in a very precise way, with each of its parts determining a part of the adult organism (Weismann, 1891).

In the 20th century, under the auspices of the Modern Synthesis, biology became dual. On the one hand, as regards evolution, the role of random mutations was recognized; but on the other hand, as regards the internal functioning of organisms, determinism was reinforced by molecular biology. In his influential book *What is Life*, Erwin Schrödinger (1944) analysed the role of randomness and drew a sharp demarcation between physics and biology. According to him, physics is ruled by an order from disorder principle. By this he meant that the regularities observed in physical systems at the macroscopic level rely on the probabilistic behaviour of atoms and molecules at the microscopic level caused by thermal agitation. This generation of order from disorder occurs because of the law of large numbers. In any physical system, since the number of particles involved is immense, variation is negligible. Schrödinger argues that in contrast to physics, the law of large numbers does not apply in biology, which is not ruled by an order from disorder principle but by an order from order principle. Biological molecules are not subjected to randomness but guided by this order principle to behave precisely at the microscopic level, in order to ensure regulation and reproducible ontogenesis. This

Towards a Theory of Development. Edited by Alessandro Minelli and Thomas Pradeu

order principle corresponds to what has later been called genetic information, after the DNA structure was elucidated. Thus, molecular biology was determinist from its foundation, and this determinism was subsequently extended to other levels of biological systems to explain the processing of genetic information. First, the so-called 'central dogma of molecular biology' stated as an inviolable law that information can only flow from DNA to proteins and phenotypes (Crick, 1958, 1970). Second, it was thought that genetic information was transferred into precise three-dimensional protein structure, making protein–protein and protein–DNA interactions specific. The concept of specific interaction excluded randomness, one molecule having only one or a very limited number of partner molecules, thus avoiding random combinatorial possibilities. Specific interactions, in turn, were supposed to cause stereospecific self-assembly processes. This concept was applied to explain the generation of viral and cellular structures (Caspar & Klug, 1962) as well as gene regulation and signal transduction (Jacob & Monod, 1961). Indeed, stereospecific interaction between a repressor protein and its target sequence in DNA lies at the root of the Jacob–Monod lactose operon regulation model. In this model, the lactose genes are either active ('on') or inactive ('off') according to whether they interact or not with the repressor protein through their operator sequence. This model was extended to explain gene regulation at the genome level. In this conception, the activity of a given gene depends on the activity of other genes by means of on/off switches, creating networks of gene regulation corresponding to the so-called genetic program; and the same mode of thinking was applied to explain signal transduction as cascades of specific interactions between biological molecules. This conception gave rise to different variations (see Britten, 1998), but the rationale on which it relied remained constant. Cell differentiation was understood as the expression at the cellular level of gene networks, and this view continued to prevail until the beginning of the 21st century (for example see Istrail et al., 2007). Randomness was taken into account by some authors but as 'developmental noise', corresponding to accidental variations in development explaining phenotypic plasticity (for example see Lewontin, 2000). In fact, this idea had been present from the beginning of genetics, since the concept of a phenotype was precisely elaborated to encompass phenotypic plasticity among individuals sharing identical genotypes. It is noteworthy here that the concept of gene networks has even been integrated into self-organization theories, although these theories were presented by their proponents as alternatives to genetic determinism (Atlan, 1979; Kaufmann, 1993). These authors also conceived randomness as noise playing a role in the expression gene networks.

Now, it has to be said that an accumulation of data obtained over the last 30 years has demonstrated that gene expression is a stochastic process. This is a major discovery, the consequences of which must be thoroughly evaluated, because it raises at least two related questions that are crucial for a theory of development. Firstly, what is the role of stochastic gene expression (SGE) during development? Secondly, is SGE simply a noise perturbing the functioning of gene networks, or is it a more profound phenomenon of cell physiology? In the former case, SGE would only be an additional parameter in the functioning of the genetic program, and the genetic programming theory of development would remain essentially valid; while in the latter, development, breaking with the determinist tradition that has prevailed until now, would have to be considered as an intrinsically stochastic phenomenon, and a new theory would be needed to explain it.

These questions are addressed in this chapter. I begin by reviewing the data as well as the mechanisms causing SGE, showing that it plays a positive role in development by creating the diversity of gene expression patterns needed for cell differentiation. I then argue that taking SGE into account leads to a new concept of development, and I put forth a model of cell differentiation based on SGE. This model underlies a new theory, named ontophylogenesis, in which ontogenesis results from an extension of natural selection working inside embryonic cell populations. I also discuss the nature of the probabilities involved in SGE, how a stochastic process can lead to reproducible development, and how ontophylogenesis fits into broader theories such as holism, reductionism, and Lamarckian and genetic determinisms.

Stochastic gene expression

Demonstration of stochastic gene expression

Discovering SGE was made possible when cell-imaging techniques using reporter genes and fluorescent proteins allowed monitoring of gene expression in single cells. Gene expression variability between isogenic cells having the same phenotype and placed in a constant environment was then observed in mammalian cells. This was first observed for glucocorticoid-inducible genes (Ko et al., 1990) and for genes induced by the transcription factor NF-AT (Fiering et al., 1990). Subsequently, a similar observation was made for the human immunodeficiency virus, the human cytomegalovirus, and the prolactin gene promoters (Ross et al., 1994; Takasuka et al., 1998; White et al., 1995). In the case of glucocorticoid-inducible genes, gene expression was shown to fit a stochastic model based on random association and dissociation between transcription factors and gene promoters (Ko, 1991). In syncytial muscle cells carrying several nuclei sharing the same cytoplasm, the expression variability of actin and troponin genes was shown to occur at the level of transcription by in situ hybridization (Newlands et al., 1998), and in rat neuroblastoma cells, heterogeneous expression of the insulin receptor gene was detected by reverse-transcription PCR in single cells (Heams & Kupiec, 2003). Heterogeneity of gene expression was also observed between the two chromosomes of a pair in diploid organisms. In situ analyses with probes specific for the different alleles of multiallelic loci revealed that it is not always the same allele that is expressed on the two chromosomes at a given time. This was shown for olfactive receptor genes (Chess et al., 1994), globin genes (Wijgerde et al., 1995), T-cell receptor and natural killer cell receptor genes (Held et al., 1999), cytokine genes (Holländer, 1999; Rivière et al., 1998), and the insulin-like growth factor 2 and the uncharacterized H19 genes (Jouvenot et al., 1999). In these studies, since the different alleles are placed in the same nuclear environment sharing the same transcriptional factors, their heterogeneous expression is a direct demonstration of SGE (Chess et al., 1994; Holländer, 1999; Wijgerde et al., 1995). SGE was proved definitively by experiments in which, in *Escherichia coli* and *Saccharomyces cerevisae*, the expression of two copies of the same gene cloned in strictly identical genetic contexts underwent major stochastic fluctuations (Elowitz et al., 2002; Raser & O'Shea, 2004) and in mammalian cells, the number of mRNA molecules produced by a gene showed stochastic variations from one cell to the other (Raj et al., 2006). Taken together, these data led to a probabilistic view of gene expression, and SGE is nowadays considered as an undeniable fact. In this conception, which departs from the classical 'on/off' view of gene regulation (see 'Introduction'), a gene in a cell has a probability of being activated at any time—a probability that varies between 0 and 1 (Balázsi et al., 2011; Fiering et al., 2000; Golubev, 1996; Hume, 2000; Kaern et al., 2005; Kupiec, 1983; 1996; McAdams & Arkin, 1997; Paldi, 2003; Spudish & Koshland, 1976). SGE is now the subject of a large body of work. Here, I shall focus mainly on aspects related to development in multicellular organisms.

The causes of stochastic gene expression

SGE is an established fact but its causes are not fully understood yet. Two main explanations have been put forward. According to one, SGE results from random fluctuations in the functioning of gene networks, commonly called 'noise', whereas according to another explanation, it is caused by random molecular events affecting the dynamic of chromatin organization.

Because of thermal agitation, there are always stochastic variations between cells in the concentration of molecules in different cellular compartments and in the speeds of biochemical reactions. These fluctuations can produce transcriptional noise when transcription factors interacting with gene promoters are present in low concentrations. They randomly fluctuate below and above their activity thresholds, leading to heterogeneity of gene expression in a cell population. This phenomenon may even be amplified if regulatory loops are present in gene networks affected by transcriptional noise (Becksei et al., 2001; Blake et al., 2003; McAdams & Arkin, 1997; Spudish & Koshland, 1976).

Thermal agitation also affects the organization of chromatin, which is known to be both highly structured and extremely dynamic. According to the second explanation, SGE is caused by stochastic molecular events occurring in chromatin. Chromosomes are organized in specific territories with genes co-expressed in one cell having a statistical tendency to be co-localized in the same place in the three-dimensional nuclear space in 'transcription factories' (Fraser & Bickmore, 2007; Misteli, 2007). However, chromatin is also extremely labile. While maintaining its overall architecture, it is continuously assembled and disassembled by a flow of molecules which associate with and dissociate from each other (Misteli, 2001). After dissociation, the diffusion of proteins and their action on another DNA site allows more or less important chromatin reorganization to occur and can lead to the expression of different genes. Owing to the stochastic nature of these biophysical processes, the modifications of gene expression they induce are themselves stochastic. An example can be given by the variegated expression of genes localized at the limit between euchromatin and heterochromatin; this expression depends on competition between proteins promoting heterochromatin and transcription factors promoting euchromatin. In each cell, the result of this competition is random, because it depends on the Brownian behaviour of molecules (Dillon & Festenstein, 2002). This phenomenon is different from transcriptional noise caused by low numbers of transcription factors. Modelling has shown that SGE can be produced by competition of transcription factors for interaction with gene promoters even when the transcription factors are present at high nuclear concentrations (Coulon et al., 2010). Stochastic gene expression can also be caused by random long-distance interactions between different chromatin regions (Dernburg et al., 1996). It was shown recently that more than 1000 long-distance interactions occur in the chromatin of different human cell lines (Sanyal et al., 2012). SGE based on Brownian behaviour of chromatin was also shown for globin and olfactory receptor genes. In these cases, it was due to local stochastic interactions between genes and a LCR (locus control region) DNA regulatory element (Chess et al., 1994; Hanscombe et al., 1991; Wijgerde et al., 1995). Moreover, it was demonstrated by experimental manipulations that SGE is affected by the chromosomal position of genes (Becksei et al., 2005). In a recent study, the extent of SGE of a transgene inserted in different locations in the genome of chicken stem cells was shown to vary according to the transgene insertion location. Using chromatin modifier agents, it was further shown that chromatin dynamics is a major player in determining the transgene transcription probability (Vinuelas et al., 2013). Taken all together, these data support a model based on two properties of chromatin molecules (DNA, histones, and all types of non-histone proteins) making chromatin dynamics stochastic. First, chromatin proteins are subjected to thermal agitation and move by Brownian diffusion until they find their DNA binding sequences (Berg & von Hippel, 1985; Halford & Marko, 2004). Second, the interactions of these molecules are not specific in the sense that a particular protein can interact with several other partner proteins or DNA binding sites. This point remained a subject of discussion for a long time, but it is now unequivocally established (see Kupiec, 2009). There are two main principles in the functioning of the model:

(i) Because interactions between chromatin molecules are not specific, there are many possible distributions of proteins (including nucleosomes) on DNA. Each of these distributions leads to a different chromatin organization and to the activation of different sets of genes.
(ii) Owing to random dissociation and reassociation of molecules, stochastic transitions between different chromatin organization states occur, causing stochastic activation of different sets of genes. The linear localization of genes along DNA and in the three-dimensional nuclear space is an important parameter determining the transition probabilities between chromatin organizations (see Figure 10.1 and Kupiec, 1983, 1997, 2009 for a more detailed presentation of the model).

A new approach to cell differentiation

Stochastic gene expression and cell differentiation

SGE raises crucial questions. Is it a source of disorganization for cells and thus a drawback that must be tightly controlled, or is it rather an integral part of their normal functioning that ensures life processes?

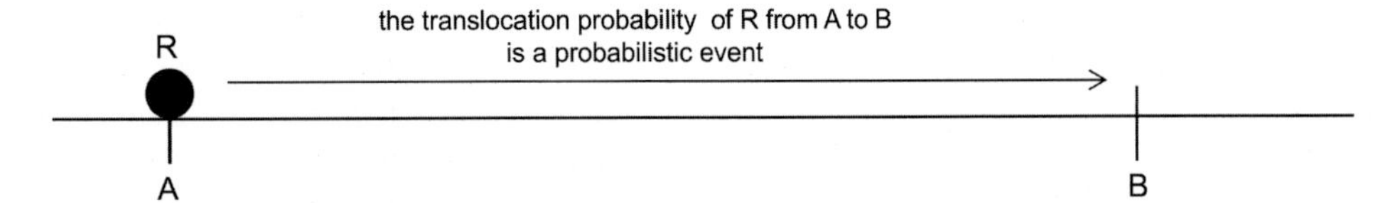

Figure 10.1 A model of stochastic gene expression based on chromatin dynamics. The most simple case of the model is shown in this diagram. One regulatory protein R can activate two genes, A and B. When R is in A, after dissociation, owing to random diffusion, it can either bind A again or move away and bind B. Its probability of translocation from A to B depends on its dissociation constant from A, its diffusion coefficient, and the distance separating B from A, either linearly along DNA or in the three-dimsensional nuclear space. In a real cell, many chromatin proteins interact with many DNA binding sites; but the rationale of the model remains the same. The transition from one distribution of chromatin proteins over DNA to another distribution is a stochastic event determined by the same parameters, allowing genes to be stochastically expressed.

An important SGE effect could be to produce variable gene expression patterns that are necessary during cell differentiation. It may even be a very simple and economical way to achieve it, since it occurs spontaneously. In fact, data showing that cell differentiation is a stochastic process existed before SGE became a demonstrated fact, when analyses of differentiation in many cell lines revealed a stochastic variability which was not compatible with a deterministic model. This variability has been observed in various experimental fashions, but with a similar rationale: if cell differentiation is a deterministic mechanism, the behaviour of all cells belonging to the same population should be homogeneous, and variability between the kinetics of differentiation of single cells is not predicted, apart from minor fluctuations. On the contrary, if cell differentiation is a stochastic phenomenon, variability is expected to occur. In practice, depending on the experimental system, in some cases each cell of a single population was shown to differentiate with a unique chronology, while in other cases the descendants of individual cells varied as regards their differentiated cell content. Such variability, which could only be modelled using stochastic models, has been observed in a variety of organisms, ex vivo and in vivo, with numerous experimental techniques and for many cell lines, normal or cancerous.

When haematopoietic stem cells divide, they either renew themselves or differentiate into blood cells that lose their capacity to multiply. The proportion of these two kinds of cells among the descendants of stem cells is highly variable from one single cloned cell to another, and this variability fits a stochastic model in which cells are assigned a probability to differentiate at each cell division (Till et al., 1964). When cultured melanoma skin cancer cells differentiate, they start producing melanin, which gives them a black pigmentation used as a differentiation marker. Dorothy Bennett filmed these differentiating cells and analysed their differentiation kinetics one by one. She found that single melanoma cells differentiate with highly variable kinetics and that this variability also fits a probabilistic model (Bennett, 1983). Cephalic neural crest embryonic cells differentiate into five different cell types (neurons, adrenergic cells, Schwann cells, melanocytes, or chondrocytes). The descendants from single clonal neural crest cells form colonies in cultures. Such colonies have been analysed using specific markers and, as is the case for haematopoietic cells, the proportion of the different cell types is highly variable from one colony to another. Statistical analysis supports a stochastic model of cell differentiation (Baroffio & Blot, 1992). Similar results have been obtained for immune system, liver, bone, and intestinal cells as well as *C. elegans* gonadal cells and mammalian blastula cells (Böhme et al., 1995; Davis et al., 1993; Godsave & Sack, 1991; Greenwald & Rubin, 1992; Lin et al., 1994; Paulus et al., 1993; van Roon et al., 1989). Finally, in addition to haematopoietic cells, stem cell differentiation has been recently shown, using cell-lineage analysis, to be stochastic in another series of stem cell lines including spermatogonial cells, intestinal epithelium cells, and epidermal cells (Simons & Clevers, 2011).

These data provide evidence for stochastic cell differentiation, but they do not give insight into its molecular mechanism. Evidence has been obtained recently that directly links SGE and cell differentiation in several cell lines. Cell-type determination of photoreceptor cells in the retina is stochastic in *Drosophila* (Bell et al., 2007) and in mammals (Gomes et al., 2011). In *Drosophila* it correlates with the stochastic expression of the transcription factor Spineless (Wernet et al., 2006). In the case of haematopoietic stem cells, Sui Huang and his team have demonstrated that SGE drives cell differentiation at the transcriptome-wide level (Chang et al., 2008). They showed that the expression level of the stem cell marker Sca-1 is highly heterogeneous in clonal populations of haematopoietic progenitor cells. Owing to SGE, cells isolated from these populations expressing either low or high levels of Sca-1 spontaneously reconstitute the parental heterogeneity of expression when left to proliferate. However, these cells express markedly different transcriptomes and show a greatly different proclivity for differentiating into either erythroid or myeloid cells, depending on their level of Sca-1 expression. SGE is also involved in the reverse differentiation process that occurs in cellular reprogramming, during which differentiated cells are reprogrammed into pluripotent stem cells. Single-cell gene expression analysis of 48 genes, including genes involved in chromatin remodelling, signal transduction, cell-cycle regulation, and a variety of pluripotency marker genes, has revealed that SGE occurs during the early phase of cellular reprogramming but diminishes in later phases (Buganim et al., 2012). Finally, and of particular interest for a theory of development as it shows that SGE is a general phenomenon occurring throughout development, it has been demonstrated that SGE also occurs during blastocyst morphogenesis in early mouse embryo. Single-cell studies show that the expression of the lineage marker proteins Cdx2 and Nanog is stochastic from the eight-cell stage to before the differentiation of the inner cell mass and trophoectoderm (Dietrich & Hiiragi, 2007). Similarly, stochastic expression of the genes encoding Nanog, Gata6, and Pdgfra has been shown to precede the segregation of the epiblast and the primitive endoderm (Chazaud et al., 2006; Plusa et al., 2008).

The selection stabilization model of cell differentiation

From induction to selection. Until now, our understanding of cell differentiation has been heavily influenced by genetic determinism and early experimental embryology work, and therefore, is predominantly deterministic. The discovery that both gene expression and cell differentiation involve stochastic mechanisms thus raises a crucial question for a theory of development: are these stochastic mechanisms compatible with the classical deterministic conception of development? The core concept of the deterministic approach is induction, as put forth by Hans Spemann (1938). He performed grafting experiments on embryos and demonstrated how cells influence each other during development: the fate of a cell in an embryo depends on the influences it receives from the other cells. According to this induction mechanism, embryonic cells are supposed to produce induction molecules which act on their neighbours to determine their fate. It should be noted, however, that Spemann's experiments did not in themselves indicate the nature of the induction mechanism (Saha, 1991). Later, the deterministic view of induction integrated knowledge from molecular biology and gave rise to the instructive model. In this model, cells differentiate because they receive 'instructions' corresponding to signals (or information) carried by proteins. These signals trigger signal transduction inside the cells, causing specific genes to be activated and thus producing cell differentiation. In the theoretical example in Figure 10.2, cell B differentiates into cell D because it receives a protein signal *d* synthesized by cell A. In the same way, cell A differentiates into cell C because it receives signal *c* synthesized by B. In this deterministic context, the differentiation of tissues during the embryogenesis of an organism is seen as a series of such elementary stages; and since each stage involves the expression of genes encoding for instructive proteins (*d* and *c* in Figure 10.2), the entire process expresses a so-called genetic program made of gene networks.

There are several versions of this inductive instructive model (IIM). Segments of differentiation pathways may be autonomous: a signal triggers a 'master gene' in a cell, and this gene in turn activates a specific program corresponding to a series of gene activations, leading a cell from one state to another without additional external signals (for example from A to C or B to D in Figure 10.2, corresponding here to terminal differentiation). If the signals are carried by membrane molecules, there

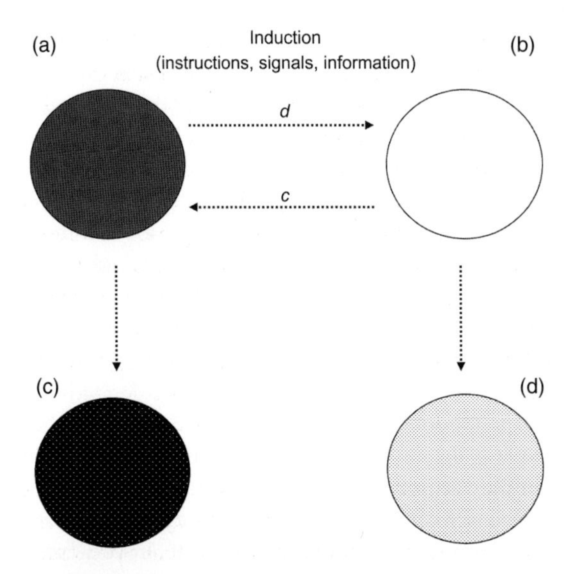

Figure 10.2 The IIM of cell differentiation. Cells are induced to differentiate according to the signals they receive. This model presumes an asymmetry between cells emitting and receiving different signals.

needs to be direct contact between the cells. If the signals are carried by diffusible molecules, they can act at a distance. In certain cases, instructive molecules may form concentration gradients and exert their specific effect only at a specific concentration. The theoretical example in Figure 10.2 involves two differentiating cells; however, other situations may exist in which a cell receives signals from several different cells or differentiates into alternative types. The signal emanating from one cell may equally correspond to a combination of several molecules; but, whatever the version of the model, its principle remains the same: cells are induced to differentiate in a way determined by the signals they receive.

This model is in agreement with all the data showing the roles of cell interactions and differential gene expression in cell differentiation. Because of its deterministic nature, it seems also relevant to explain a reproducible phenomenon such as cell differentiation. It presents, however, some weaknesses which must lead us to challenge it.

The first weakness is precisely linked to its deterministic nature. Because of this, the IIM does not account for all the data, showing the importance of randomness in gene expression and cell differentiation. In fact, the IIM is refuted by these data. The second weakness is theoretical: the IIM does not explain how different cell lines are produced from a single cell. The production of cell diversity during cell differentiation has to be explained by a theory of development. The IIM fails to do so, because it posits an initial asymmetry at the beginning of cell differentiation. In Figure 10. 2, the two cells A and B are already different from the start, since they synthesize different signals. The model therefore presumes a diversity of cells, the appearance of which it is supposed to explain. The usual practice for resolving this contradiction is to evoke the effect of morphogenetic gradients pre-existing in the egg. Owing to their heterogeneous distribution, there is said to be unequal distribution in each daughter cell of the molecules present in the egg after each of its cleavage steps. This mechanism is supposed to create initial differentiation of the cells, which would set differentiation in motion according to the IIM. Although morphogenetic gradients are an indisputable reality and play a role in embryogenesis, this explanation does not really resolve the problem. It bases all embryogenesis on the egg's initial gradients, whereas classical experiments and data demonstrate that regulatory mechanisms must exist to create heterogeneity of cell types in the course of embryogenesis or later in the adult during stem cell differentiation.

The problems of the IIM can be eliminated by integrating SGE and stochastic cell behaviour into a new approach to cell differentiation. As previously mentioned, SGE could be a simple way to produce the variable gene expression patterns underlying cell differentiation. This is the starting point of the selection stabilization model (SSM) in which interactions between cells control this process of SGE-driven cell differentiation (Kupiec, 1983). The word 'model' is used here to denote schematically the way cell differentiation usually occurs. The simplest theoretical case is depicted in Figure 10.3. The two cells A and A′ are identical. One of the two genes, *b* or *c*, may be randomly activated in a cell, and depending on which of these events occurs, the cell differentiates into type B or C. Interactions between cells do play an important role in this model, but their function is not to induce differentiation, as is the case in the IIM. They select or stabilize the cells, which differentiate randomly. In Figure 10.3, the interaction

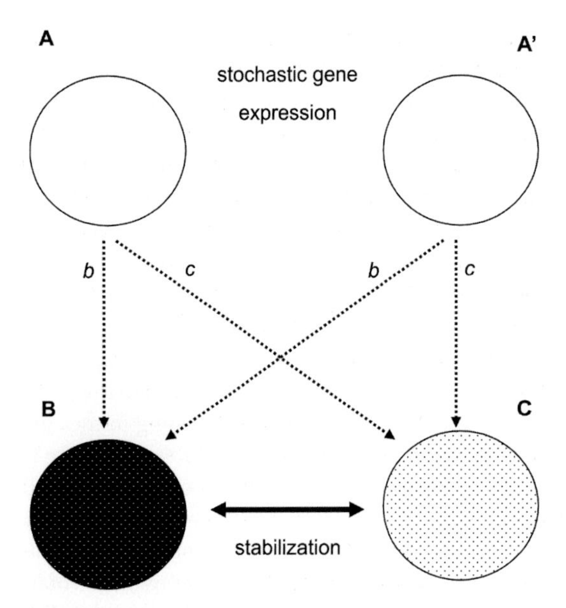

Figure 10.3 The SSM of cell differentiation. A and A′ cells are identical. In each of them either gene *b* or gene *c* can be randomly activated, leading to differentiation into a B or C cell phenotype. Only the combination of B and C cells leads to the stabilization of cell phenotypes. Otherwise gene expression continues to be variable and cell phenotypes remain unstable. If, by chance, B and C are not produced together, unstabilized SGE can eventually lead to cell death. (Adapted from Kupiec, 1983, 1997.)

between cells B and C leads to selection or stabilization of B by C and of C by B, and therefore to their coordinated differentiation. In this model, only the combination of B and C cell phenotypes leads to stabilization of SGE and thus of cell phenotypes. The other combinations (B/B or C/C) do not lead to stabilization. If, by chance, the 'right' combination of phenotypes, that which stabilizes SGE, is not produced, cells may eventually die. Unlike the IIM, in this model the precision of embryonic development is not based on molecular deterministic events controlling gene expression. Two factors play complementary roles, each of them being more or less important according to the various situations occurring in organismal development. First, there is a population effect. While gene expression and cell differentiation are random, they are nevertheless reproduced statistically in a cell population. This occurs with mean frequencies, subject to variations depending on the size of the population. In accordance with the law of large numbers, the larger the size of the cell population, the less variability there will be. Second, cell interactions also impose coordinated differentiation between cell lines. A cell cannot differentiate independently from the other cells. In the example given in Figure 10.3, cells B and C will necessarily be produced together because this is the only combination of cell phenotypes leading to SGE stabilization. In the SSM, the environment of a cell inside an embryo, which consists predominantly of other cells, is a constraint guiding its gradual change during development by stabilization. In the IIM, the cell's environment guides cell differentiation as well, but by induction.

The model depicted in Figure 10.3 is intended to explain the rationale of cell differentiation with a simple theoretical case involving a bifurcation between two cell lines only. At some stages of development, more cell lines can differentiate. In vertebrate development, much of the substantive differentiation takes place very early on, and once the core tissues are in place—the germ layers and their immediate derivatives—cells are most often engaged in binary choices. However, this is not a general rule. For example, somites have three choices, and cephalic neural crest cells, six. At later stages, most differentiation is specialization, with again more than two cell lines at some steps. It should be stressed that the general rationale depicted in Figure 10.3 can apply to all these cases involving multiple differentiating cell lines with minor modifications: multiple gene expression patterns can be produced by SGE, and each committed cell can be stabilized by its microenvironment made of signals emanating from other differentiating cell lines, whatever the number (see also 'Natural selection inside the organism' in 'Discussion and conclusion').

The SSM actually eliminates the weaknesses of the IIM. It fully integrates the data demonstrating the existence of SGE and randomness in cell differentiation, and it does not need to presume cell diversity. Cells A and A′ of Figure 10.3 may be identical, but owing to SGE, they still have different potential fates. However, the SSM in turn raises a question, because it inverts the problem of differentiation as it is usually posed. It implies that, due to inherent randomness, cells can change state without induction. This prediction agrees with data which has shown that cells are spontaneously transformed when they are cultured outside the internal environment of the organism (Rubin, 1990). Because of this well-known phenomenon, in vitro cultivated cells must be cloned regularly so that they keep their original characteristics; otherwise, they transform and change phenotype beyond control. It is usually thought that this phenomenon is of no significance, that it is an artifact connected with cultivating cells, and does not contribute any relevant information to differentiation produced in vivo in the organism. In contrast, for the SSM it assumes essential significance. It reveals the inherent probabilistic and spontaneous nature of cell differentiation. Having escaped from the constraint that the internal environment exerts in vivo, cells transform spontaneously because of SGE which is no longer controlled. In this context, the problem is not so much to explain why two cells are different from each other, but rather to understand how, despite this inherent tendency to differentiate, homogeneous tissues of identical cells can form in organisms. To resolve this problem, the cell stabilization process must be more precisely explained.

Making sense of epigenetic chromatin modifications. Cellular interactions trigger signal transduction, which leads to epigenetic modifications of DNA and chromatin proteins and so affects gene expression (Delcuve et al., 2009; Suganuma & Workman, 2011). It remains to be explained, however, how this process can direct the expression of specific genes in differentiating cells in the context of the IIM, since the enzymes responsible for these epigenetic modifications have a very wide range of potential substrates in chromatin. In the context of the SSM, in contrast, a simple mechanism can explain the effect of epigenetic modifications. It has been shown that phosphorylation or dephosphorylation of transcription factors drastically modifies their dissociation constants with DNA (for example: Bourbon et al., 1995; Li et al., 1994; Takenaka et al., 1995; Xu et al., 1994; and references therein). Due to this effect on DNA binding, it will also significantly modify their probabilities of displacement in chromatin (see Figure 10.1), and thus the expression probabilities of the genes they regulate. If the dissociation constant of a transcription factor is low, its probability of moving to another gene will be low, and the expression of genes it activates will be stabilized. Through this mechanism, cellular interactions that trigger protein kinases or phosphatases could stabilize or destabilize gene expression (Kupiec, 1997), and the same effect could be achieved with all chromatin epigenetic modifications (Paldi, 2003). In the context of this model, the role of epigenetic modifications is to stabilize chromatin states that are randomly produced. In support of this model, it was recently demonstrated using chromatin-modifying agents that the level of variability in gene expression depends on histone acetylation and DNA methylation (Viñuelas et al., 2012).

Evidence for the selection stabilization model. Many data support the SSM in its essential aspects.

The SSM predicts that differentiation is associated with chromatin and the stabilization of gene expression. This has been shown to be the case during stem cell differentiation and cellular reprogramming. A wide and variable range of genes is expressed in embryonic stem cells (Efroni et al., 2008), notably, the expression of stem cell markers such as Nanog, REX1, PECAM1, SSEA1, or Stella varies from one single cell to another (Koh et al., 2010). This unstable expression pattern correlates with a widely open chromatin structure (Mattout & Meshorer, 2010). Chromatin is highly dynamic in these embryonic stem cells with unstable protein–protein and protein–DNA interactions but is stabilized when they differentiate (Meshorer et al., 2006). While the characteristic open structure of their chromatin becomes 'closed' during differentiation, there appears to be a generalized repression of gene expression, with the expression level of most genes being lowered (Efroni et al., 2009). Similarly, gene expression is stochastic during the early phase of cellular reprograming of differentiated cells into pluripotent cells but becomes stabilized in the later phase (Buganim et al., 2012).

The SSM also predicts that perturbing cell interactions during cell differentiation should affect the level of variation in gene expression. If cell interactions are prevented, gene expression should be destabilized, whereas if they are stimulated, gene expression stabilization should be facilitated. These predictions have been experimentally confirmed. Prolactin expression is homogeneous in pituitary tissue but variable when pituitary cells are dissociated by enzymatic means thus preventing cell interactions (Harper et al., 2010). CDX2 protein expression is random in early mouse cloned embryos. These embryos are deficient in cell interactions because they have fewer cells than normal fertilized embryos. If a normal number of cells is restored by aggregating several embryos, thus restoring a normal level of cell interactions, CDX2 expression is stabilized as predicted by the SSM (Balbach et al., 2010).

By using a combination of experimental and modelling methods, the phenotypic diversification of clonal mammalian cell populations in vitro has been shown to conform to the SSM. It relies on both cell intrinsic SGE and influences from the local cell microenvironment built up by cells themselves (Neildez-Nguyen et al., 2008; Stockholm et al., 2007, 2010).

In the *Drosophila* embryo, maintenance of the expression of the gene *engrailed* depends on the protein Wingless (the equivalent of Wnt in this organism). In the absence of Wingless, *engrailed* expression can be initiated, but it is soon interrupted instead of being maintained (Martinez-Arias & Hayward, 2006). In line with the SSM, this shows that Wingless does not play the role of inducer but of SGE stabilizer. In the vertebrate hindbrain, the interaction between the Eph receptor and its ephrin ligand plays a crucial role in sorting cell populations at the boundaries of rhombomeres (hindbrain segments). Also in line with the SSM, mosaic activation of Eph receptors and ephrins in distinct rhombomeres ends in cell population homogenization (Xu et al., 1999).

Another prediction of the SSM is that cell death occurs during development if the 'right' combination of cell phenotypes is not produced (cells not adjusted to their local internal environment die). In agreement with this prediction, cell death is a widespread phenomenon during development occurring in the genesis of numerous tissues (Michaelson, 1993; Penaloza et al., 2006), and it has been shown to happen if a competent tissue fails to receive the signal it needs to stimulate development (Koseki, 1993). Cell death caused by cell competition has been also demonstrated in *Drosophila* wing development (Diaz & Moreno, 2005) and in the early mammalian embryo (Claveria et al., 2013).

Finally, evidence for the SSM has been provided by numerical simulations. Cellular automata functioning according to the SSM possess the main property expected for cell differentiation: they reproducibly give rise to differentiated cell types forming organized tissue patterns. Thus, randomness is not an obstacle to cell differentiation and tissue organization. Various tissue patterns could be produced in these numerical simulations by quantitatively and qualitatively varying the parameters of the SSM, showing its potential for a theory of development. Another important finding concerned the role of morphogen gradients. It was found that molecules involved in stabilization do form gradients in the simulated growing cell populations and that cells are stabilized in a particular phenotype according to their position within these gradients. Thus, there is no contradiction between the existence of morphogen gradients in embryos and the SSM (Laforge et al., 2005).

Discussion and conclusion

What kind of probability is involved in development?

The use of probability in a theory of development may be a source of misunderstanding. It should be clear that probabilistic does not mean absolute absence of determination (Zernicka-Goertz & Huang, 2010). In a determinist process, the probability of an event is either 0 or 1, whereas in a probabilistic process, the probability of an event is between 0 and 1; however, both determinist and probabilistic processes are determined by the material conditions in which they occur. When we play dice or flip a coin, the frequencies obtained (1/6 and 1/2, respectively) are determined by the structure of dice and coins, and this implies statistical reproducibility described by mean and variance. Similarly, SGE is determined by a series of parameters, for example, the diffusion coefficients of transcriptional regulators or the dissociation constants between these regulators and target sequences in DNA. Thus, the expression probability of a given gene is defined by the biophysical properties of proteins and DNA involved in its expression; and this probability in turn determines statistically the copy number of the protein coded by this gene. The difference between a determinist and a probabilistic gene expression process concerns the variations of the protein copy number. In a determinist process, the protein copy number will be constant, whereas it will be subject to random variations in a probabilistic one (it will vary in different cells and according to time). Another question raised by a probabilistic theory of development concerns the nature of the probability involved in gene expression. In physics the status of probability in statistical physics differs greatly to that in quantum physics. In the former, atoms and molecules are subjected to Newton's deterministic laws, but because of the complexity caused by the immense number of these particles, their movements can only be analysed using a probabilistic formalism; whereas in the latter, there is inherent indetermination in the behaviour of particles. Thus, probability appears to be subjective or epistemic in statistical physics, whereas it is objective or ontological in quantum physics. Although quantum physics might also have some effects in biological systems, the probability involved in SGE is caused by the complexity of random diffusion processes depending on thermal agitation and is thus kindred to statistical physics probability. In this sense, SGE probability can be said to be subjective or epistemic, similar to statistical physics probability.

There are, however, particularities in biology that must be introduced in the analysis of SGE. As exposed in 'The causes of stochastic gene expression', two explanations have been given to SGE, both linked to thermal agitation. According to the first one, SGE is noise that perturbs the functioning of gene networks, which still govern development. Thus, in this frame the genetic programming theory remains essentially valid. According to the second explanation, SGE is a consequence of probabilistic chromatin dynamics caused by competitive and combinatorial interactions between chromatin molecules. In this frame development is not driven by gene networks but by this intrinsic probabilistic process. Thus, the genetic programming theory of development is no longer valid.

Order from disorder?

Although evidence for SGE and stochastic cell differentiation is now accepted, it has not yet been integrated into a theory that accepts development as an intrinsically probabilistic process. I have exposed such a theory in this chapter. It comprises a model explaining SGE and another model explaining cell differentiation. Although it is based on data, it raises a question that needs to be discussed: how can an ordered and reproducible phenomenon such as development rely on an intrinsically probabilistic mechanism? In fact, a probabilistic process is not per se irreproducible. It depends on the conditions in which it occurs. Two conditions can make it reproducible: the law of large numbers and the presence of constraints.

Does the law of large numbers apply to biological systems? Schrödinger (1944) described many physical processes that are stochastic at the molecular level and determinist at the macroscopic one. This generation of order from disorder is a consequence of the law of large numbers (see 'Introduction'). This law states that when the number of probabilistic events of a process tends towards the infinite, the variability of the process as a whole tends towards 0. In practice, when the number of events is very high, the variability is negligible. A paradigmatic example is diffusion. Although atoms and molecules move by random walk, diffusion can be described by Fick's determinist laws at the macroscopic level. Could the same phenomenon occur during the generation of order in biological systems? Could the law of large numbers also apply there? Recent data concerning chromatin allow us to revisit this question. Using fluorescence imaging techniques, it is now possible to determine the parameters of protein interaction and diffusion in living cells. When these techniques were applied to chromatin, it was found that protein–protein and protein–DNA interactions are very brief, just a few seconds in the large majority of cases (Misteli, 2001). This result implies that chromatin is a continuously assembling and disassembling structure. It is not a static entity but an ensemble of interacting proteins and DNA sequences. This apparent lability, stochastic in essence, could seem paradoxical because chromatin stability is expected in order to maintain gene expression and phenotype stability. In fact, it is the contrary. This result shows that there is a high number of interactions in chromatin and thus that the law of large numbers could apply to some extent. Because of the high number of stochastic molecular interactions, chromatin is both dynamical and relatively stable.

It is in an intermediate state between total disorder and absolute stability. It is sufficiently stable to ensure the stability of cell phenotypes but also subject to stochastic variations allowing cell differentiation.

Development as a stochastic phenomenon under constraints. In addition to the law of large numbers, the presence of constraints can render a stochastic process reproducible. To understand this point, a thought experiment might be useful. Let us consider the trajectory of a small ball moving randomly on a surface. When it moves freely, the ball's trajectory will be totally unpredictable and irreproducible (Figure 10.4a). If the ball's movement is constrained, for example by the presence of two walls, the ball's trajectory becomes predictable and reproducible to a certain extent: the ball will always move between the two walls (Figure 10.4b). Now, the extent of reproducibility depends on the strength of the constraints. If the distance between the walls is just above the ball diameter, the ball will move according to a one-dimensional random walk between the walls (Figure 10.4c). In this case the ball's trajectory is predictable and reproducible. The ball will always move in a straight line, although it will go forwards and backwards randomly. If the shape of the constraints changes, for example if the two walls form concentric circles separated by the ball diameter (not shown), the ball's trajectory will be a circle; and if the experiment is repeated several times, according to the type of constraints applied on the ball, the same form will be reproduced. This example shows that a constrained stochastic process can lead to form reproducibility. Could a similar process be at work in biological systems? Cells are highly compartmentalized structures, and for this reason, proteins do not diffuse by a three-dimensional random walk as in a glass of water. Instead, cell structure acts as a strong constraint on diffusion processes and optimizes the probabilities of stochastic cellular processes. Diffusion in chromatin is an example of constrained diffusion, with important consequences for gene expression. Chromatin proteins find their DNA target sequence much faster than if they were moving in a three-dimensional random walk. This occurs because of the constraints exerted by DNA molecules and nuclear spatial organization, which make the proteins slide along the DNA in a one-dimensional random walk (Berg & von Hippel, 1985; Halford & Marko, 2004). Thanks to these constraints, the efficiency of protein target-finding is much improved and, although still being a stochastic process, it is reproducible, just as the ball's trajectory in the previous thought experiment. Development, although more complex, is such a constrained stochastic process. Indeed, during cell differentiation, in addition to cell structure constraints, cell nuclei receive signals coming from neighbouring cells, leading to epigenetic chromatin modifications which, in turn, modify the biophysical properties of proteins and DNA. Thus, these signals also act as constraints on chromatin stability controlling SGE. According to the SSM, both cellular organization and multicellular interactions optimize the probability of development to the point that it is a reproducible process. This raises the question of the inner nature of these constraints.

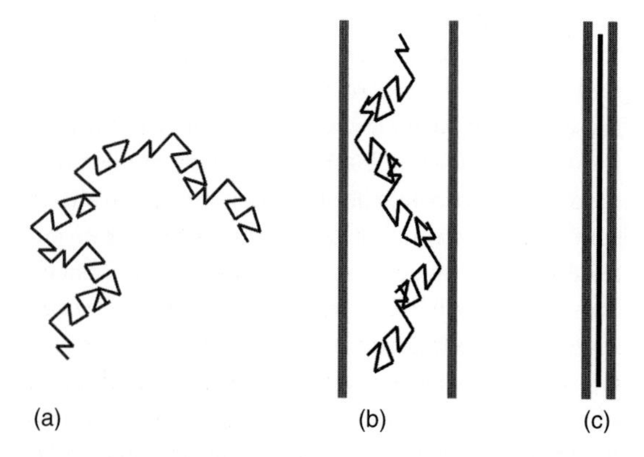

Figure 10.4 The effect of constraints on a stochastic process. Black lines represent the ball's trajectory, walls are in grey. The ball's movement, although stochastic, is constrained by the walls. The reproducibility of the ball's trajectory depends on the strength of this constraint.

Natural selection inside the organism

The SSM is based on a combination of molecular randomness (SGE) and selection or stabilization at the cellular level. Natural selection is also based on a combination of molecular randomness (DNA mutations) and selection of organisms at the ecosystem level. In this sense, the SSM is analogous to natural selection occurring among embryonic cells and similar to other already known Darwinian mechanisms operating at the cellular level, such as the clonal selection of lymphocytes for antibody production (Jerne, 1955) and the selective stabilization of synapses during neurogenesis (Changeux et al., 1973). In the 19th century Roux proposed a theory of functional adaptation based on internal selection (see 'Introduction'), and more recently Leo Buss put forth a theory based on competition between cell lines to explain the evolution of multicellularity (Buss, 1987). In this final discussion section, I want to suggest that development is a real extension of natural selection inside cell populations of living organisms.

In the context of molecular biology, ontogenesis is conceived as a unidirectional bottom–top process stemming from DNA genetic information, itself propagated to macromolecular structures to build organisms (see 'Introduction'). For various reasons, it is nowadays rejected by the biologists who tend to adopt a more holistic view of life. (On this subject, the reader can refer, for example, to Gilbert & Sarkar, 2000; Noble, 2006; Sarkar, 1998; Soto & Sonnenschein, 2006). I shall only recall two well-known factual arguments concerning two major steps of ontogenesis. They demonstrate that it is a bidirectional bottom–top and top–bottom process, in which gene action is selectively constrained by cell and multicellular structures (see also Noble, 2008 for a discussion of the importance of top–bottom processes).

It is now known that the genome is not self-sufficient in determining its own functioning. Gene expression is controlled by chromatin epigenetic modifications (Delcuve et al., 2009; Suganuma & Workman, 2011). As explained above, the chromatin is a cell structure whose state depends on the signals it receives from within the cell and from the cell's environment through cell interactions. Genome functioning is therefore controlled by effects caused at the cellular and multicellular levels. Genetic information was previously thought to be transferred to a protein's three-dimensional structure, to make proteins interact stereospecifically with each other or with DNA sequences, in a one-to-one fashion, to build either gene and protein networks or cell structures (see 'Introduction'). Although in this conception, combinatorial possibilities in protein interactions that could be a source of randomness were excluded, it has now been demonstrated that the connectivity of global protein networks is very high, with typically more than 10% proteins interacting with more than 100 other proteins. As a consequence, there are huge combinatorial possibilities, and all signalling pathways or gene networks are interconnected, with numerous contact points between them (Albert, 2005; Barabási & Oltvai, 2004; Bork, 2004). A series of mechanisms have been documented to explain how only specific subparts of global networks are active when cells need to respond specifically to a signal or to ensure differentiation of cell lines (Dumont et al., 2002; Komarova et al., 2005; Schwartz & Madhani, 2004). The most common mechanism is spatial compartmentalization, whereby proteins are distributed specifically in different cell compartments, with unwanted interactions being thus avoided. This mechanism of protein-binding restriction implies bidirectional ontogenesis, because it presupposes the existence and the agency of an organized cell to explain the specificity of molecular interactions. On the one hand, DNA codes for protein synthesis; and on the other hand, the global cell structure acts as a selective constraint, sorting protein interactions. I cannot review here the other mechanisms suggested for sorting protein interactions; but, as with cell compartmentalization, they all presuppose the existence of an organized cell, and they all imply bidirectional ontogenesis (see Kupiec, 2009).

The SSM fits in with a bidirectional scheme. According to it, ontogenesis is a dual mechanism with, on the one hand, probabilistic effects caused at the molecular level, gene expression probabilities being determined by the genome structure and random combinatorial molecular interactions in chromatin; and on the other hand, effects caused at the cell and multicellular levels, chromatin epigenetic states and gene expression being constrained by either the egg-cell's initial structure (by its chromatin state), which allows pluripotency gene expression, or by multicellular interactions in embryos, which selectively stabilize the expression of genes involved in differentiation. But, more importantly, bidirectional ontogenesis also implies a causal interconnection between ontogenesis and phylogenesis. If cell and multicellular structures constrain SGE, being themselves sorted by natural selection, there is a chain of causality starting from the organism's environment and

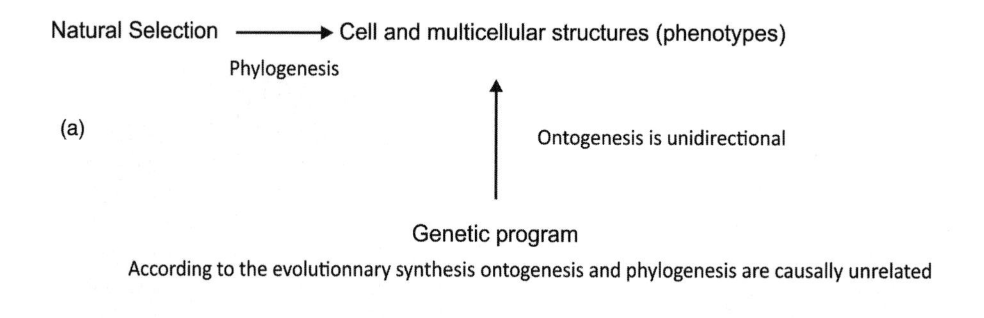

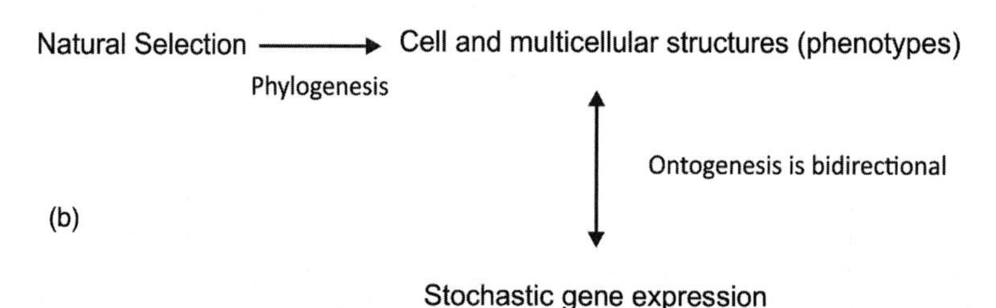

Figure 10.5 The extension of natural selection inside the organism. (a) Within the classical evolutionary synthesis, cells and multicellular organisms are produced by genetic programs and sorted by natural selection. These two processes do not interfere with each other. Ontogenesis and phylogenesis are causally unrelated. Their only relation is through mutations, which are selected along with phenotypes; but otherwise, natural selection does not intervene in the course of ontogenesis. (b) In the context of the SSM, SGE is selectively stabilized by cells or multicellular structures (by combinations of interacting cells) giving rise to adult phenotypes. Since cells and multicellular structures are selected by natural selection, the two processes of ontogenesis and phylogenesis are causally related, with natural selection being the ultimate cause of the internal selective process.

entering into the organism, with natural selection being the ultimate cause determining which genes are expressed during ontogenesis (see Figure 10.5). The SSM is therefore a real extension of natural selection inside organisms. To better understand this point, one may consider a cell population growing either on a solid medium or in a liquid medium. Some cells at the interface with the external environment have direct access to substrates and signals from the outside, whereas internal cells have only access to substrates and signals coming from their internal environment, which is made primarily of other cells. As the cell population continues to grow, the internal environment inside the cell population becomes more differentiated, and each cell is placed in a different internal environment (Figure 10.6). According to the SSM, cell differentiation is the adaptation of cells to their differentiated internal environment. The key point here is that the way the internal environment differentiates is in part dependent on the external environment. For example, the way substrates and signals are distributed inside the cell population depends on their spatial and quantitative distribution in the external environment, which, in turn, constrains the way the cell population differentiates as a whole. Of course, the external environment is only one of several factors influencing development. The structures of the initial egg and genome play a primordial role, notably in determining the gene expression probabilities that actually drive development.

This conception of the relation between the internal and external environments is both similar to and different from Claude Bernard's internal environment concept and from Paul Weiss's morphogenetic field theory (Weiss, 1973). For Claude Bernard, the internal environment was not merely a way to

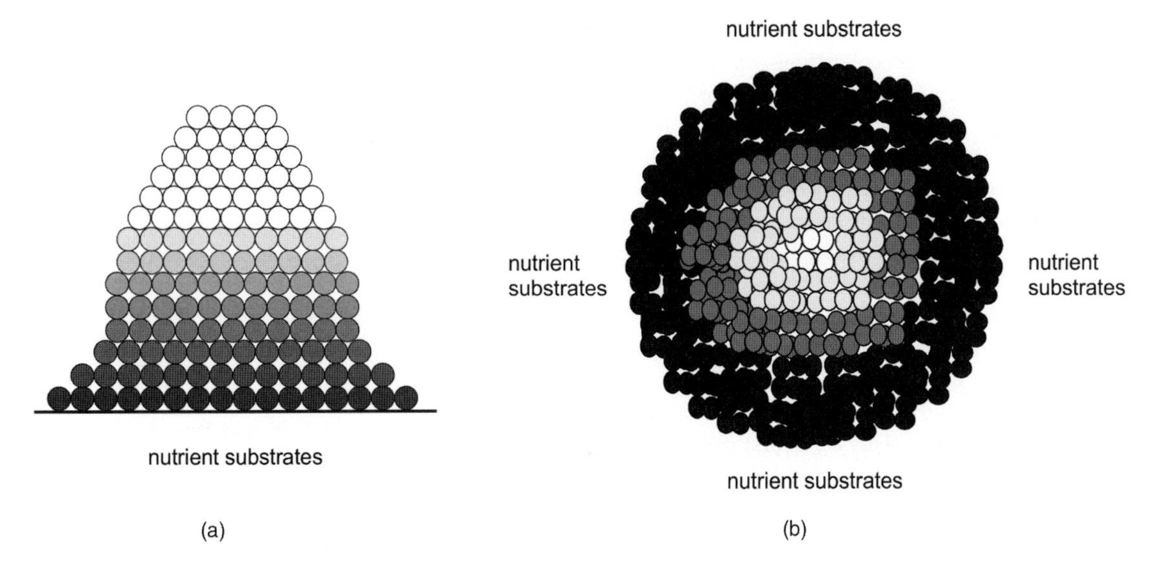

Figure 10.6 The external environment constrains the internal environment. The varying shades of cells in the diagrams indicate gradients of nutrients, signals, and metabolites, either coming from the external environment or produced by the activity of cells. The distribution of these substances inside growing cell populations initially depends predominantly on their distribution in the external environment and then on the activity of cells, which further differentiates the internal environment. As growth proceeds, the embryo becomes more autonomous in relation to the external environment. Two examples are given showing how the external environment can shape the internal environment differentially: (a) the cell population growing on a solid substrate; and (b) the cell population growing in a liquid medium.

maintain homeostasis, as it is often believed; it was primarily a way to mediate the influences exerted by the external environment on the parts of organisms (Bernard, 1878). Paul Weiss also thought that morphogenetic fields in embryos are dependent on the relation of embryos with the environment. However, both of them were determinist. The probabilistic and selective aspects of the SSM mark a crucial difference with previous theories. This opens a new way to the understanding of the role of environment in ontogenesis and sheds new light on the nature vs nurture debate.

Indeed, like Lamarck, the proponents of a major role of the environment in phenotype building usually think of it as determinist and instructive. This generates a problem: in the current debate, it is difficult to understand how environmental and genetic determinisms can be integrated together to account for the functioning of organisms. Usually, it is assumed that there is a superposition of the two, as the term epigenetic precisely denotes, but it remains unclear how they articulate with each other. This difficulty does not occur with the theory presented here, as it departs from both environmental and genetic determinisms. In it, the role of the environment is selective rather than instructive. Thus, it does not compete with gene action, and development is still directed by the genome, although in a probabilistic fashion (gene expression probabilities are optimized by natural selection, thus ensuring proper development).

The final point I would like to make concerns the traditional separation between ontogenesis and phylogenesis. I point out that they are not separate phenomena, but on the contrary form a single phenomenon, *ontophylogenesis* (Kupiec, 1986, 1997, 2009) The relativity of the individual organism concept has already been stressed by several authors (Ludwig & Pradeu, 2008; Wilson, 1999). Alessandro Minelli has given many examples showing that the origin and the end of ontogenesis are arbitrary. For this reason, he has defined development as an open-ended segment of life (Minelli, 2011). In the continuous process of life, consisting in multicellular organisms of an alternation of unicellular and multicellular phases, we are the ones who isolate segments going from the unicellular to the multicellular phases, and *we* call it development. Darwin also underlined that species are arbitrary entities, which he defined as collections of individuals sharing the same ancestor (Darwin, 1859). In this perspective, phylogenesis is a collection of longer segments corresponding to genealogical lines, which *we* also isolate in the process of life. Thus, ontogenesis and phylogenesis are subjective concepts. In both, the continuous alternation of unicellular and multicellular phases that make up both the generation of organisms and genealogical lines is the sole real underlying phenomenon, on which *we* isolate different portions. I have named it ontophylogenesis to highlight the unity between ontogenesis and phylogenesis. This is not just playing with words. If there is only one ontophylogenetic phenomenon, then there is also only one mechanism to produce it, and, as shown in this chapter, a single natural selection theory is sufficient to explain it.

Acknowledgments

I thank Jonathan Bard, Alessandro Minelli, and Thomas Pradeu for very helpful comments after reading the first draft of this chapter.

References

Albert, R. (2005). Scale-free networks in cell biology. *Journal of Cell Science*, **118**, 4947–57.

Atlan, H. (1979). *Entre le cristal et la fumée*. Seuil, Paris.

Balázsi, G., van Oudenaarden, A., and Collins, J.J. (2011). Cellular decision making and biological noise: from microbes to mammals. *Cell*, **144**, 910–25.

Balbach, S.T., Esteves, T.C., Brink, T., et al. (2010). Governing cell lineage formation in cloned mouse embryos. *Developmental Biology*, **343**, 71–83.

Barabási, A.L., and Oltvai, Z.N. (2004). Network biology: understanding the cell's functional organization. *Nature Reviews Genetics*, **5**, 101–13.

Baroffio, A., and Blot, M. (1992). Statistical evidence for a random commitment of pluripotent cephalic neural crest cells. *Journal of Cell Science*, **103**, 581–7.

Becskei, A., Kaufmann, B.B., and van Oudenaarden, A. (2005). Contributions of low molecule number and chromosomal positioning to stochastic gene expression. *Nature Genetics*, **37**, 937–44.

Becskei, A., Séraphin, B., and Serrano, L. (2001). Positive feedback in eukaryotic gene networks: cell differentiation by graded to binary response conversion. *EMBO Journal*, **20**, 2528–35.

Bell, M.L., Earl, J.B., and Britt, S.G. (2007). Two types of *Drosophila* R7 photoreceptor cells are arranged randomly: a model for stochastic cell-fate determination. *Journal of Comparative Neurology*, **502**, 75–85.

Bennett, D.C. (1983). Differentiation in mouse melanoma cells: initial reversibility and an on-off stochastic model. *Cell*, **34**, 445–53.

Berg, O.G., and von Hippel, P.H. (1985). Diffusion-controlled macromolecular interactions. *Annual Review of Biophysics and Biophysical Biochemistry*, **4**, 131–60.

Bernard, C. (1878). *Leçons sur les phénomènes de la vie*. Baillière, Paris.

Blake, W.J., Kaern, M., Cantor, C.R., et al. (2003). Noise in eukaryotic gene expression. *Nature*, **422**, 633–7.

Böhme, K., Winterhalter, K.H., and Bruckner, P. (1995). Terminal differentiation of chondrocytes in culture is a spontaneous process and is arrested by transforming growth factor-β 2 and basic fibroblast growth factor in synergy. *Experimental Cell Research*, **216**, 191–8.

Bork, P., Jensen, L.J., von Mering, C., et al. (2004). Protein interaction networks from yeast to human. *Current Opinion in Structural Biology*, **14**, 292–9.

Bourbon, H.M., Martin-Blanco, E., Rosen, D., et al. (1995). Phosphorylation of the *Drosophila* engrailed protein at a site outside its homeodomain enhances DNA binding. *Journal of Biological Chemistry*, **270**, 11130–9.

Britten, R.J. (1998). Underlying assumptions of developmental models. *Proceedings of the National Academy of Sciences USA*, **95**, 9372–7.

Buganim, Y., Faddah, D.A., Cheng, A.W., et al. (2012). Single-cell expression analyses during cellular reprogramming reveal an early stochastic and a late hierarchic phase. *Cell*, **150**, 1209–22.

Buss, L.W. (1987), *The Evolution of Individuality*. Princeton University Press, Princeton.

Caspar, D.L.D., and Klug, A. (1962). Physical principles in the construction of regular viruses. *Cold Spring Harbor Symposia on Quantitative Biology*, **27**, 1–24.

Chang, H.H., Hemberg, M., Barahona, M., et al. (2008). Transcriptome-wide noise controls lineage choice in mammalian progenitor cells. *Nature*, **453**, 544–7.

Changeux, J.P., Courrège, P., and Danchin, A. (1973). A theory of the epigenesis of neuronal networks by selective stabilization of synapses. *Proceedings of the National Academy of Sciences USA*, **70**, 2974–8.

Chazaud, C., Yamanaka, Y., Pawson, T., et al. (2006). Early lineage segregation between epiblast and primitive endoderm in mouse blastocysts through the Grb2-MAPK pathway. *Developmental Cell*, **10**, 615–24.

Chess, A., Simon, I., Cedar, H., et al. (1994). Allelic inactivation regulates olfactory receptor gene expression. *Cell*, **78**, 823–34.

Claveria, C., Giovinazzo, G., Sierra, R., et al. (2013). Myc-driven endogenous cell competition in the early mammalian embryo. *Nature*, **500**, 39–44.

Coulon, A., Gandrillon, O., and Beslon, G. (2010). On the spontaneous stochastic dynamics of a single gene: complexity of the molecular interplay at the promoter. *BMC Systems Biology*, **4**, 2.

Crick, F. (1958). On protein synthesis. *Symposia of the Society for Experimental Biology*, **12**, 139–63.

Crick, F. (1970). Central dogma of molecular biology. *Nature*, **227**, 561–3.

Darwin, C. (1859). *On the Origin of Species*. Murray, London.

Davis, C.B., Killeen, N., Crooks, M.E., et al. (1993). Evidence for a stochastic mechanism in the differentiation of mature subsets of T lymphocytes. *Cell*, **73**, 237–47.

Delcuve, G.P., Rastegar, M., and Davie, J.R. (2009). Epigenetic control. *Journal of Cell Physiology*, **219**, 243–50.

Dernburg, A.F., Broman, K.W., Fung, J.C., et al. (1996). Perturbation of nuclear architecture by long-distance chromosome interactions. *Cell*, **85**, 745–59.

Díaz, B., and Moreno, E. (2005). The competitive nature of cells. *Experimental Cell Research*, **306**, 317–22.

Dietrich, J.E., and Hiiragi, T. (2007). Stochastic patterning in the mouse pre-implantation embryo. *Development*, **134**, 4219–31.

Dillon, N., and Festenstein, R. (2002). Unravelling heterochromatin: competition between positive and negative factors regulates accessibility. *Trends in Genetics*, **18**, 252–8.

Dumont, J.E., Dremier, S., Pirson, I., et al. (2002). Cross signaling, cell specificity, and physiology. *American Journal of Physiology (Cell Physiology)*, **283**, C2–28.

Efroni, S., Duttagupta, R., Cheng, J., et al. (2008). Global transcription in pluripotent embryonic stem cells. *Cell Stem Cell*, **2**, 437–47.

Efroni, S., Melcer, S., Nissim-Rafinia, M., et al. (2009). Stem cells do play with dice: a statistical physics view of transcription. *Cell Cycle*, **8**, 43–8.

Elowitz, M.B., Levine, A.J., Siggia, E.D., et al. (2002). Stochastic gene expression in a single cell. *Science*, **297**, 1183–6.

Fiering, S., Northrop, J.P., Nolan, G.P., et al. (1990). Single cell assay of a transcription factor reveals a threshold in transcription activated by signals emanating from the T-cell antigen receptor. *Genes and Development*, **4**, 1823–34.

Fiering, S., Whitelaw, E., and Martin, D.I. (2000). To be or not to be active: the stochastic nature of enhancer action. *BioEssays*, **22**, 381–7.

Fraser, P., and Bickmore, W. (2007). Nuclear organization of the genome and the potential for gene regulation. *Nature*, **447**, 413–7.

Gilbert, S.F., and Sarkar, S. (2000). Embracing complexity: organicism for the 21st century. *Developmental Dynamics*, **219**, 1–9.

Golubev, A.G. (1996). Random necessity, transcription initiation, induction of differentiation and need for randomness. *Biokhimiia*, **61**, 1303–19.

Godsave, S.F., and Slack, J.M. (1991). Single cell analysis of mesoderm formation in the *Xenopus* embryo. *Development*, **111**, 523–30.

Gomes, F.L., Zhang, G., Carbonell, F., et al. (2011). Reconstruction of rat retinal progenitor cell lineages in vitro reveals a surprising degree of stochasticity in cell fate decisions. *Development*, **138**, 227–35.

Greenwald, I., and Rubin, G.M. (1992). Making a difference: the role of cell-cell interactions in establishing separate identities for equivalent cells. *Cell*, **68**, 271–81.

Halford, S.E., and Marko, J.F. (2004). How do site-specific DNA-binding proteins find their targets? *Nucleic Acids Reearch*, **32**, 3040–52.

Hanscombe, O., Whyatt, D., Fraser, P., et al. (1991). Importance of globin gene order for correct developmental expression. *Genes and Development*, **5**, 1387–94.

Harper, C.V., Featherstone, K., Semprini, S., et al. (2010). Dynamic organization of prolactin gene expression in living pituitary tissue. *Journal of Cell Science*, **123**, 424–30.

Heams, T., and Kupiec J.J. (2003). Modified 3′-end amplification PCR for gene expression analysis in single cells. *Biotechniques*, **34**, 712–6.

Held, W., Kunz, B., Ioannidis, V., et al. (1999). Mono-allelic Ly49 NK cell receptor expression. *Seminars in Immunology*, **11**, 349–55.

Holländer, G.A. (1999). On the stochastic regulation of interleukin-2 transcription. *Seminars in Immunology*, **11**, 357–67.

Hume, D.A. (2000). Probability in transcriptional regulation and its implication for leukocyte differentiation and inducible gene expression. *Blood*, **96**, 2323–8.

Istrail, S., De-Leon, S.B., and Davidson, E.H. (2007). The regulatory genome and the computer. *Developmental Biology*, **310**, 187–95.

Jacob, F., and Monod, J. (1961). Genetic regulatory mechanisms in the synthesis of proteins. *Journal of Molecular Biology*, **3**, 318–56.

Jerne, N.K. (1955). The natural-selection theory of antibody formation. *Proceedings of the National Academy of Sciences USA*, **41**, 849–57.

Jouvenot, Y., Poirier, F., Jami, J., et al. (1999). Biallelic transcription of Igf2 and H19 in individual cells suggests a post-transcriptional contribution to genomic imprinting. *Current Biology*, **9**, 1199–1202.

Kaern, M., Elston, T.C., Blake, W.J., et al. (2005). Stochasticity in gene expression: from theories to phenotypes. *Nature Reviews Genetics*, **6**, 451–64.

Kauffman, S.A. (1993). *The Origins of Order: Self-organization and Selection in Evolution*. Oxford University Press, Oxford.

Ko, M.S. (1991). A stochastic model for gene induction. *Journal of Theoretical Biology*, **153**, 181–94.

Ko, M.S., Nakauchi, H., and Takahashi, N. (1990). The dose dependence of glucocorticoid-inducible gene expression results from changes in the number of transcriptionally active templates. *EMBO Journal*, **9**, 2835–42.

Koh, F.M., Sachs, M., Guzman-Ayala, M., et al. (2010). Parallel gateways to pluripotency: open chromatin in stem cells and development. *Current Opinion in Genetics and Development*, **20**, 492–9.

Komarova, N.L., Zou, X., Nie, Q., et al. (2005). A theoretical framework for specificity in cell signaling. *Molecular Systems Biology*, **1**, 2005.0023.

Koseki, C. (1993). Cell death programmed in uninduced metanephric mesenchymal cells. *Pediatric Nephrology*, **7**, 609–11.

Kupiec, J.J. (1983). A probabilist theory for cell differentiation, embryonic mortality and DNA C-value paradox. *Speculation in Science and Technology*, **6**, 471–8.

Kupiec, J.J. (1986). A probabilist theory for cell differentiation: the extension of Darwinian principles to embryogenesis. *Speculation in Science and Technology*, **9**, 19–22.

Kupiec, J.J. (1996). A chance-selection model for cell differentiation. *Cell Death and Differentiation*, **3**, 385–90.

Kupiec, J.J. (1997). A Darwinian theory for the origin of cellular differentiation. *Molecular and General Genetics*, **255**, 201–8.

Kupiec, J.J. (2009). *The Origin of Individuals*. World Scientific, Singapore.

Laforge, B., Guez, D., Martinez, M., et al. (2005). Modeling embryogenesis and cancer: an approach based on an equilibrium between the autostabilization of stochastic gene expression and the interdependence of cells for proliferation. *Progress in Biophysics and Molecular Biology*, **89**, 93–120.

Lewontin, R. (2000). *The Triple Helix*. Harvard University Press.

Li, C.C., Dai, R.M., Chen, E., et al. (1994). Phosphorylation of NF-KB1-p50 is involved in NF-κB activation and stable DNA binding. *Journal of Biological Chemistry*, **269**, 30089–92.

Lin, Z., Lu, M.H., Schultheiss, T., et al. (1994). Sequential appearance of muscle-specific proteins in myoblasts as a function of time after cell division: evidence for a conserved myoblast differentiation program in skeletal muscle. *Cell Motility and the Cytoskeleton*, **29**, 1–19.

Ludwig, P., and Pradeu, T. (2008). *L'individu. Perspectives contemporaines*. Vrin, Paris.

Martinez Arias, A.M., and Hayward, P. (2006). Filtering transcriptional noise during development: concepts and mechanisms. *Nature Reviews Genetics*, **7**, 34–44.

Mattout, A., and Meshorer, E. (2010). Chromatin plasticity and genome organization in pluripotent embryonic stem cells. *Current Opinion in Cell Biology*, **22**, 334–41.

McAdams, H.H., and Arkin, A. (1997). Stochastic mechanisms in gene expression. *Proceedings of the National Academy of Sciences USA*, **94**, 814–9.

Meshorer, E., Yellajoshula, D., George, E., et al. (2006). Hyperdynamic plasticity of chromatin proteins in pluripotent embryonic stem cells. *Developmental Cell*, **10**, 105–16.

Michaelson, J. (1993). Cellular selection in the genesis of multicellular organization. *Laboratory Investigations*, **69**, 136–51.

Minelli, A. (2011). Animal development, an open-ended segment of life. *Biological Theory*, **6**, 4–15.

Misteli, T. (2001). Protein dynamics: implications for nuclear architecture and gene expression. *Science*, **291**, 843–7.

Misteli, T. (2007). Beyond the sequence: cellular organization of genome function. *Cell*, **128**, 787–800.

Neildez-Nguyen, T.M., Parisot, A., Vignal, C., et al. (2008). Epigenetic gene expression noise and phenotypic diversification of clonal cell populations. *Differentiation*, **76**, 33–40.

Newlands, S., Levitt, L.K., Robinson, C.S., et al. (1998). Transcription occurs in pulses in muscle fibers. *Genes and Development*, **12**, 2748–58.

Noble, D. (2006). *The Music of Life*. Oxford University Press, Oxford.

Noble, D. (2008). Claude Bernard, the first systems biologist, and the future of physiology. *Experimental Physiology*, **93**, 16–26.

Paldi, A. (2003). Stochastic gene expression during cell differentiation: order from disorder? *Cellular and Molecular Life Sciences*, **60**, 1775–8.

Paulus, U., Loeffler, M., Zeidler, J., et al. (1993). The differentiation and lineage development of goblet cells in the murine small intestinal crypt: experimental and modelling studies. *Journal of Cell Science*, **106**, 473–83.

Penaloza, C., Lin, L., Lockshin, R.A., et al. (2006). Cell death in development: shaping the embryo. *Histochemistry and Cell Biology*, **126**, 149–58.

Plusa, B., Piliszek, A., Frankenberg, S., et al. (2008). Distinct sequential cell behaviours direct primitive endoderm formation in the mouse blastocyst. *Development*, **135**, 3081–91.

Raj, A., Peskin, C.S., Tranchina, D., et al. (2006). Stochastic mRNA synthesis in mammalian cells. *PLoS Biology*, **4**: e309.

Raser, J.M., and O'Shea, E.K. (2004). Control of stochasticity in eukaryotic gene expression. *Science*, **30**, 1811–14.

Rivière, I., Sunshine, M.J., and Littman, D.R. (1998). Regulation of IL-4 expression by activation of individual alleles. *Immunity*, **9**, 217–28.

Ross, I.L., Browne, C.M., and Hume, D.A. (1994). Transcription of individual genes in eukaryotic cells occurs randomly and infrequently. *Immunology and Cellular Biology*, **72**, 177–85.

Roux, W. (1881). *Der Kampf der Teile im Organismus*. W. Engelmann, Leipzig; repr., Esther von Krosigk, ed. (2007). VDM, Saarbrücken.

Roux, W. (1895). *Gesammelte Abhandlungen über Entwickelungsmechanik der Organismen*. Engelmann, Leipzig.

Rubin, H. (1990). On the nature of enduring modifications induced in cells and organisms. *American Journal of Physiology*, **258**, L19–L24.

Saha, M. (1991). Spemann seen through a lens. In S.F. Gilbert, ed., *A Conceptual History of Modern Embryology*. Plenum Press, New York, pp. 91–108.

Sanyal, A., Lajoie, B.R., Jain, G., et al. (2012). The long-range interaction landscape of gene promoters. *Nature*, **489**, 109–13.

Sarkar, S. (1998). *Genetics and Reductionism*. Cambridge University Press, Cambridge.

Schrödinger E. (1944). *What is Life?* Cambridge University Press, Cambridge.

Schwartz, M.A., and Madhani, H.D. (2004). Principles of MAP kinase signaling specificity in *Saccharomyces cerevisiae*. *Annual Reviews in Genetics*, **38**, 725–48.

Simons, B.D., and Clevers, H. (2011). Strategies for homeostatic stem cell self-renewal in adult tissues. *Cell*, **145**, 851–62.

Soto, A.M., and Sonnenschein, C. (2006). Emergentism by default: a view from the bench. *Synthese*, **151**, 361–76.

Spemann, H. (1938). *Embryonic Development and Induction*. Yale University Press, New Haven.

Spudish, J.L., and Koshland, D.E. (1976). Non genetic individuality: chance in the single cell. *Nature*, **262**, 467–71.

Stockholm, D., Benchaouir, R., Picot, J., et al. (2007). The origin of phenotypic heterogeneity in a clonal cell population in vitro. *PLoS One*, **2**:e394.

Stockholm, D., Edom-Vovard, F., Coutant, S., et al. (2010). Bistable cell fate specification as a result of stochastic fluctuations and collective spatial cell behaviour. *PLoS One*, 5: e14441.

Suganuma, T. and Workman, J.L. (2011). Signals and combinatorial functions of histone modifications. *Annual Review of Biochemistry*, **80**, 473–99.

Takasuka, N., White, M.R., Wood, C.D., et al. (1998). Dynamic changes in prolactin promoter activation in individual living lactotrophic cells. *Endocrinology*, **139**, 1361–8.

Takenaka, I., Morin, F., Seizinger, B.R., et al. (1995). Regulation of the sequence-specific DNA binding function of p53 by protein kinase C and protein phosphatases. *Journal of Biological Chemistry*, **270**, 5405–11.

Till, J.E., McCulloch, E.A., and Siminovitch, L. (1964). A stochastic model of stem cell proliferation, based on the growth of spleen colony-forming cells. *Proceedings of the National Academy of Sciences USA*, **51**, 29–36.

van Roon, M.A., Aten, J.A., van Oven, C.H., et al. (1989). The initiation of hepatocyte-specific gene expression within embryonic hepatocytes is a stochastic event. *Developmental Biology*, **136**, 508–16.

Viñuelas, J., Kaneko, G., Coulon, A., et al. (2012). Towards experimental manipulation of stochasticity in gene expression. *Progress in Biophysics and Molecular Biology*, **110**, 44–53.

Viñuelas, J., Kaneko, G., Coulon, A., et al. (2013). Quantifying the contribution of chromatin dynamics to stochastic gene expression reveals long, locus-dependent periods between transcriptional bursts. *BMC Biology*, **11**, 15.

Weiss, P.A. (1973). *The Science of Life*. Futura, Mt. Kisko.

Weismann, A. (1891). *Essay on Heredity and Kindred Biological Problems*. Clarendon Press, Oxford.

Wernet, M.F., Mazzoni, E.O., Celik, A., et al. (2006). Stochastic *spineless* expression creates the retinal mosaic for colour vision. *Nature*, **440**, 174–80.

White, M.R., Masuko, M., Amet, L., et al. (1995). Real-time analysis of the transcriptional regulation of HIV and hCMV promoters in single mammalian cells. *Journal of Cell Science*, **108**, 441–55.

Wijgerde, M., Grosveld, F., and Fraser, P. (1995). Transcription complex stability and chromatin dynamics in vivo. *Nature*, **377**, 209–13.

Wilson, J. (1999). *Biological Individuality*. Cambridge University Press, Cambridge.

Xu, M., Sheppard, K.A., Peng, C.Y., et al. (1994). Cyclin A/CDK2 binds directly to E2F-1 and inhibits the DNA-binding activity of E2F-1/DP-1 by phosphorylation. *Molecular and Cellular Biology*, **14**, 8420–31.

Xu, Q., Melitzer, G., Robinson, V., et al. (1999). In vivo cell sorting in complementary segmental domains mediated by Eph receptors and ephrins. *Nature*, **399**, 267–71.

Zernicka-Goertz, M., and Huang, S. (2010). Stochasticity versus determinism in development: a false dichotomy? *Nature Reviews Genetics*, **11**, 743–44.

CHAPTER 11

From genes to gene regulatory networks: the progressive historical construction of a genetic theory of development and evolution

Michel Morange

Introduction

How to explain embryological development? This question has received different answers since the renaissance of epigenesis in the middle of the 18th century. A majority of present approaches gives a central place to genes. In some of them, development is explained by the action of regulatory genes organized in networks (gene regulatory networks—GRNs). This emphasizes the importance of the genome (as an ensemble of genes), and of the temporal functional deployment of GRNs during development. Development is described as the successive activation or inhibition of genes of the GRNs in space (the different cells of the organism) and time (of development) (see for instance Davidson, 2006, 2010). This description of development is that of causal chains running from genes to proteins and from proteins to genes. These chains are chemical, but they consist more of interactions through the formation of weak bonds than of traditional chemical reactions. Central to these networks are regulatory sequences found upstream of the genes: the activity (or not) of the genes they control is the result of a combination of transcription factors binding to these sequences. Evolution is identified with modifications of these networks. These modifications can affect any component, any node of these networks. Nevertheless, the nodes do not evolve at the same rate, and their modifications do not have the same impact. Modification of the upstream regulatory sequences may have important effects—permitting the expression of a gene in a tissue or organ in which it was not previously expressed—without altering the expression of the same gene in other tissues. This is why many developmental biologists ascribe to these mutations a major role in the evolution of morphology.

The GRN models are not the only ones existing in extant developmental biology. But they are the most ambitious, and for this reason, the most prone to highlight the main issues raised by the growth of evo-devo. I will focus my presentation on these models, describe their origins, and discuss some of these issues.

GRNs have their origin in a long embryological tradition emphasizing the existence of a *Bauplan* guiding the development of organisms. A second root of the notion of GRNs was in genetics. Nevertheless, the path was long from the early evidence that genes were involved in development to the elaboration of the first GRNs (Morange, 2000a, 2011). I will describe this path in 'The genetic origin and the progressive construction of the GRN models'. The fact that the Modern Synthesis was the product of the encounter between evolutionary biology and genetics did not prevent the existence of a permanent conflict between the works and hypotheses underlying the conception of GRNs, and

Towards a Theory of Development. Edited by Alessandro Minelli and Thomas Pradeu

the Modern Synthesis. The nature of this conflict, which is still alive today, will be examined in 'The conflicts with the Modern Synthesis': the relation between the models provided by evolutionary biologists and the description of GRNs and their modifications remains unclear. An interesting current issue concerns the existence (or not) of an abstract theory of innovation, which would be additive to the description of GRNs. As discussed in 'Something more than GRNs? The search for a theory of innovation', perhaps the solution that will emerge is simply the coexistence of an abstract evolutionary theory and a precise description of the historical, more or less contingent modifications of the GRNs that were responsible for the evolutionary transformations of organisms.

The genetic origin and the progressive construction of the GRN models

The problem of the relation between particles inherited by offspring from their parents, and the development of these offspring emerged at the end of the 19th century, when a material and corpuscular theory of inheritance was progressively elaborated by August Weismann, Hugo de Vries, and other biologists.

The model proposed by Theodor Boveri of an unequal distribution of the hereditary material between differentiated cells was abandoned when it was shown that, in most organisms, differentiated cells had the same number and structure of chromosomes as the fertilized egg.

Whereas most embryologists considered that the major part of development did not require the action of genes but was the consequence of the organization of the egg, geneticists were convinced, as early as the 1920s, from studies on *Drosophila* and the isolation of mutations affecting development, that genes were involved at each step of development.

The model proposed by Thomas Morgan in his book *Embryology and Genetics* published in 1934 (Morgan, 1934) and in his Nobel lecture (Morgan, 1935) posited that development was the result of the differential activity of genes in the different embryonic cells. This differential activity results from the control of genes by the cellular cytoplasm. The initial trigger to development may be the heterogeneity of the cytoplasm of the egg, leading to differential gene activation in the cells resulting from its division. This differential activity generates new characteristics of the cytoplasm in these cells. It promotes a new pattern of gene activity, and the process repeats itself throughout development. The cytoplasm can also be modified by signals coming from the environment, or from surrounding cells.

Trained in the tradition of experimental embryology, Conrad Waddington considered nevertheless that genes had a major role in development from its earliest stages. He adopted Morgan's model, but tried to picture the role of genes in development more precisely (Waddington, 1940). He hypothesized that numerous genes determined what he called the 'epigenetic landscape' in which cells progressively differentiate.

Morgan's model was widely accepted. It reduced the problem of development to a question of gene regulation. Development was equivalent to the process of differentiation affecting the cells forming the organism. In his influential 1968 book *Gene Activity in Early Development*, Eric Davidson described the first results of applying molecular techniques to the study of embryological development, and explicitly stated that his early results confirmed 'the variable gene activity theory of cell differentiation' and 'the cytoplasmic localization of morphogenetic potential' (Davidson, 1968: Ch. 1).

The success of the model proposed by Morgan sharply contrasted with the total ignorance of the mechanisms explaining how genes could be regulated by molecules coming from the cytoplasm. This explains why the first model of gene regulation, the operon model, was so rapidly accepted, and immediately used to explain development, even though it had been discovered in bacteria, organisms not known for the complexity of their developmental processes!

In the conclusion of the Cold Spring Harbor Symposium of 1961, Monod and Jacob imagined different combinations or modifications of the operon model to account for the stable genetic activity of differentiated cells (Monod & Jacob, 1961).

Some of the characteristics of the operon model were rapidly shown not to be general: grouping of the genes on the chromosomes, and their transcription into one unit, and the dominant role of negative regulation. But what was most important in the model of Monod and Jacob was the distinction introduced in 1959 between two categories of genes: structural genes encoding the components of the cells, and regulatory genes controlling the expression of the former (Jacob & Monod, 1959; Morange, 2000a). Monod and Jacob hypothesized the existence of a small ensemble of regulatory genes controlling development: the latter could be explained through the isolation and characterization of this small group of genes controlling it.

Remarkably, some years later, Jacob turned to the study of the early development of mice. He focused on the study of the *T*-complex, considered at that time to encode an ensemble of proteins successively expressed at the surface of cell membranes during development (Morange, 2000b). The model he had in mind to explain development was identical to that of Morgan. Regulatory genes controlled the expression of these membrane proteins. By modifying the interactions between cells, these proteins led to changes in the characteristics of the cytoplasm, and consequently to the activation or inhibition of other regulatory genes, generating the expression of new membrane proteins. And the process repeated itself throughout early development.

A similar program of research was immediately initiated by geneticists working on *Drosophila* development. The first was Antonio Garcia-Bellido. He identified the selector genes controlling the formation of cellular compartments during insect development with regulatory genes (Garcia-Bellido et al., 1973). Francis Crick and Peter Lawrence made explicit and publicized this new vision of the genetic control of development (Crick & Lawrence, 1975). These ideas were well accepted by most *Drosophila* geneticists, convinced that 'a genetic framework for *Drosophila* development' had been discovered (Baker, 1978). The systematic search for early developmental genes in *Drosophila* initiated by Christiane Nüsslein-Volhard and Eric Wieschaus in their laboratory at EMBL and their classification of these genes (Nüsslein-Volhard & Wieschaus, 1980) found their justification in this new vision of development.

In parallel and independently, Allan Wilson immediately seized on the importance of regulatory genes for an understanding of evolution. He was also impressed by the model proposed by Roy Britten and Eric Davidson in 1969 and 1971 (Britten & Davidson, 1969, 1971). He spent some years characterizing regulatory genes in prokaryotes and then turned his attention to vertebrates and mammals. The famous 1975 paper written with Mary-Claire King in which they compared proteins from humans and chimpanzees is considered today as the first demonstration of the low genetic distance separating these two species (King & Wilson, 1975). King and Wilson's conclusion was oriented in a different direction: their work demonstrated that the huge phenotypic differences between humans and chimpanzees were due, not to numerous mutations in 'ordinary' genes, but to mutations affecting a small group of regulatory genes controlling development, genes that had not yet been characterized. Wilson emphasized the consequences of the existence of regulatory genes for evolutionary theory, which led him to a firm opposition to the Modern Synthesis (see 'The conflicts with the Modern Synthesis').

A similar attitude was shared by Stephen Jay Gould. In 1977 he published a book entitled *Ontogeny and Phylogeny*, in which he criticized the principle of recapitulation advocated by Ernst Haeckel and its wide use outside biology, in particular in support of racist theories (Gould, 1977). He suggested that the observations behind the principle were partially right but misinterpreted: they could be explained by the existence of regulatory genes that control development and the mutation of which leads to heterochrony in one or other developmental process.

What triggered the rapid expansion of work on developmental regulatory genes was their isolation, made possible by the new tools of genetic engineering, and the discovery in the mid-1980s that these genes had been conserved during evolution. An emblematic example was the homeobox-containing genes. This was a huge surprise: what was common between the development of a *Drosophila* and that of a mouse? Jacob acknowledged some years later that, when he introduced the notion of tinkering in 1977—the hypothesis that evolution recombines the same pieces to build new devices (Jacob, 1977)—he did not consider that it applied to the master genes controlling development. He thought, as did most biologists, including Ed Lewis the specialist of homeotic genes, that these genes had progressively appeared during evolution, in parallel with the developmental processes they controlled. This firm belief was in agreement with the presupposed higher number of genes in the genomes of humans compared, for instance, with lower species such as *Drosophila* (Jacob, 1994). The complexity of structures such as the brain required the action of new and numerous regulatory genes.

The first developmental genes to be characterized were regulatory genes, such as *engrailed* or the homeobox-containing genes previously mentioned, encoding transcription factors controlling the activity (expression) of other genes (Gehring, 1998). But as other developmental genes were progressively characterized, some were shown to be different: they had been as extensively conserved during evolution as the regulatory genes, but they encoded components of the cell signalling pathways. The description of these signalling pathways was parallel to the demonstration of their involvement in the development of highly different organisms (see Morange, 2001 for a discussion). Transcription factors and components of the signalling pathways are the main actors in GRNs.

The conflicts with the Modern Synthesis

From its beginnings, the new genetic model of development was in conflict with the conceptions of evolutionary biologists, as expressed within the framework of the Modern Synthesis elaborated in the 1930s and 1940s. The latter privileged, as Darwin did, the role of 'small' adaptive variations in evolution, i.e. an uniformitarian view. But the attitudes of the proponents of the new genetic conception of development were highly varied, with some dismissing the problem while others openly challenging the Modern Synthesis.

Jacob and Monod adopted the first attitude. The evolutionary consequences of the distinction between regulatory and structural genes were briefly discussed in a cryptic publication of the Pontifical Academy of Sciences (Jacob & Monod, 1962). Jacob was personally convinced that the explanation of evolution was to be found in the properties of the master control genes of development and that evolution operated through huge leaps, in contrast to the uniformitarian view of Darwin and the Modern Synthesis. He acknowledged in private his admiration for Richard Goldschmidt, the German-born geneticist who introduced the distinction between micro- and macromutations, micro- and macro-evolution, and outlined the role of 'hopeful monsters' in evolution. But he never loudly addressed the issue of the incompatibility between the Modern Synthesis and the new genetic conception of development. One reason was that Jacob had no scientific facts in favour of his vision. A long time was necessary to adapt and extend the tools used on microorganisms to be able to describe the mechanisms involved in the development of insects and vertebrates. This cautious attitude also finds its explanation in the close links established by molecular biologists with evolutionary biologists. The experiments of Max Delbrück and Salvador Luria, the founders of the American Phage Group, in the 1940s on the origin of mutations in bacteria expelled Lamarckism from its last field of application—microbiology. Molecular biologists and evolutionary biologists shared the same opposition to traditional biology, to embryology, which they considered to be impregnated with metaphysical flavours, and also to all biologists who still supported the Lamarckian or neo-Lamarckian theories. The latter were particularly active in France, which probably explains the cautious attitude of Monod and Jacob (Loison, 2011). They did not want to appear, in one way or another, as opponents to Darwinism, and on the same side as neo-Lamarckians. The behaviours of Garcia-Bellido and other *Drosophila* geneticists were similar. Focused on the description of the genetic mechanisms of *Drosophila* development, they did not address the evolutionary issue.

In contrast, Gould emphasized that heterochronic mutations were exactly the kind of macromutations imagined by Goldschmidt. He directed a new edition of the fundamental book of Goldschmidt *The Material Basis of Evolution* (Goldschmidt, 1940) and wrote a preface for it (Gould, 1982). The 1970s to 1980s were the culmination of Gould's partially disorganized and confused attacks on the Modern Synthesis. His critique of panadaptationism and his support for punctuated equilibria are well known. The interpretations of the latter—the existence during evolution of long stases followed by rapid changes—were diverse, and an extensive debate followed the publications of Niles Eldredge and Gould (Eldredge & Gould, 1972). One of the simplest and most direct interpretations of punctuated equilibria is that they simply reflect the existence of leaps in evolution.

The major enemy within the Modern Synthesis for Gould was obviously uniformitarianism. He opposed it in numerous ways throughout his career. For instance, he emphasized the role of catastrophic events, such as extinctions, in the evolution of life (Gould, 2002).

Allan Wilson also opposed the uniformitarian view of the Modern Synthesis. Not only did he suggest that the modification of a limited number of regulatory genes played a major role in evolution, but he also went further and argued that these regulatory mutations were not point mutations but rearrangements of the genome (Wilson et al., 1974). Such a statement was in agreement with the ideas of Goldschmidt for whom macromutations were of a different physical nature than simple mutations.

Eric Davidson also rapidly addressed the evolutionary issues raised by the new genetic models of development. In sharp contrast with Monod and Jacob, only two years separated the publication by Britten and Davidson of their model of the genetic control of development (Britten & Davidson, 1969) and the publication of an article dealing with the evolutionary consequences of such a model (Britten & Davidson, 1971). The perspectives were clearly fully at odds with the evolutionary vision of the Modern Synthesis, and the opposition of Davidson to Darwinism became more and more open in the following years and decades.

What are the main aspects of this opposition between the new genetic vision of development and the Modern Synthesis?

The first is the focus placed on a small group of genes designated as the main players in development, and consequently, evolution. In the traditional models of evolutionary genetics, all genes participated on an equal footing in the evolutionary transformations of organisms. Adaptation, the motor of evolution for neo-Darwinians, may concern every character of an organism. From Monod and Jacob to Davidson and the four-dimensional regulatory genome, the genome is no longer seen as a collection of equal genes, but as a hierarchical structure in which each gene has different functions and roles. Both studies and explanations have to take into account this hierarchy. All genes are important, but some genes are more important than others!

The second source of conflict is uniformitarianism. The new genetic model of development implied that some mutations may have a dramatic effect on the development and structure of organisms. This was a consequence of the models pointed out discretely but clearly by Jacob and Monod in 1962 (Jacob & Monod, 1962): a mutation in a regulatory gene does not have the same consequences as a mutation in a structural gene. In one step, it affects the expression of numerous genes. Wilson and Gould shared the same view: a simple heterochronic mutation can have a dramatic effect. Within a GRN, as demonstrated by Davidson, a mutation in a gene coding for one of the components of a central (kernel) subcircuit of the network will have dramatic consequences, whereas a mutation affecting a component of a subcircuit controlling a battery of differentiation genes will have a far more limited impact (Erwin & Davidson, 2009). It is necessary to distinguish micro- and macromutations, and the consequence is that the rhythm of evolution is irregular. As argued by Gould, Darwinians have wrongly conflated two different statements in uniformitarianism (Gould, 2002). The first, which remains unquestionable, is that to explain the facts of the past, one must not appeal to extraordinary causes—causes that are unknown. It is possible that certain causes no longer operate in the present world, but reasons for their disappearance have to be found. The second statement is that natural causes have always operated in a regular way, with the same (limited) strengths. The result of this second dimension of uniformitarianism for evolution is that all mutations are considered to have roughly the same consequences. For Gould, this second component of uniformitarianism imposes limits that are too strong and unjustified on the range of possible explanations of natural phenomena.

The last conflict between evolutionary biology and this new genetic vision of development originates when the tree of life is no longer seen as the result of a progressive divergence of organisms by adaptation but as the direct consequence of the nature of the variations that have occurred in evolution and have produced it. By considering the nature of variations, it is possible to anticipate whether they will generate a new variety or species, or a new phylum (Davidson & Erwin, 2006). If a mutation affects one of the components of a battery of differentiation genes, its consequences will be minor, whereas if it modifies a kernel subcircuit, it will lead to a dramatic change in the development and structures of organisms. Such a vision of evolution is totally at odds with the single theoretical diagram describing the divergence of organisms by variations proposed by Darwin in *On the Origin of Species*, and also with the evolutionary Modern Synthesis. The driver of evolution is adaptation for the supporters of the Modern Synthesis, whereas it is the nature of the GRNs and of their variations for Davidson.

Is it possible to reconcile these antagonistic views on evolution? It is fair to recognize that both sides—evolutionary biologists and developmental geneticists—make little effort to integrate their own views with those of others. One argument in favour of a regular rhythm of evolution—by variations of a small amplitude at the phenotypic level—is that such variations are less likely to be detrimental and therefore eliminated by natural selection. In addition to the fact that catastrophic events could be too easily identified with miraculous events, the preference for small variations and progressive evolution was probably also a belated answer from Darwin to Georges Cuvier and his demonstration that the different parts of an organism are related to one another and consequently that no one part can be altered without modifying the others. To see evolution as the consequence of 'small' variations was the only way to reconcile the existence of evolution with the results of Cuvier.

The necessity for variations to have limited phenotypic amplitudes and the dramatic consequences of a mutation in a regulatory gene controlling the rhythm of development, or in a gene encoding a component of a kernel subcircuit, can be reconciled if there exist within organisms mechanisms buffering the initial consequences of mutations. Plenty of such mechanisms have been described: the existence of genetic (by gene duplication; Ohno, 1970) and functional redundancy, the action of chaperones attenuating the effects of mutations on protein structures and functions (Rutherford & Lindquist, 1998), etc.

Interestingly, even Davidson has admitted that while mutations in the genes encoding components of the kernel subcircuits are impossible today, such modifications were possible in the past, in particular in the early Cambrian period when there was an explosion of animal forms (Davidson, 2011). A further step would have been to admit that some buffering mechanisms were also present at these periods and allowed such dramatic transformations to occur in a soft way. But, as we have already underlined, the protagonists do not make serious efforts to solve the conflicts!

Something more than GRNs? The search for a theory of innovation

The characterization of GRNs is intended not only to provide a precise description of the development of organisms and to explain the mechanisms of the evolutionary transformations that occurred in the past but also to describe the space of possibles in which future evolution will occur and in this way, to anticipate it.

The limits of the present descriptions of GRNs are obvious. They are restricted to few organisms and will have to be enriched by the incorporation in the present schemes of additional components, microRNAs, and epigenetic modifications among others. The precise quantitative description of the GRNs in four dimensions remains more an objective than a reality!

Nevertheless, despite these present limitations, it is already possible to raise two fundamental and closely linked issues concerning the place and value of GRNs. The first concerns the relations between evolutionary theory and the description of GRNs and their variations. Is the latter sufficient to explain development and evolution, and the first, evolutionary theory—a framework simply recalling that natural selection has a (limited) place in the process? Or, conversely, does a new theory need to be built by rejuvenating evolutionary theory through the integration within it of the results accumulated on GRNs? The origin of the problem is seen by most of those in the field as the following: abstract evolutionary theory was built at a time when nothing was known of the mechanisms involved in the developmental construction of organisms. Now that these mechanisms have been described, it is obvious that evolutionary theory has to be modified. But what will be the amplitude of these modifications (Morange, 2012)?

The second issue concerns the eventual necessity to add something to the present theories and models. Will the explanation of development and evolution coincide with the precise description of GRNs, or is something else required to account for evolution? This 'something else' might be a more abstract description at a higher level of organization than that of genes and components. But it might also be the enunciation of some general principles that have emerged during evolution and are responsible for its present course. Some innovations have transformed the course of organismal evolution such as, for instance, the formation of a central nervous system; and the elaboration of a theory of innovation would be required besides the present evolutionary theory and beyond the simple description of GRNs.

Such a conviction is shared by many researchers working in evo-devo. But despite this apparent unanimity, the nature of this theory of innovation is still debated. For John Maynard-Smith and Eörs Szathmáry, the important innovations concern the emergence of information and of its treatment (Maynard-Smith & Szathmáry, 1999). For Marc Kirschner and John Gerhart, innovations consist of the emergence of global functionalities: exploratory behaviours, the formation of weak linkages, and compartmentation (Kirschner & Gerhart, 2005). For Andreas Wagner, the capacity to innovate is to be found in the particular nature of the relation between the space of genotypes and the space of phenotypes (Wagner, 2011). The same phenotype can be generated by a network of different genotypes. In the whole space of genotypes, genotype networks giving rise to different phenotypes can nevertheless be close. It is this meta-architecture of genotype networks that explains the capacity of organisms to evolve and to innovate.

Today two representations of evolution coexist: on the one side abstract evolutionary theory; on the other, a precise description of the molecular modifications that occurred during evolution. Between the two, something is missing that could establish a bridge between the description of GRNs and of their transformations and evolutionary theory: maybe a theory of innovation.

Historians abandoned in the middle of the 20th century any hope of drawing up a theory of historical transformations for human societies. Biologists have not yet included in their conception of evolution its full historical dimension. This delay probably has its roots in the complex and tumultuous history of the theory of evolution. The first to propose such a theory was Lamarck, but his vision was not historical. He proposed a theory explaining the permanent transformations of organisms, and the general trend towards complexity. But it was not a history, since there was no beginning; it looked more like a conveyor belt, permanently transforming the organisms in the same direction. As noted by Ernst Mayr (2004), Darwin simultaneously did two different things. He collected facts in favour of a transformation of organisms, and he proposed a mechanism (variation and natural selection) to explain it. By so doing, he reduced the role attributed to historical contingency in the evolution of organisms. The absolute necessity to provide a mechanism of evolution in addition to the proofs of its existence probably came from the fact that the biologist who convinced his contemporaries that there had been a history of life on Earth, Cuvier, was paradoxically an opponent to transformism!

The contingent dimension of the history of life was progressively reintroduced, first through the importance given to genetic drift in the Modern Synthesis, and later by Stephen Jay Gould and others in the second half of the 20th century.

Is it possible to imagine a theory of innovation that would not abolish the historical dimension of life? What would be its structure? The answers are not obvious. Maybe the present situation, with a general abstract theory of evolution and an increasingly precise description of GRNs and of their transformations during evolution, is the last word that has to be said on the evolution of organisms.

Conclusion

What are the main drivers of evolution? Variations or the action of natural selection? Has an abstract theory of innovation to be added to the description of GRNs, and what would be its content?

It is too early to answer these questions. But the description of development as the result of the action of GRNs provides one way to address them experimentally.

There is no doubt that advances will be made in the description of GRNs in the years to come. More importantly, scenarios of the modifications of GRNs during evolution will be proposed, and these scenarios will be tested through synthetic experimental evolution.

Answers to the previous questions will not be immediately forthcoming. Nevertheless, the strength of the present description and explanations of development is that they can be tested, and falsified. For instance, the occurrence of large variations at the phenotypic level can be experimentally shown to be the result of the action of natural selection on minor variations. Only the accumulation of descriptions of GRNs and of their transformations will justify (or not) the existence of a theory of innovation. The time of fruitless speculations is over!

Acknowledgments

I am indebted to Dr. David Marsh for critical reading of the manuscript.

References

Baker, A. (1978). A genetic framework for *Drosophila* development. *Annual Review of Genetics*, **12**, 451–70.

Britten, R.J., and Davidson, E.H. (1969). Gene regulation for higher organisms: a theory. *Science*, **165**, 349–57.

Britten, R.J., and Davidson, E.H. (1971). Repetitive and non-repetitive DNA sequences and a speculation on the origins of evolutionary novelty. *Quarterly Review of Biology*, **46**, 111–33.

Crick, F.H.C., and Lawrence, P. (1975). Compartments and polyclones in insect development. *Science*, **189**, 340–47.

Davidson, E.H. (1968). *Gene Activity in Early Development*. Academic Press, New York.

Davidson, E.H. (2006). *The Regulatory Genome: Gene Regulatory Networks in Development and Evolution*. Academic Press, Burlington.

Davidson, E.H. (2010). Emerging properties of animal gene regulatory networks. *Nature*, **468**, 911–20.

Davidson, E.H. (2011). Evolutionary bioscience as regulatory systems biology. *Developmental Biology*, **357**, 35–40.

Davidson, E.H., and Erwin, D.H. (2006). Gene regulatory networks and the evolution of animal body plans. *Science*, **311**, 796–800.

Eldredge, N., and Gould, S.J. (1972). Punctuated equilibria: An alternative to phyletic gradualism. In T.J.M. Schopf, ed., *Models in Paleobiology*. Freeman, San Francisco, pp. 82–115.

Erwin, D.H., and Davidson, E.H. (2009). The evolution of hierarchical gene regulatory networks. *Nature Reviews Genetics*, **10**, 141–8.

Garcia-Bellido, A., Ripoll, P., and Morata, G. (1973). Developmental compartmentalisation of the wing disk of *Drosophila*. *Nature New Biology*, **245**, 251–3.

Gehring, W.J. (1998). *Master control genes in development and evolution*. Yale University Press, New Haven.

Goldschmidt, R. (1940). *The Material Basis of Evolution*. Yale University Press, New Haven.

Gould, S.J. (1977). *Ontogeny and Phylogeny*. The Belknap Press of Harvard University Press, Cambridge.

Gould, S.J. (1982). The uses of heresy: an introduction to Richard Goldschmidt's *The Material Basis of Evolution*. In R. Goldschmidt, *The Material Basis of Evolution*. Yale University Press, New Haven, pp. xiii–xvii.

Gould, S.J. (2002). *The Structure of Evolutionary Theory*. The Belknap Press of Harvard University Press, Cambridge.

Jacob, F. (1977). Evolution and tinkering. *Science*, **196**, 1161–6.

Jacob, F. (1994). L'irrésistible ascension des gènes *Hox*. *Médecine/Sciences*, **10**, 145–8.

Jacob, F., and Monod, J. (1959). Gènes de structure et gènes de régulation dans la biosynthèse des protéines. *Comptes-rendus de l'Académie des sciences de Paris*, **249**, 1282–4.

Jacob, F., and Monod, J. (1962). Sur le mode d'action des gènes et leur régulation. *Comptes rendus de l'Académie Pontificale des sciences*, **22**, 85–95.

King, M.C., and Wilson, A.C. (1975). Evolution at two levels in humans and chimpanzees. *Science*, **188**, 107–16.

Kirschner, M.W., and Gerhart, J.C. (2005). *The Plausibility of Life: Resolving Darwin's Dilemma*. Yale University Press, New Haven.

Loison, L. (2011). French roots of French neo-lamarckisms, 1879–1985. *Journal of the History of Biology*, **44**, 713–44.

Maynard-Smith, J., and Szathmáry, E. (1999). *The Origin of Life: From the Birth of Life to the Origin of Language*. Oxford University Press, Oxford.

Mayr, E. (2004). *What Makes Biology Unique?* Cambridge University Press, Cambridge.

Monod, J., and Jacob, F. (1961). General conclusions: Teleonomic mechanisms in cellular metabolism, growth and differentiation. *Cold Spring Harbor Symposia on Quantitative Biology*, **26**, 389–401.

Morange, M. (2000a). The developmental gene concept: History and limits. In P. Beurton, R. Falk and H.-J. Rheinberger, eds., *The Concept of the Gene in Development and Evolution: Historical and Epistemological Perspectives*. Cambridge University Press, Cambridge, pp. 193–215.

Morange, M. (2000b). François Jacob's Lab in the seventies: the T-complex and the mouse developmental genetic program. *History and Philosophy of the Life Sciences*, **22**, 397–411.

Morange, M. (2001). *The Misunderstood Gene*. Harvard University Press, Cambridge.

Morange, M. (2011). Evolutionary developmental biology: its roots and characteristics. *Developmental Biology*, **357**, 13–6.

Morange, M. (2012). What might be a new 'view of evolution'? *Studies in History and Philosophy of Biological and Biomedical Sciences*, **43**, 578–81.

Morgan, T.H. (1934). *Embryology and Genetics*. Columbia University Press, New York.

Morgan, T. H. (1935). The relation of genetics to physiology and medicine. *Scientific Monthly*, **41**, 5–18.

Nüsslein-Volhard, C., and Wieschaus, E. (1980). Mutations affecting segment number and polarity in *Drosophila*. *Nature*, **287**, 795–801.

Ohno, S. (1970). *Evolution by Gene Duplication*. Springer, Berlin.

Rutherford, S.L., and Lindquist, S. (1998). Hsp90 as a capacitor for morphological evolution. *Nature*, **396**, 336–42.

Waddington, C.H. (1940). *Organisers and Genes*. Cambridge University Press, Cambridge.

Wagner, A. (2011). *The Origin of Evolutionary Innovation: A Theory of Transformative Change in Living Systems*. Oxford University Press, Oxford.

Wilson, A.C., Sarich, V.M., and Maxson L.R. (1974). The importance of gene rearrangement in evolution: evidence from studies on rate of chromosomal, protein, and anatomical evolution. *Proceedings of the National Academy of Sciences USA*, **71**, 3028–30.

CHAPTER 12

Reproduction and scaffolded developmental processes: an integrated evolutionary perspective

James Griesemer

Introduction

Traditional concepts of development concern the embryogenesis of organisms from single-celled zygote or egg to multicellular differentiated adult. The goal of this chapter is to probe the conceptual boundaries of traditional concepts of development and their relations to heredity and evolution by refining a concept of development designed for integration into a general account of units of evolution (Griesemer, 2000a, b, c, 2005, 2006a). The concept of development has already been extended beyond embryogenesis in the 20th century transition from embryology to developmental biology, though the implications are still being worked out (e.g. *Biological Theory* 2011 special issue, *The Boundaries of Development*). Extended concepts of reproduction, heredity, and selection are needed as well if an integrated theory of evolution is to be achieved.

That project is too ambitious for one chapter. Here I consider the concept of development in the context of an account of reproduction designed to serve some general principles of ecological, evolutionary developmental biology and the place of concepts in an inclusive view of biological theories. If theory is taken to include several kinds of theoretical components rather than just one, it is plausible to think that developmental biology already has and operates with many theories, even if practitioners rarely articulate or present theories of development as a goal of research. However, because there seems to be no single, widely accepted 'grand' or 'overarching' theory of development (Love, this volume) that 'organizes' a whole domain of knowledge (cf. Waters, 2008a) in the way that Darwin's principles (Darwin, 1859) are seen to organize evolutionary knowledge (see Lewontin, 1970) or Newton's principles organize and even define classical mechanics (see Giere, 1988), it is worthwhile to speculate about what an articulated theory of development might look like. I propose that (mature) scientific theories be conceptualized as comprising three kinds of components: a set of core principles, at least one family of models, and a theoretical perspective. This set is sufficient to make a theory potentially empirical. The empirical content of a theory depends in addition on 'theoretical hypotheses' (Giere, 1988) linking principles or models to 'nature' or constructed phenomena and stable 'effects' (Hacking, 1983).

After locating the concept of development I favour within a very abstract general account of reproduction, I consider the complex life cycle of malaria parasites to explore the concept of development in cellular contexts where development in complex life cycles is best not considered separately from reproduction. I then extend the concept of development beyond cellular life to consider an example of molecular development: the replication of HIV-1 retroviruses. Understanding retrovirus replication as a developmental process requires considering the whole life cycle, multiple levels of biological organization, and the fact that at some

Towards a Theory of Development. Edited by Alessandro Minelli and Thomas Pradeu

stages the virus is a molecule while at others it is a complex, membrane-bound entity. At all stages, the virus per se lacks metabolism but has development. Metabolism is the province of a hybrid entity—the virus–host fusion—while virus developmental capacities are carried through this 'hybrid state' by virus molecular entities through several 'generations' that develop in ecologically different molecular environments of the host. While it is tempting to treat this description as mere analogy, I argue that pushing the ecological-evolutionary-developmental biology of viral infection to the molecular level is a heuristically useful extension of the concept of development to the molecular level. I return at the end of the chapter to consider whether the proposed conceptual change advances the pursuit of a *theory* of development in the expanded sense presented in this chapter.

Conceptual change has been a common, but not the sole, means of theory development in biology. It is widely recognized already, for example, that the units of selection are not limited to within-generation changes in frequencies of organism phenotypes, so Darwinian theory can be extended by conceptualizing units of selection or 'individuals' more broadly than just 'organisms'. Evolution may also proceed by genetic drift rather than or in combination with natural or sexual selection, so evolutionary theory can also be extended by the addition of evolutionary forces to the list of Darwinian principles. But more importantly, hereditary relations between parents and offspring depend on the developmental processes linking offspring phenotypic states to adult/parent phenotypic states. Thus, to extend evolutionary theory by considering novel mechanisms (and levels) of heredity, such as trans-generational epigenetic inheritance and epigenetic heritability (e.g. Tal et al., 2010), theories of inheritance must be articulated with a theory of development. A theory of development is therefore a requirement for extending evolutionary theory, even if it were not needed for the conduct of inquiry into development per se. Griesemer (2013) argues that the role of theory in inquiry is in part through 'formalization', a process by which exact theory rather than inexact empirical conduct and modelling is taken to delimit and control the domain of study. Developmental biology can be practiced as an 'inexact science' without a theory in this sense. However, the integration of developmental biology with evolutionary theory may well require a formalized, exact theory of development.

Expansion of evolution and inheritance theories to include development calls not only for adding developmental principles to Darwin's principles (see Wimsatt, 2001) but also for a careful consideration of how concepts, principles, and models of development articulate with concepts, principles, and models of heredity and evolution. Moreover, good conceptual practice calls for probing the boundaries or limits of concepts central to the core principles of theories by stretching them beyond their original domains of application. This is an 'engineering' conception of philosophy *for* science (Griesemer, 2006b). It is a heuristic practice of 'conceptual mechanics' seeking to learn how to build better, more robust concepts, models, and theories. The method is to stretch concepts to breaking points by application to potentially inappropriate subject matter, such as the application of developmental concepts to the molecular level, as considered here in the replication cycle of HIV-1 and malarial parasites, so as to study how they fail in the hope of understanding why they fail (see Wimsatt, 2007). Conceptual interventions of this kind, when practiced in the sciences, are usually coordinated with material experimental practices and the deliberate construction of false models as means to 'truer' theories (Wimsatt, 1987, 1992, 2007).

Expanding the conceptual boundaries of development

This chapter employs such a heuristic strategy. The goal is a robust concept of development, articulated with concepts of heredity and evolution, that does not presuppose any particular, historically contingent, evolved mechanism for its operation. The aim is not conceptual *analysis*, but rather to explore how revised, articulated concepts might transform the ways models and theories are built and how the conduct of empirical and theoretical research can be envisioned, carried out, and evaluated. The aim is thus empirical and practical, even though the work here is conceptual and theoretical.

A concept of units of selection based on properties of modern genes, for example, cannot provide the conceptual resources to explain the evolutionary *origin* of genes (Griesemer, 2000c, 2005). Analogously, I am in sympathy with Minelli, who calls for a theory of development that does not depend on any 'adultocentric' periodization of the developmental process manifested by any particular (typically animal) taxa (Minelli, 2011b: 5). I am also in sympathy with Laplane, who rejects the idea that either reproduction or death, treated as events in the life cycle, are adequate delimiters of the temporal boundaries of development (Laplane, 2011). I go farther, however, in calling for an account of the process of development that is not delimited by a conceptual boundary accepted by Minelli, Laplane, and most other analysts of concepts of biological development: that development is necessarily a *cellular* process. While it may be empirically true that life is inherently cellular, it is nevertheless problematic to presuppose cellularity in characterizing or defining development (Griesemer & Szathmáry, 2009: 499).

Minelli proposes that 'a comprehensive theory of development should start with a zero principle of "developmental inertia," corresponding to an indeterminate local self-perpetuation of cell-level dynamics' (Minelli, 2011b: 4; see also Minelli, 2011a). This kind of proposal is a step towards effectively overcoming human, vertebrate, or animal biases by conceptualizing development comprehensively, without regard to any particular mode or taxon, in terms of a process of self-maintaining cell dynamics rather than in terms of a developmental goal, fate, or evolutionary consequence. Laplane (2011) likewise resists traditional approaches to the concept of development as biased towards certain forms of organization. She argues that a consideration of stem cells rather than eggs or zygotes leads to a more useful and more comprehensive account of developmental capacities that is not distorted by delimitation in terms of the ultimate developmental capacity: reproductive capacity. She argues for a temporal ordering of development in terms of the differing proximate developmental capacities of totipotent, pluripotent, multipotent, and unipotent stem cells. Laplane's proposal provides excellent guidance towards a less biased conception of development. Minelli and Laplane argue convincingly that the temporal boundaries of development in various kinds of organisms are better described with these expanded concepts than with traditional ones.

Minelli usefully frames the problem of searching for a comprehensive theory of development in terms of what he says development is not: 'development (1) is not restricted to the multi-cellular organisms, (2) does not necessarily start from an egg, (3) does not necessarily start from a single cell, (4) does not necessarily imply an increase in structural complexity, and (5) does not necessarily end with the achievement of sexual maturity' (Minelli, 2011b: 5).

Minelli's and Laplane's concepts are, however, both *cellular* concepts of development. Minelli's third claim rests on the observation that development may start from multicellular bodies such as buds rather than single cells, not that development applies to non-cellular processes. Yet if we want to deploy such a concept of development in an account of mechanisms of evolutionary transition, the evolutionary origin of new levels of biological organization (Maynard Smith & Szathmáry, 1995), this limitation to cellular development is too restrictive. Cellular life could not *evolve* from non-cellular life on the presuppositions that development is an *inherently* cellular process and that development is necessary for evolution. Moreover, if evolution is a cellular process (because development is), evolutionary change above the level of multicellular organisms, e.g. among groups or populations, would not be appropriately understood as evolutionary processes because groups and populations are not cellular, even if their organism *parts* are composed of cells; so the 'development' of groups and populations would not be cellular processes, except derivatively on a reductive account of group and population evolution in terms of cells.

I seek a concept of development that does not conceptually rule out the *evolutionary* origin and elaboration of cellular life or adaptive group, social, or cultural change by presupposing the conceptual boundary of development is to be drawn around cellular forms of organization. I seek to probe that boundary, though I have no stake in whether evolution below the cell or above the group occurs. I do insist that it is an empirical question to be answered by empirical investigation and that my project is not to be taken as one of conceptual *analysis* of heredity, development, and reproduction, but rather as one of *modelling* concepts for the sake of an extended, integrated theory (see Griesemer, 2014). Rather than defining concepts, the conceptual modeller represents possible meanings that function as empirical hypotheses: suppose biological development were not a cellular concept tied to certain structural *levels* of organization, but rather a processual concept tied to certain functional *modes* of organization, could we then extend development to phenomena not typically investigated developmentally? If so, that discovery would reflect on our understanding of cellular as well as extended processes of development. It would also generalize the significance of theoretical principles involving the concept by abstracting them from structural constraints (Griesemer, 2005).

If extension of the concept of development beyond cellular life is successful, we might then gain conceptual and theoretical resources to turn back to traditional exemplars of cellular development with fresh eyes and research questions, about the nature of cellularity, for example. Is the key concept of cellularity boundedness by a lipid membrane, a mode of autonomy grounded in metabolism, or something else? We might also gain new tools with which to explore the chemical and human cultural domains in which 'non-biological' analogues of processes of development, heredity, and evolution may occur. If the extended concept breaks down, then we will perhaps learn something about the unique developmental significance of cellular organization we didn't understand before.

A reproducer perspective on development

From an evolutionary perspective, the full phylogenetic distribution of material modes of reproduction is an empirical question depending not only on DNA genealogy, but also in part on the ways and extent to which 'exogenetic' environments involving a wide range of developmental resources beyond protein-coding genes (see Griffiths & Stotz, 2013) contribute to inheritance across the extended genealogical nexus. That in turn depends on the roles environments play as contributors to reproduction. To what extent and in what ways are environments reliable enough to deliver developmental capacities to offspring in the form of *nutrition* (material, energy) and *environmental triggers* (reviewed in Gilbert & Epel, 2009), *scaffolding* (physical interactions formed between developers and scaffolds to facilitate development; discussed in Caporael et al., 2014), and *prostheses* (organized parts added to a developing system that enhance or substitute for developed parts, e.g. hermit crab houses, nests and nest sites, knowledge recorded in books, scientific instruments that enhance perception)? To the extent material modes of reproduction processes involve environmental contributors, developing systems may forego making or managing developmental components on their own or receiving them ready-made from parents, thus making genealogies more a matter of organized developmental environments or niches and less a matter of uniquely *genetic* inheritance (Griesemer, 2014; Griffiths & Stotz, 2013). Niche construction (Odling-Smee et al., 2003), ecological engineering (e.g. Sterelny, 2001), and developmental systems theory (Griffiths & Gray, 1994; Oyama, 1985; Oyama et al., 2001) all integrate developmental concepts into evolutionary thinking in the form of extended concepts, principles, and models which are ingredients for an integrated and extended theory of ecological developmental evolution. How such efforts relate to a theory of development remains to be seen.

Moreover, to address problems of evolutionary transition, characterized as transitions to new levels of reproduction, in which 'entities that were capable of independent reproduction before the transition can reproduce only as parts of a larger whole after it' (Griesemer, 2000c: 79) and for philosophical reasons explored more fully elsewhere, I articulated a view of development in relation to reproduction and inheritance (Griesemer, 2000a, b, c, 2005, 2006a, 2014; Griesemer & Szathmáry, 2009; Wimsatt & Griesemer, 2007). Here I refine the account in line with some aspects of Laplane's (2011) and Minelli's (2011b) proposals, but abstract from cells as units of development to a more general concept suitable for the heuristic purposes mentioned above. These refinements require considering ecological *context* as part of the concept of development.

Contemporary developmental biology is 'going eco', i.e. recognizing that normal development is plastic in ways that depend on specific kinds of environmental inputs and feedback (Gilbert, 2001; Gilbert & Bolker, 2003; Gilbert & Epel, 2009; West-Eberhard, 2003). Environmental components must be included in the concept of development. The concept of development is thus being extended to developmental *systems* (Oyama, 1985; Oyama et al., 2001) of quite heterogeneous kinds of elements from developing *organisms* organized around traditional notions of gene and cell lineages. I focus here on certain of the environmental inputs to development which function as developmental 'scaffolding'.

Concepts of heredity, heritability, and inheritance carry with them implicit conceptions of development by presupposing the operation of developmental processes leading to the repeated dynamic realization of stable phenotypes. Accounts of genotype/phenotype relations for example typically assume a cellular environment in which gene expression depends on developmental mechanisms transmissible between cell generations. Bringing such conceptual dependencies to light may reveal hidden theoretical presuppositions and improve understanding of well-entrenched models and theories (Griesemer, 2007).

For these and other reasons (see Griesemer, 2005), I proposed the core elements of the following account of evolution (see Griesemer, 2000a, b, c). Evolution is descent with modification of a population of reproducers. Reproducers are entities that have the capacity to make more reproducers, i.e. generate descent relations among entities organized by reproduction into lineages that are genealogically connected, such that offspring have relations of material overlap with their parents. Material overlap means that at least some material parts of the offspring were formerly material parts of the parents. Thus, reproduction involves a bond of material continuity, not merely one of resemblance nor merely one of formal transmission of information (see Griesemer, 2014 for a recent statement and Griesemer, 2005 for arguments that reproduction understood in terms of copying or merely formal relations are problematic as stand-alone concepts). Most importantly, at least some materially overlapping parts convey developmental capacities to offspring via the transfer of parts, i.e. a material 'progeneration' (propagule generation) or propagation with material overlap (see also Griesemer, 2000a). Development is the *recursive* acquisition (over a compositional hierarchy of parts and wholes) of a capacity to reproduce. Recursion bottoms out in 'null development' in which progenerated entities are born 'ready-made' with a capacity to develop, rather than having to *acquire* a capacity to develop. Development in the special case of multicellular organisms requires reproduction of cells, reproduction of cells requires development of cellular capacities, and cell development requires autocatalysis (null development) of cellular constituents.

An evolutionary reason biological reproduction may depend on material overlap is that it might improve the robust, reliable propagation of organized mechanisms of development over what an unstable, uncertain, less complexly organized environment can deliver in suitable temporal order and spatial configuration to a developing offspring. The argument rests on a developmental version of Simon's (1962) evolutionary argument for the hierarchical architecture of complex systems. Development of systems from previously developed material modules in organized spatio-temporal patterns conferring novel capacities on the modules, rather than from de novo assembly of the entire collection of unorganized 'adult' parts, is dynamically more effective and more stable and reliable. Material propagules in reproduction, like material modules in hierarchical assembly processes, can more effectively preserve and propagate developmental order because they organize and preserve complex developmental mechanisms that carry capacities from parental to offspring context.

A material product or part spatially isolated from the reproducer(s) that progenerated it and which is lacking in developmental capacity should not really be considered an offspring. Thus, all (biological) descent relations involve relations of material overlap, but not all relations of material overlap constitute (biological) descent relations. (Biological) descent relations hold only for reproducers. Developmental capacity is, in the most expansive sense, a capacity to *acquire* the capacity to reproduce. Reproductive capacity is the capacity to *realize* the progeneration of developmental capacity in new reproducers. A sterile 'offspring' can be said to be an offspring in the restricted sense that it inherits an incomplete developmental capacity. If developmental biologists consider a sterile but otherwise somatically matured organism to have fully developed, this fact (if it is a fact) indicates that developmental biologists demarcate development and reproduction according to within and among generation processes and capacities differently than does my extended concept. While my proposal may seem counter-intuitive and even to offend ordinary usage, the traditional demarcation leads to equally counter-intuitive results. A fertile adult male moth incapable of feeding but which is capable of reproducing would on traditional views be considered fully developed, yet 'maturation' or 'adulthood' in this case *incorporates* the aspect of reproductive capacity in order to discount the inability to feed as not counting against 'full' development (compared, say, to a female of the same species). The traditional demarcation of development and reproduction does not explain why reproduction is part of the concept of 'full development' in some cases but not others.

A population of reproducers is the minimal unit of evolution (see Millstein, 2009, 2010 on a concept of population that can be generalized to serve present purposes). Reproducers in such a population can vary, can produce offspring differentially, and can 'transmit' properties by conveying developmental capacities to offspring, so they are capable of fulfilling Darwin's principles for evolution by natural selection (Darwin, 1859; Lewontin, 1970). Such populations are capable of fulfilling Godfrey-Smith's criteria for Darwinian populations: 'collections of things that vary, reproduce at different rates, and inherit some of this variation' (Godfrey-Smith, 2009: 107). Godfrey-Smith (2009: Ch. 4) argues that my account of reproduction is sufficient but not necessary for units of evolution, on the grounds that material overlap is not necessary for reproduction. Reproduction might be 'formal' rather than material, i.e. parents can be causally responsible for the determination of offspring form but make no material contribution to offspring, such that parent–offspring correlations are produced without material transfer of parts (Godfrey-Smith, 2009: 79–81). Godfrey-Smith argues that formal reproduction actually occurs in a few kinds of cases: retroviruses, prions, and some transposons. I address his criticism below by arguing that cases which appear not to involve material overlap between parents and offspring, such as retrovirus replication, can be understood as cases of development in complex *molecular* life cycles, in which an offspring virion escaping from a host cell is not the offspring of a parent virion infecting that host but the grand-offspring of several molecular generations intervening between the infecting parent and the escaping offspring virions. Each *molecular* generation involves material overlap and movement to a new molecular or sub-cellular 'host' environment *within the host cell* between each pair of molecular generations. Further aspects of my response to this kind of challenge to the reproducer account are beyond the scope of this essay (but see Griesemer, 2014).

In the abstract account, relations of heredity are formed at minimum between parents and offspring simply because they share parts; hence there will automatically be some parent–offspring correlations ('heritability') in the properties of shared parts (cf. Darwin, 1868: 397). This abstract account can also support further concepts more apposite for contemporary biology by considering *evolved* forms of the transfer of developmental capacities, such as the coding system of DNA replication and transcription/translation (see Griesemer, 2000a).

Systems of heredity in which evolved mechanisms insure that particular forms, structures, or organizations are repeatedly assembled in development may be called inheritance systems (Griesemer, 2000a; on repeated assembly see Caporael, 2003; Caporael et al., 2014). Recent work on epigenetic inheritance points to the possibility of various kinds of inheritance systems evolved from more basic systems of reproduction, including genetic, epigenetic, behavioural, and symbolic inheritance systems (Jablonka & Lamb, 2005). More advanced inheritance systems involve particular mechanisms for controlling the transfer and developmental articulation of offspring structures. One type of advanced inheritance system includes coding mechanisms fundamental to modern genetic systems. Thus reproductive, inheritance, and genetic systems can be interpreted as 'grades' of evolved developmental organization covering a range of reproducers that belong to the kinds of populations known to biologists to constitute potential units of evolution.

Three features of the abstract concept of development introduced above are important to note. First, development is characterized in terms of capacities that can realize reproductive capacity. In this way, the concept is made not to depend on any particular kind, level, scale, or grade of material, structural organization such as a cell or a self-replicating nucleotide polymer. A huge diversity of modes, patterns, and processes of apparent development are already well known to developmental biologists and embryologists. Any of these that cannot, given a range of environments and population contexts, contribute to the reproductive potential of some reproducer or other, is not really a mode of biological *development*. In saying this, I do not exclude from development the elaboration or emergence of 'post-reproductive' (i.e. post-progenerative) traits, since a trait in one reproducer may serve a scaffolding function for another. Traits such as post-reproductive cancers are dead-ends in a developmental sense as well as in Dawkins' (1982) sense of dead-end rather than germ-line replicators.

On this broad characterization of the concept of development abstracted from any particular cellular or other mechanistic mode, one can model development in non-cellular processes such as proto-cell evolution (see Gánti et al., 2003; Griesemer, 2008; Griesemer & Szathmáry, 2009; Szathmáry, 2006). Cultural development resulting from social interactions among biological (and other) constituents can also be modelled as possibly involving reproducers that convey developmental capacities at a cultural level (Caporael et al., 2014; Griesemer, 2014; Wimsatt & Griesemer, 2007).

Second, as noted above, development is recursively linked to the more-making process of progeneration, in which material parts of parents become material parts of offspring. The recursion is an important feature of the view because it conceptualizes the notion of levels of organization in terms of the process of reproduction at a level. For multicellular organisms to reproduce, they must typically develop, but their development typically involves the reproduction of their (cellular) parts, which in turn involves (cellular) development, and so on down to a chemical level at which developmental capacity is an automatic consequence of autocatalytic cyclic chemical synthesis, as in intermediary cellular metabolism, rather than an acquired biological capacity. Note that the recursive feature of development as characterized here does not exclude recursion *up* levels of compositional organization: organisms might reproduce only as their scaffolding contexts develop, which may involve the reproduction of parts of those contexts, and so on (see Caporael et al., 2014 and 'Development of hybrids').

Third, and most importantly for present purposes, although the concept of development is articulated in relation to reproduction, the concept proposed here does not entail the temporal boundary of development rejected by Laplane (2011), nor is it a violation of Minelli's point 5 that development does not necessarily end with the achievement of sexual maturity. The realization of a *capacity* (or skill) does not entail the end of the *process* realizing it nor the beginning of another process exercising it. Capacities and skills may be refined after they are acquired and may even require their exercise/realization in order for refinement or elaboration to occur. Taking death, reproduction, or sexual maturity as traditionally understood as a temporal boundary focusses too much on conditions of the persistence of cellular entities and time of first appearance of traits which, from the reproducer perspective, are better understood as conventions for modelling capacities rather than basic to the concept of development.

The process by which I learn to ride a bicycle need not end at the point when I can balance while moving forward: my skill at riding increases well beyond that point with practice. Those refinements may have consequences for other downstream developmental processes such as hand, foot, and eye coordination, ability to shift attention on short time scales, ability to ride safely in traffic, and even ability to teach others to ride. In other words, the *refinement* of a capacity beyond its acquisition, initial exercise, and first and often only realization may be continuing aspects of the process that generates it.

Nor does the concept proposed here entail the kind of 'teleological' dependency of development on evolution rejected by Minelli (2011b) in favour of an account of development for its own sake.

The proposed concepts of development and reproduction were designed to address limitations in treatments of units of heredity and evolution that presupposed evolved characteristics of genes (e.g. in Dawkins' and Hull's replicator concepts; see Dawkins, 1982; Hull, 1988; cf. Griesemer, 2005). As such, the concepts of development and reproduction proposed here are, by design, conceptually *prior* to a concept and principle of evolution. On the reproducer account, development and reproduction are characterized completely independently of evolution and of any *particular* material basis or mechanism for reproduction by incorporating a *general* material condition for the propagation of developmental capacities. There is no restriction on the mechanistic means through which the exercise of developmental capacities achieves reproductive capacity, save that it must occur through material transfer of parts, nor in terms of the evolutionary potential, fitness, or fate of varying reproducers in populations. More specifically, developmental capacities can realize other developmental capacities, which thus can form causal sequences, the full realization of which results in a reproductive capacity. Development and reproduction are continuous and coincident *aspects* of a process that unfolds over a whole life trajectory. In what follows, I discuss complex multi-'host' or multi-site life cycles in order to clarify the refinements sketched above and to address issues raised by Minelli (2011b) and Godfrey-Smith (2009).

Development in complex life cycles

The view of development as the acquisition (and refinement) of a capacity to reproduce described above assumed a simple life cycle in which there is a straightforward progression of progenerating and becoming a reproducer. A developmental capacity, D1, is transferred through material propagules from parent to offspring reproducers that is then either exercised in the offspring to realize reproductive capacity, R, or whose exercise results in a cascade of developmental capacities: realization of D1 produces D2, realization of D2 produces D3, . . . , realization of Dn produces R. Further developmental capacities may be produced and realized even after R by refining R. (The recursive aspect of acquiring a developmental capacity Dj from the exercise of a developmental capacity Di involving the reproduction of components of the sub- (or super-) system that has or carries Di had been set aside for purposes of this discussion.) The entity or process bearing each capacity was presumed to persist through each successive capacity.

Complex life cycles challenge this framing of a concept of development in terms of the acquisition of capacities, because capacities are dispositions that depend on context. Development in complex life cycles involves *multiple* contexts. How we understand and interpret developmental capacities also depends on the entities to which capacities are attributed via a concept of development: the system as a whole or only some among its parts. The key conceptual question is whether to think of developmental capacities as properties of life cycles or of 'organisms' in lineages of reproducing entities. In complex life cycles, multiple 'organism' trajectories or generations are recognized within the span of a single life cycle that returns the *lineage* to a given developmental or reproductive state.

Here I focus specifically on the developmental status of the 'hybrid' entities formed in reproduction processes spanning more than one 'organism' generation across complex life cycles involving multiple developmental contexts. 'Organism' is in quotation marks because I seek a general, integrated theory of units of evolution, not one limited to the evolved properties of the organism grade of organization delimited traditionally from egg or zygote to adult, or in terms of mitotic or meiotic events. 'Hybrid' is also in quotation marks because I seek a general theory of units of development, not one limited to cellularity or other traditionally delimited grades of developmental organization requiring specific mechanisms of hybridization, such as the combining of genetically distinct chromosomes inherited from sexually reproducing parents.

Since 'development from the hybrids' was the problem Mendel claimed his theory sought to address, resolution of questions about the concept of development have direct bearing on concepts of hereditary factors as well (see Griesemer, 2007). Some of the implications of this linkage between heredity and development are explored below in terms of two widely described examples: the reproduction and development of malaria-causing *Plasmodium* species, a group of apicomplexan cellular parasites with complex life cycles, and HIV-1, a retrovirus that I suggest has an equally complex life cycle, viewed at a molecular or sub-cellular level.

Development of hybrids

In this section, I revise the general account of reproduction to accommodate complex life cycles leading to consideration of the central role of states of hybrid, scaffolded offspring for understanding development. Here, the concept of a 'hybrid' is meant in the broad sense of biological systems incorporating parts of different provenance (lineage, genealogy) rather than hybrid specifically in genetic state. The goal is to understand hybridization (progenerative fusion and scaffolding events) as delimiting generations of hybrid individuals in complex life cycles rather than as mere transitional phases between genetic parents and offspring completing a life cycle (see also Griesemer, 2007, 2014; Wimsatt & Griesemer, 2007).

I briefly discuss the life cycle of apicomplexans such as *Plasmodium falciparum* (one of the malaria-causing species) to explore the implications of complex life cycles for conceptualizing development in familiar cellular terms. I introduce a further concept of scaffolded development to distinguish several ways in which material overlap relations can figure in reproduction. Then, I argue that extension of the refined concept to non-cellular cases, such as the *molecular* development of the retrovirus HIV-1, reveals how material overlap relations between parents and offspring can be understood to hold in cases where it appears there is none. The argument reveals a conceptual trade-off between invoking formal reproduction and thus downplaying the hybrid formation and scaffolded development of offspring that complete a life cycle vs retaining material reproduction and thus driving a heuristic search for modes of reproduction that involve non-traditional, sometimes non-cellular modes of scaffolded 'hybridization'.

Virion particles are the closest thing to a cellular stage in the life cycle of HIV. I argue that escaping virion particles in HIV replication are, in a complex life cycle, not the offspring, but the grand-offspring of invading virions; thus, one should not expect to find material overlap relations between invading and escaping virions. (The material overlap relation is reflexive, symmetric, and not necessarily transitive, so even if there is material overlap between parent and offspring, there need not be one between grandparent and grand-offspring for material overlap relations in a lineage to constitute the relevant bond of 'material continuity'.) The mere fact that the material propagules produced in the steps in the process taking place inside the host don't resemble virions is no more important to the analysis of retrovirus development and reproduction than the fact that winged forms of insect species or spring wing-colour patterns in seasonally polyphenic butterflies don't resemble their unwinged or autumn-coloured parents (reviewed in Gilbert & Epel, 2009), nor the fact that human sperm and eggs don't resemble their adult parents in outward morphology. More centrally, tracking the sources of the raw materials that form a hybrid is only one way to track material overlap relations. Another is to track the sources of developmental capacities that are propagated through scaffolding interactions that form material hybrids.

On the reproducer view, enhanced by general concepts of hybrid systems and scaffolded development, most life cycles turn out to be at least somewhat 'complex', and thus there may be offspring generations in between what are conventionally called the parent and offspring generations of a single reproductive cycle. We may say that human parents do not materially overlap 'their offspring' as fairly as retroviral parent RNA does not materially overlap 'its offspring' retroviral RNA, because in both cases, there is an intermediate generation of progenerants: haploid gametes in the case of human (cellular) reproducers which then fuse to form the *gametes'* offspring, a zygote; and double-stranded DNA helixes and single-stranded RNA helixes in the case of HIV virus parents and grandparents of the offspring RNA. Differently put, if we mark generations of reproducers according to material overlap relations conveying salient developmental capacities rather than according to generations of (cellular) organisms or mitotic or meiotic cycles, then it turns out all traditional life cycles are complex in the sense that there is typically at least one substantial change of developmental context or niche, with multiple generations of progenerants, before a life cycle has been completed.

The development of hybrids is central not only to a theory of development, but also to the whole theoretical project of modern genetics (Griesemer, 2007). Mendel's question was: what law can explain patterns of plant development of, and from, the (multicellular) pea hybrids formed from crosses of pure-breeding lines? There are many sorts of articulations of lineages in which the system hasn't quite resolved into trackable hybrid individuals. This is not just a matter of scientists lacking tools for tracking the kinds of progeny produced in breeding experiments but also a matter of the system not having resolved relations among its parts into (modular) mechanisms that can function as traceable individuals, despite or because of their hybridity. Here, I generalize hybridity in the context of complex life cycles to generations of progenerants of whatever morphology—cellular or not—produced by the articulation of multiple lineages of parents conveying developmental capacities of whatever morphology—again, cellular or not. Assuming cells are delimited by lipid membranes, I interpret HIV as a non-cellular, molecular parasite that forms multiple molecular generations of molecular progenerants that develop in complex cytoplasmic and nuclear molecular host environments situated between virion generations.

The complication of complex life cycles

As Minelli (2011b: 5) points out, many uni- and multicellular organisms have complex life cycles that involve multiple hosts. In some cases, such as the several malaria-causing *Plasmodium* parasites, the organisms in a given reproductive lineage develop different morphologies and can reproduce by different means in each host environment. It is critical to bear in mind that life cycles may span more than one organism or cell generation. We should not be misled by the *characterization* of familiar life cycles as involving exactly one organism, cell, or more generally, one reproducer generation, especially when tied to a criterion of cell or organism generations such as mitosis/meiosis events in the cell cycle or the kinds of criteria Minelli and Laplane have criticized. In general, the life trajectory of a single developer may not encompass an entire life cycle, even if cellular boundaries around development are relaxed.

Minelli seeks a concept that serves a theory of development for its own sake focussed on developmental processes as the subject matter of developmental biology, regardless of outcomes specified by other disciplines such as evolutionary biology or genetics. My account appears to treat development in the instrumental, evolution-dependent way criticized by Minelli. Here, I consider several complications in developmental processes pointed out by Minelli that challenge the generality of the relationship I characterized in terms of simple life cycles.

Metamorphosis. Minelli (2011b: 7) regards talk of set-aside cells in some modes of metamorphic development (*Drosophila* cell lineages fated to produce imaginal discs, for example) to be 'adultocentric' because it characterizes development in terms of the prospective fate of the set-aside cells in contrast to other cells that constitute the part of the soma fated not to persist beyond the metamorphic transition.

Non-disc larval parts and characters can, however, serve indirectly in the acquisition of developmental capacities by 'set-aside' cells, e.g. in larval feeding that not only maintains larval metabolism but acquires and delivers enough of the right kinds of nutrition and structural scaffolding to fuel the cell and tissue growth and development necessary for imaginal discs to form, develop, and metamorphose. 'House-keeping' sensory functions might also transduce sensory cues via neuroendocrine pathways into *developmental* signals (see Gilbert & Epel, 2009). Somatic characters functioning to trigger, signal, or scaffold development should therefore be counted as developmental components characterized not in terms of their own very different cellular reproductive fates but rather in terms of different developmental capacity-forming functions. The idea is that specific forms of 'housekeeping' behaviour (metabolic, sensory, predator-avoidance, mate-seeking) may have either direct or scaffolding developmental functions as well, so there is no need for the reproducer account to distinguish types of cellular function in terms of reproductive fate rather than present developmental function.

A key question raised by metamorphosis for the reproducer account of development concerns the developmental role of the non-'set-asides'. The notion of scaffolding helps to differentiate two kinds of developmental capacities among cells. Imaginal disc cells participate in development by directly giving rise to the somatic cells of the adult animal. Cells in lineages excluded from imaginal discs may be understood to have developmental roles as scaffolding for the disc cells, i.e. serving as environments that facilitate acquisition by the discs of the developmental capacities that can produce adult structures (presumably by setting positional, transcriptional regulatory contexts for cascades of gene expression).

This interpretation of the acquisition of developmental capacities in terms of relations between scaffolding and scaffolded tissues or cell lineages effectively pushes the 'eco-devo' perspective inside the body of the fly. We can see that the mere fact that a cell does not contribute its own material parts to adult *cellular* structure does not mean that it lacks a developmental role or capacity from the reproducer perspective. As scaffolds, the non-disc cells and tissues may facilitate disc development, and, since they bear material overlap relationships back down their cell lineages to the zygote, the propagation of developmental capacity as described above still holds. It is a refinement of the reproducer perspective nevertheless to recognize that there are two kinds of developmental capacity in metamorphic development: 'direct'—providing for further developmental capacity acquisition through propagation of cells via material overlap to offspring cells forming adult structures—and 'scaffolding'—providing facilitation of development via the formation of hybrid disc/non-disc intermediates that enable the cells with direct developmental capacities to pass through metamorphosis and form adult tissues.

Alternation of (progenerant) generations. Alternation of morphologic states across stages of a life cycle seems to pose a different kind of problem for adultocentric conceptions of development and for the account of reproduction described above. Although a certain set of components are configured to produce a certain sort of adult body (a wingless body, say), those same (genetic and cellular) components in a similar starting configuration may yet produce a different sort of adult body (winged, say) if exposed to a different external environment (e.g. a different seasonal, light, or temperature regime). Many sorts of polyphenism induce quite different developmental outcomes in similar sets of starting developmental materials (see Gilbert & Epel, 2009 for a review containing many examples). The entire explanation of these developmental outcomes must be in terms of the external, environmental differences, because these are what make the developmental difference as the environment triggers direct development down one path or another. Identifying the triggering events, however, fails to explain much of the differences in developmental process or outcome, because most of the working parts of the developmental mechanism are inside the organism. The eco-devo solution expands the notion of developmental system beyond the organism to include the triggering environment, but it does not trace the consequences for the concept of reproduction of developmental systems.

Alternation of morphological generations looks problematic from the reproducer perspective because developmental capacities that realize a particular mode of reproductive capacity are outside the developer, i.e. beyond the system's spatial boundaries. Eco-devo's (and developmental systems theory's) solution is to expand the notion of a developer or developmental system to include those 'environmental' mechanisms that provide developmental feedback but again without clarifying how the expanded mechanisms figure as part of a system that must not only develop but reproduce. The reproducer perspective recognizes this expansion as well as analogous and other kinds of feedback relationships *within* the organism by identifying certain kinds of somatic structures as internal developmental scaffolds and uses that insight to extend the view to reproductive systems that potentially include non-cellular, non-organism 'parents' beyond traditional cell or organism reproducers.

Within a multicellular embryo, a tissue can provide a material substrate with specific cell surface molecules that interact with an informational gradient that both instructs cells within the tissue to acquire various different developmental fates and provides positional signals to cells migrating outside the tissue. For example, during gastrulation in the vertebrate embryo, the ectoderm expresses bone morphogenic protein (BMP) receptors. These receptors interact with a gradient of BMP activity that is established in the ectoderm by BMP antagonists secreted by the dorsal mesoderm; as a result, the ectoderm responds to this gradient by forming neural plate at regions of low BMP activity, neural crest cells at regions of intermediate BMP activity, and non-neural ectoderm at regions of high BMP activity. The BMP activity gradient within the ectoderm also functions to guide neural crest cells as they migrate to positions where they can differentiate into a variety of vertebrate-specific tissues (Huang & Saint-Jeannet, 2004). The mesoderm thus functions as 'internal-' or 'self-' scaffolding for the development of neural plate, neural crest cells, and non-neural ectoderm from the point of view of the multicellular organism as a developmental system, or as 'external' scaffolding from the point of view of the neural crest cells as reproducers forming cell lines within the organism in an environment of developing tissues.

The developmental capacities of an organism of a given generation appear to serve not only the acquisition of developmental and reproductive capacities of the material body that carries them, but also the developmental and reproductive capacities of offspring that develop very different morphologies. A seasonally polyphenic grasshopper is a spring grasshopper because its parent was an autumn grasshopper, which raises the probability that its offspring develops in spring. Acquisition of an adult form depends on the environments in which their parents found themselves and the developmental capacities (including the probability of experiencing a different developmental environment than they did) that their parents passed on to them.

Alternation of generations only looks like a problem for the reproducer perspective by assuming that a single life cycle must be that of a single developer, whose temporal boundaries are set by birth and death (or reproduction) events in that life cycle. As argued above, the reproducer perspective does not endorse these boundaries nor the interpretation of reproduction as an event in the life cycle. The fact that different developmental capacities are conveyed to offspring who grow up in a different environment than their parents is not precluded by the reproducer account. Moreover, the reproducer perspective supports a separation of continuous reproduction processes into generations by tracking material overlap relations that convey developmental capacities to progenerant material 'offspring'. From this perspective, many of the transformations traditionally identified as 'developmental' *rather than* reproductive may need to be rethought.

Parasite development in complex life cycles

Although they are special cases, parasitic life cycles epitomize the complexity of development and its dependency on environmental feedback. The many detailed studies of parasites which threaten human health have revealed fundamental aspects of development that otherwise tend to go unnoticed; medical researchers are intent on finding out about every point in the life cycle where parasites might be stopped.

According to the reproducer account, development of any biological complexity tends to be internally scaffolded because external environments, traditionally conceived, tend to lack enough temporal and spatial organization to assemble developing systems from simple raw materials. Biological reproduction starts from highly organized propagules, but even so, whenever agents or structures in the environment can act or operate to reliably organize developmental processes, there is an evolutionary benefit in taking advantage of them as scaffolds.

The term 'scaffolding' refers to facilitation of a process that (1) would otherwise be more difficult or costly without it (Bickhard, 1992, 2005, 2007); (2) would be acquired with lower quality, fidelity, or reliability without it; and (3) tends to be temporary: an element of a maintenance, growth, assembly, development, or construction process that fades away, is removed, or becomes 'invisible' even if it remains structurally integral to the product (Caporael et al., 2014; Griesemer, 2014). Progenerated developmental organization transferred to offspring by means of material propagules together with a scaffolding context jointly ensure (or raise the probability) that offspring have or acquire a salient capacity to develop in a specific environment (or range of environments).

Apicomplexan reproduction. Malaria-causing parasites such as *Plasmodium falciparum* pass through two very different kinds of host organisms to complete their life cycle: humans and mosquitoes (Centers for Disease Control 2009; National Institute of Allergy and Infectious Diseases, 2012; Wiser, 2000). I review this complex life cycle briefly to point out that there are multiple cellular parasite organism generations per life cycle, each associated with a different developmental environment. Many of these environments scaffold parasite development by providing a context in which the parasite can gain a developmental capacity and reproduce that it would not so easily acquire in the otherwise lethal environment of the host.

Most life cycle narratives for parasites begin with a free living or adult form that infects a host; but this victim-centred idea of a unique starting point is problematic in describing a cycle, and more so as the number of hosts included in a complete life cycle of the parasite increases. Female *Anopheles* mosquitoes carrying a malaria-causing parasite feed on a human, injecting sporozoites into the bloodstream. These invade liver cells within seconds or minutes, or they are attacked by the human immune system. As they invade the protective environment of the liver, the sporozoites shed the apical structures that permit them to invade liver cells and form a vacuole in which to undergo schizogonic development, shedding organelles and performing multiple rounds of nuclear division; they then segment into asexually reproduced, separate cells that differentiate, a process involving other changes of organelle structure, into merozoites. The merozoites exit into the blood stream, where they invade blood cells and develop the ability to feed (on host cytoplasm and haemoglobin); at this point they are called trophozoites. Upon invading blood cells, again forming a membrane-bound vacuole, the parasites cause changes in the host cell membrane and cytoskeleton that facilitate parasite feeding on (mostly) haemoglobin.

The parasite acts to increase host membrane permeability so that digested haemoglobin not needed for metabolism by the parasite can be excreted rather than remaining internal to the host and causing osmotic host cell bursting prior to the completion of parasite development (Wiser, 2000). The parasite undergoes multiple rounds of asexual reproduction to yield merozoites that can invade new blood cells, forming a 'sub-cycle' of the overall life cycle, as well as a sexual phase of 'gametocytogenesis' that produces male and female gametocytes.

Gametocytes break out of blood cells and bind to the human blood glycoprotein Factor H, wearing it like a hat to prevent digestion of the gametocyte as it circulates in the host blood awaiting ingestion by a female *Anopheles* mosquito during a blood meal. Parasite gametogenesis and fertilization is completed in the mosquito gut to form a diploid zygote, which develops into a mobile form (ookinete) that invades the insect midgut wall, where it completes the life cycle by forming an oocyst. The oocyst then undergoes asexual reproduction to form multiple haploid sporozoites that migrate to the mosquito salivary glands so as to infect a human host once more.

Depending on how one counts, there are around six generations of sporozoite, merozoite, gameotcyte, and zygote cellular parasite organisms in each complete life cycle. Human bloodstream, liver cells, and blood cells (and the special vacuoles inside these cells), along with mosquito saliva, haemolymph, gut wall, and salivary glands each constitute a distinct developmental environment or niche that scaffolds one of these parasite generations to acquire specific developmental capacities that facilitate movement to the next stage (generation). Most of these scaffolding interactions involve modifications of both host and parasite. All of them involve transfers of material parts conveying developmental capacities from the scaffolded parent to the offspring (or scaffolding parents, if one is willing to count the scaffolding host parts as contributors to offspring host–parasite hybrids; see 'Scaffolded development').

The vacuole membranes formed when the parasite invades human liver and blood cells may even be a hybrid sub-cellular structure including both host and parasite components (see Wiser, 1999). Malaria parasite lineages passing through a series of hosts presenting very different environments for development resembles a cell-level version of the molecular HIV life cycle.

HIV-1 reproduction. I applied the reproducer perspective to the HIV-1 replication cycle elsewhere

(Griesemer, 2014). Here, I summarize a few points to emphasize the value of treating the virus 'life' cycle as complex and scaffolded ('life' in quotation marks to signal that viruses are at a dubious 'edge' of the living state).

The standard narrative of HIV replication begins with an invading virion ('mature' virus containing an RNA genome) binding a CD4 receptor and G-protein co-receptor of a host cell, e.g. a human helper T-cell (see e.g. Scherer et al., 2007; discussed in Griesemer, 2014). The virion fuses with the host cell membrane. The virus nucleocapsid is inserted into the host cell cytoplasm where it is uncoated. The RNA genome is reverse transcribed to double-stranded DNA with host nucleotides by a viral enzyme that came in with the virus genome. The parent virus matrix protein, integrase enzyme, and DNA genome are assembled into a pre-integration complex that is transported to the host cell nucleus. The 'provirus' (DNA genome) is integrated into the host cell genome. There, the DNA virus/genome is transcribed to RNA, some of which serves as mRNA coding for proteins, such as a protease, that aggregate at the host cell surface, along with some of the new RNA serving as genomes. The synthesized virus parts assemble at host cell membrane raft structures recruited by transmembrane virus envelope proteins and are then budded out of the host cell as new virion particles. The newly budded particles are not yet infectious, as the protease must cleave a polyprotein inside the virion to prepare it for infection.

At each of the 'stages' of replication described above, a new, structured object is formed in a different region of the host cell: at the host plasma membrane, in the cytoplasm, in the host nucleus, again in the host cytoplasm, and again at the host plasma membrane. At each stage (i.e. passage through a distinct developmental environment), the new structured object ('offspring') is formed that bears a relation of material overlap to the object(s) from which it was formed. In some cases, the material overlap is merely the relation of a (former) whole to a subset of (subsequent) parts. For example, some objects are formed by stripping off some infecting virus parts that make it inside the host cell (and which are degraded) leaving others exposed to the cytoplasm (early stages). Some objects are synthesized hybrids of infecting viral genome and host nucleotides (RNA–DNA hybrids formed by templating, which is a form of scaffolding). Some are synthesized hybrids of a daughter 'virus' DNA strand from the RNA–DNA hybrid and more host nucleotides that form double-stranded DNA 'virus'. (HIV is called an RNA virus; yet at one stage of 'its' life cycle, it is comprised of DNA, not RNA.) Some are synthesized hybrid DNA–RNA in the host nucleus at the host genome via transcription to form 'offspring' 'virus' mRNA and 'virus' RNA genome, by assembling at the host cell membrane and taking a bit of host membrane with the rest as they bud to form a new virion outside the host cell.

'Virus' is in quotation marks because at each of these stages, the new objects formed are scaffolded by material in distinct regions of the host cell: (1) the cytoplasmic milieu in which the virus nucleocapsid can be uncoated to expose the RNA genome; (2) the RNA genome and host cytoplasm scaffolding the synthesis of a complementary DNA strand that originates as a part of an RNA–DNA hybrid; (3) the DNA strand and host cytoplasm scaffolding the synthesis of a complementary DNA strand forming a DNA–DNA hybrid; (4) the DNA and nucleus scaffolding mRNA synthesis via a DNA–RNA hybrid; and so on. Each point of formation of a new molecular 'offspring' is also a place where the scaffolded development of ancestral parts are hybridized with parts from the host cell to form 'offspring' progenerants that then have new developmental capacities: to move (or be moved) to a new host cell region and to undergo scaffolded development with different hybridizing material facilitating the production of still more 'offspring' progenerants.

Differently put, each scaffolded stage in the process initiated by virion infection results in the articulation of a new generation—provided generation is recognized, not as *genetic* ancestry in particular but as demarcated by material overlap relations in which the overlapping parts acquire a new *developmental* capacity. In so far as developmental capacities can form the core of different kinds of 'inheritance system' (Jablonka & Lamb, 2005), the abstract reproducer account of development characterizes its structure and identifies it for evolutionary investigation as a potential member of a population that forms a unit of evolution.

HIV retrovirus replication is sometimes claimed as an example of 'formal' rather than 'material' reproduction on the grounds that the formation of an RNA–DNA hybrid by reverse transcription shows that there is a barrier to material overlap between parent virus and offspring virus (see Godfrey-Smith, 2009: Ch. 4). This interpretation, however, assumes that the DNA strand is not virus while the RNA strand is. There is no disagreement that the host is the source of the DNA nucleotides (see Griesemer, 2014). But the interpretation of a barrier step to material overlap amounts to a form of polynucleotide essentialism which treats the virus as having an RNA genome essentially and the host cell as having a DNA genome essentially, thus treating the RNA–DNA hybrid as a nonentity without a developmental status, a mere representational figure expressing the interaction of two essentially different entities. The reproducer perspective rejects such an essentialist, genetic-information-centred view of the matter. RNA–DNA hybrids have their own chemistry, their own developmental capacities and niches, and their own evolutionary fates. The fact that this hybrid, like transition complexes in many chemical reactions, persists on a short timescale relative to virion particles is not a good reason to discount it as a material offspring of a generation intermediate between infecting virion and escaping virion.

Instead, each new hybrid in which materially overlapping parts carry and realize new developmental capacities is recognized from the reproducer perspective as a progenerant entity of a new *developmental* generation. Instead of mitosis or meiosis, or cell membrane partition as a general criterion marking generations, distinct developmental capacities carried by material propagules distinguish generations of progenerant individuals. In a life *cycle*, any developmental stage can be taken as a 'starting' point (see Griesemer, 2014 for further discussion). The narrative above treated the mature virion just before infection as the life cycle's starting point, making the RNA–DNA hybrid appear to be a mere transitional state between 'mature' or 'adult' virions. From that narrative perspective, a host cell is a virion's way of making another virion. Instead, consider the RNA–DNA two strand hybrid as the narrative starting point. From that perspective, the virus *and* the host cell are jointly the hybrid's way of making another hybrid. From the latter perspective, it seems clear that there *is* a material overlap relation between its DNA part and the DNA integrated into the host genome and also a material overlap relation between the parental virion's RNA genome and the RNA–DNA hybrid. Thus, the RNA–DNA hybrid is in a developmental generation of its own, situated between single-stranded RNA and double-stranded DNA generations.

Scaffolded development

Application of the reproducer perspective on development, as prompted by consideration of the examples of complex parasitic life cycles described above, can be summarized as follows.

Ecological developmental biology expands the scope of developmental biology to consider the role(s) of environmental inputs to normal development. Some inputs are mere 'triggers', having little structure of their own that can guide or facilitate development, so their developmental effects depend mainly on the structure and organization of the triggered developmental system. Regulatory molecular cascades triggered by environmental sensory or metabolic inputs depend for their developmental consequences on transduction of the triggering signal by the developer's system (reviewed in Gilbert & Epel, 2009: Ch. 2).

Other environmental inputs are 'scaffolds', material environmental inputs with organizations that are sensitive and responsive to the developmental state of the developmental system being scaffolded. I here distinguish organization from structure. A structure is an arrangement or configuration of parts that does not change on some timescale of interest. An organization is a structure that changes (or is in dynamic equilibrium) on a timescale of interest. Organizations of scientific interest tend to be ones that maintain or change structure dynamically in some particular respect. A dissipative structure, such as a stream of water out of a tap for example, is an organization that maintains a structure at a gross level by a continual flow of matter and energy through the system, by continually reorganizing the system on microlevels.

A developing system may have evolved mechanisms that adapt it to scaffolding developmental environments, as malaria-causing parasites typically do, which allow them to modify the host cell to be more responsive to the parasite's developmental needs, e.g. the vacuole membrane, which might be a hybrid structure of host and parasite parts, that facilitates parasite growth and development and at the same time prevents host cell death that would otherwise result from the parasite's presence. Some scaffolds are so key to development that they are 'internalized' parts of the developing system itself. Internal scaffolds, such as the template polynucleotide systems of all cellular life, have specific structures relevant to their developmental effects on the templated developmental systems, which may or may not themselves be responsive to the developmental state of the system. Increasingly, genomes are viewed as responsive systems (e.g. Griffiths & Stotz, 2013). DNA sequence per se is (mostly) not responsive to development in the sense that, in most cases, there appears not to be genome-level DNA sequence editing, but it may be used as a scaffolding 'infrastructure' by the developmentally responsive transcriptome. HIV RNA replication is hypervariable, due to the inaccuracy of the reverse transcriptase, which might count as a developmental response system in so far as that variability lowers the probability of an effective host immune response.

Complex life cycles, as traditionally understood, are ones in which multiple environments of development figure in the developmental processes of reproducers which must pass through more than one 'organism' generation to complete an entire life cycle. The offspring of *Plasmodium* organisms that developed in mosquito inherit developmental capacities facilitating their development in humans, e.g. the specific sensory apical apparatus that allows them to get out of the dangerous human bloodstream and into the shelter of a liver cell and the ability of offspring escaping blood cells to put on a hat consisting of host protein to protect them from immune system attack while they wait in the human bloodstream for uptake in a mosquito's blood meal.

The reproducer perspective suggests an expansive notion of 'pro-generations' when considering lineages of reproducers, the production of entities bearing material overlap relations carrying developmental capacities to their progeny. HIV, I suggested above, is a molecular parasite as much as a virion particle parasite because many of the 'stages' (progenerant generations) in the HIV life cycle are molecular rather than virion or quasi-cellular in character. The reproducer perspective on generations contrasts with the traditional notion of 'organism' and 'genetic parents', in which one or two parent organisms are considered the genetic ancestors of each offspring organism because they provide a specific internal-templating scaffold which other 'environmental' inputs to development do not. The traditional view is narrower because genes are a special kind of evolved class of scaffolding developmental mechanisms that exploit template-coding mechanisms to carry and transmit developmental capacities. On the reproducer perspective, 'developmental parents' provide a structured developmental environment in which material parts of the parents become the material (or material parts) of offspring, which embody mechanisms of development that carry the capacity to reproduce (perhaps in virtue of carrying capacities to develop the capacity to reproduce). Complex life cycles involving multiple developmental environments include multiple developmental 'pro-generations' of progenerant entities related by material overlap, which is a broader criterion than material overlap by genetic ancestry.

Progeneration of parasites, especially those that invade host bodies, are cases where it is easy to identify scaffolding processes and relations through which the host provides a structured environment that facilitates the parasite's development and reproduction. The identification of scaffolding processes in development is salient to the expanded reproducer perspective because scaffolds not only act to facilitate development but do so typically by the formation of a temporarily hybrid material entity composed of the scaffold and the developing system. In traditional accounts, scaffold and developing systems are both cells or organisms whose identities are held separate and distinct as units of development.

Using material overlap of parts carrying developmental capacities, or 'progeneration', as a criterion delimiting generations of progenerant entities, hybrids can be seen to be the 'offspring' of an environmental scaffold and a developer, which in turn are the 'developmental parents' of the hybrid. Although it seems a step too far to treat the hybrid combinations of *Plasmodium* and human erythrocyte cells or HIV virion particles and human T-cells as 'offspring' and the *Plasmodium* and erythrocyte or virion and T-cell as 'parents', this expansion does point the way towards consideration of reproductive systems that are neither tracked nor organized by DNA- or RNA-template replication systems, cells, or organisms (Griesemer, 2014). As I argued at the outset, this move is needed conceptually in order to explain origins and transitions of these advanced grades of contemporary life without begging the question of whether they evolved because concepts of evolution, reproduction, or development are delimited in terms of those advanced, evolved grades of organization.

Scaffolds operate to facilitate development at points in life cycles where such hybrids form, by definition, since scaffolding is a form of hybridization with developmental implications for the constituent entities or for the new, fused hybrid entity. On the broader notion of generations delimited by material overlap relations rather than by more specific genetic overlap relations, we might attempt to recognize potential lineages of reproducers of kinds that may be vastly different from familiar forms of life. Life before cells (Griesemer, 2008; Griesemer & Szathmáry, 2009), life after symbiogenesis, life as a community of micro- and macroorganisms, all must look radically different than the paradigms of 'organismal' life with which we are most familiar. In addition to describing a perspective that can accommodate all of the familiar forms of life, we need a concept and a theoretical perspective on development and reproduction that can accommodate potential forms of life still more radically different, because we want to know if the forms of life which have been and are being evolved truly are endless. We also want to know if there are forms beyond the realm of biology as we know it, most wonderful in the sense that they reproduce, even if not most beautiful in the sense that they lack the compact, coherent, integrated, autonomous characteristics of organisms. Did life precede cellularity? Does life extend to complex organizations comprised of what we now recognize as whole 'organisms' as parts?

Conclusion

I proposed at the outset that theories be thought of as comprised of three kinds of components: a set of core principles, a family of models, and a theoretical perspective. In this chapter, I characterized a very abstract concept and general principle that development is the acquisition of the capacity to reproduce, where reproduction involves material propagation of developmental capacities from parents to offspring. I further proposed that development often involves scaffolding interactions in which components of 'the environment' that form a developmental niche for a developing system hybridize (form a material complex) at least temporarily with it, generating entities with novel developmental possibilities due to structures and organizations that are typically quite different from those of any of the contributing parents. The developmental potentialities of a zygote are quite different than those of gametes. The developmental potentialities of an HIV provirus in DNA form integrated into a host T-cell genome are quite different than those of an HIV pre-integration complex or a free virion. I took the perspective of the hybrid rather than that of the invading parasite or invaded host in narrating the complex parasitic life cycles, including the cellular malaria parasite *Plasmodium falciparum* and the molecular parasite HIV-1.

Concepts and principles, however, only take us part way towards a theory of development. On analogy with Darwinian evolutionary theory, Darwin's principles provide a framework for developing a theory of evolution, but only in a very abstract formal sense could Darwin's principles be taken to *be* Darwin's theory. The development of the neo-Darwinian modern synthetic theory of evolution, for example, required the integration of Darwin's principles with principles of genetics, the articulation of families of models of genetic transmission, population genetics, adaptive and non-adaptive change, speciation, and more (Mayr & Provine, 1980; Provine, 1971). I think that a further complementing element is also needed for a theory to be *empirical*, since models per se need not involve any empirical claim or hypothesis. Darwin's principles of natural selection are stated conditionally: 'if it be in any degree profitable to an individual of any species'; the principles do not make the further empirical claim that any natural system actually conforms to them. Theories gain empirical content through theoretical hypotheses asserting that natural systems bear the right relation to some model included as part of the theory (Giere, 1988).

To be *potentially* empirical, a theory also needs a 'theoretical perspective' that can coordinate empirical investigative strategies (Waters, 2008a, b) with theoretical modelling strategies so that models and phenomena can be meaningfully compared (Griesemer, 2000a). Theories, through their theoretical perspectives, function to guide empirical research as well as to serve descriptions and explanations of nature. Theoretical principles do not provide normative or directive force by themselves. There is no 'should' or 'ought' or heuristic directive in Darwin's principles, Newton's principles, or the core principles of any other scientific theory. Put differently, 'theory is as theory does' (Love, 2013). Theories should be understood in terms of what they do, not (only) in terms of formal models of their logical structure. Theories in action have complex structures in virtue of their multiple roles in guiding empirical inquiry, description, explanation, confirmation, prediction, and control. Philosophers such as Giere (1988) suggest generic criteria for comparison and evaluation of the fit of models to empirical phenomena needed for prediction, confirmation, and hypothesis testing in terms of 'degrees and respects of fit'; but clearly a perspective on what *counts* as empirical phenomena to which core principles apply and what *counts* as adequate degrees and relevant respects of fit of models of a family to phenomena requires more specificity than that. The dual commitment to model Darwinian 'individuals' as *organisms* and to track organisms in empirical inquiry is part of one theoretical perspective. The dual commitment to model Darwinian 'individuals' as any of a wider range of entities such as genes, cells, kin groups, breeding groups, or species and to track them in empirical inquiry is part of another theoretical perspective.

Principles usually admit of multiple interpretations, and each interpretation suggests a somewhat different domain of application and possibly a different theoretical perspective on how to coordinate modelling with investigative strategies. If Darwin's principle of fitness or 'profit' is interpreted in light of a Malthusian assumption of universal competition, Darwinian theory looks quite different than if it is interpreted in light of Kropotkin's assumption that cooperation is basic. To make different empirical theories, there must be a commitment to look for competitive or cooperative arrangements in nature and at the same time to model Darwinian processes using principles of competition or cooperation. A conceptual difference is not enough. A concept of development as cellular supports a theoretical perspective in which the replication of HIV is looked on as the formal transmission of virus information via the mediation of a co-opted host cell genome and transcription mechanisms. The molecular phenomena of reverse transcription appear to be 'owned' by either the parent virion (a cell-like entity) or the host cell. That theoretical commitment is manifest in models of the mechanism of virus replication as a barrier to the transfer of parent virus *material* bridged by a flow of genetic information but not the transfer of developmental material.

A change in the concept of development from a cellular concept to a more expansive one based on collectives of materials functionally organized as reproducers does not by itself introduce a new theory of development, but the juxtaposition of novel with traditional concepts (and perspectives) points to new possibilities for modelling and for empirical investigation (e.g. above the level of organisms by considering scaffolded development of social and cultural entities; see Caporael et al., 2014; Wimsatt & Griesemer, 2007). Moreover, it suggests that the 'problem' with developmental biology may not be that it lacks a grand, overarching set of core principles but rather that there are many theoretical perspectives that can guide research. The plurality of theories in the expanded sense proposed in this chapter, however, is only a difficulty if the lack of theoretical consensus is an obstacle to research. It is not at all obvious that a science needs a formalized, exact theory to be successful (Griesemer, 2013).

Acknowledgments

I thank Xóchitl Arteaga Villamil, Bert Baumgärtner, Linnda Caporael, Roberta Millstein, and Michael Trestman as well as Sandro Minelli, Thomas Pradeu, and Michel Morange for comments on the manuscript. Elizabeth Farrell provided much appreciated copy editing. A Herbert A. Young Society Fellowship, UC Davis, 2011–2014 provided financial support.

References

Bickhard, M.H. (1992). Scaffolding and self-scaffolding: central aspects of development. In L.T. Winegar and J. Valsiner, eds., *Children's Development within Social Contexts: Volume 2. Research and Methodology*. Erlbaum, Hillsdale, pp. 33–52.

Bickhard, M.H. (2005). Functional scaffolding and self-scaffolding. *New Ideas in Psychology*, **23**, 166–73.

Bickhard, M.H. (2007). Learning is scaffolded construction. In D.W. Kritt and L.T. Winegar, eds., *Education and Technology*. Rowman & Littlefield, New York, pp. 73–88.

Caporael, L.R. (2003). Repeated assembly: prospects for saying what we mean. In S.J. Scher and F. Rauscher, eds., *Evolutionary Psychology: Alternative Approaches*. Kluwer Academic, Norwell, pp. 71–89.

Caporael, L., Griesemer, J., and Wimsatt, W., eds. (2014). *Developing Scaffolds in Evolution, Culture, and Cognition*. MIT Press, Cambridge.

Centers for Disease Control, Laboratory Identification of Parasites of Public Health Concern, Malaria. URL: http://www.dpd.cdc.gov/dpdx/HTML/Malaria.htm. Last updated 20 July 2009. Consulted 25 April 2013.

Darwin, C. (1859). *On the Origin of Species*. John Murray, London.

Darwin, C. (1868). *Variation of Animals and Plants under Domestication*. John Murray, London.

Dawkins, R. (1982). *The Extended Phenotype: The Gene as the Unit of Selection*. Oxford University Press, New York.

Gánti, T., Griesemer, J., and Szathmáry, E. (2003). *The Principles of Life, with a Commentary by James Griesemer and Eörs Szathmáry* (trans. Vekerdi, L., Czaran, E., and Muller, V.). Oxford University Press, Oxford.

Giere, R.N. (1988). *Explaining Science: A Cognitive Approach*. University of Chicago Press, Chicago.

Gilbert, S. (2001). Ecological developmental biology: Developmental biology meets the real world. *Developmental Biology*, **233**, 1–12.

Gilbert, S., and Bolker, J. (2003). Ecological developmental biology: preface to the symposium. *Evolution and Development*, **5**, 3–8.

Gilbert, S., and Epel, D. (2009). *Ecological Developmental Biology*. Sinauer Associates, Sunderland.

Godfrey-Smith, P. (2009). *Darwinian Populations and Natural Selection*. Oxford University Press, New York.

Griesemer, J. (2000a). Development, culture and the units of inheritance. *Philosophy of Science*, **67**, S348–S368.

Griesemer, J. (2000b). Reproduction and the reduction of genetics. In P. Beurton, R. Falk, and H-J. Rheinberger, eds., *The Concept of the Gene in Development and Evolution, Historical and Epistemological Perspectives*. Cambridge University Press, New York, pp. 240–285.

Griesemer, J. (2000c). The units of evolutionary transition. *Selection*, **1**, 67–80.

Griesemer, J. (2005). The informational gene and the substantial body: on the generalization of evolutionary theory by abstraction. In M.R. Jones and N. Cartwright, eds., *Idealization XII: Correcting the Model, Idealization and Abstraction in the Sciences*. Rodopi, Amsterdam, pp. 59–115.

Griesemer, J. (2006a). Genetics from an evolutionary process perspective. In E.M. Neumann-Held and C. Rehmann-Sutter, eds., *Genes in Development*. Duke University Press, Durham, pp. 199–237.

Griesemer, J. (2006b). Theoretical integration, cooperation, and theories as tracking devices. *Biological Theory*, **1**, 4–7.

Griesemer, J. (2007). Tracking organic processes: representations and research styles in classical embryology and genetics. In M.D. Laubichler and J. Maienschein, eds., *From Embryology to Evo-Devo: A History of Developmental Evolution*. MIT Press, Cambridge, pp. 375–433.

Griesemer, J. (2008). Origins of life studies. In M. Ruse, ed., *The Oxford Handbook of Philosophy of Biology*. Oxford University Press, New York, pp. 263–290.

Griesemer, J. (2013). Formalization and the meaning of 'theory' in the inexact biological sciences. *Biological Theory*, **7**, 298–310.

Griesemer, J. (2014). Reproduction and the scaffolded development of hybrids. In L. Caporael, J. Griesemer and W. Wimsatt, eds., *Developing Scaffolds in Evolution, Culture, and Cognition*. MIT Press, Cambridge, pp. 23–55.

Griesemer, J., and Szathmáry, E. (2009). Gánti's chemoton model and life criteria. In S. Rasmussen, M.A. Bedau, L.-h. Chen, D. Deamer, D.C. Krakauer and P.F. Stadler, eds., *Protocells: Bridging Nonliving and Living Matter*. MIT Press, Cambridge, pp. 481–512.

Griffiths, P.E., and Gray, R.D. (1994). Developmental systems and evolutionary explanation. *Journal of Philosophy*, **91**, 277–304.

Griffiths, P.E., and Stotz, K. (2013). *Genetics and Philosophy: An Introduction*. Cambridge University Press, New York.

Hacking, I. (1983). *Representing and Intervening: Introductory Topics in the Philosophy of Science*. Cambridge University Press, New York.

Huang, X., and Saint-Jeannet, J-P. (2004). Induction of the neural crest and the opportunities of life on the edge. *Developmental Biology*, **275**, 1–11.

Hull, D.L. (1988). *Science as a Process*. University of Chicago Press, Chicago.

Jablonka, E. and Lamb, M. (2005). *Evolution in Four Dimensions*. MIT Press, Cambridge.

Laplane, L. (2011). Stem cells and the temporal boundaries of development: toward a species-dependent view. *Biological Theory*, **6**, 48–58.

Lewontin, R.C. (1970). The units of selection. *Annual Review of Ecology and Systematics*, **1**, 1–17.

Love, A.C. (2013). Theory is as theory does: scientific practice and theory structure in biology. *Biological Theory*, **7**, 325–37.

Maynard Smith, J., and Szathmáry, E. (1995). *The Major Transitions in Evolution*. W.H. Freeman Spektrum, Oxford.

Mayr, E., and Provine, W., eds. (1980). *The Evolutionary Synthesis: Perspectives on the Unification of Biology*. Harvard University Press, Cambridge.

Millstein, R.L. (2009). Populations as individuals. *Biological Theory*, **4**, 267–73.

Millstein, R.L. (2010). The concepts of population and metapopulation in evolutionary biology and ecology. In M.A. Bell, D.J. Futuyma, W.F. Eanes and J.S. Levinton, eds., *Evolution Since Darwin: The First 150 Years*. Sinauer, Sunderland, pp. 61–86.

Minelli, A. (2011a). A principle of developmental inertia. In B. Hallgrímsson and B.K. Hall, eds., *Epigenetics: Linking Genotype and Phenotype in Development and Evolution*. University of California Press, San Francisco, pp. 116–133.

Minelli, A. (2011b). Animal development, an open-ended segment of life. *Biological Theory*, **6**, 4–15.

National Institute of Allergy and Infectious Diseases (2012). Life cycle of the malaria parasite. http://www.niaid.nih.gov/topics/malaria/pages/lifecycle.aspx. Last updated 03 April 2012. Accessed 30 May 2013.

Odling-Smee, F.J., Laland, K.N., and Feldman, M.W. (2003). *Niche Construction: The Neglected Process in Evolution*. Princeton University Press, Princeton.

Oyama, S. (1985). *The Ontogeny of Information*, 2nd ed. 2000. Duke University Press, Durham.

Oyama, S., Griffiths, P.E., and Gray, R.D., eds., (2001). *Cycles of Contingency: Developmental Systems and Evolution*. MIT Press, Cambridge.

Provine, W. (1971). *Origins of Theoretical Population Genetics*. University of Chicago Press, Chicago.

Scherer, L., Rossi, J., and Weinberg, M. (2007). Progress and prospects: RNA-based therapies for treatment of HIV infection. *Gene Therapy*, **14**, 1057–64.

Simon, H.A. (1962). The architecture of complexity. *Proceedings of the American Philosophical Society*, **106**, 467–82.

Sterelny, K. (2001). Niche construction, developmental systems, and the extended replicator. In S. Oyama, P. Griffiths, and R. Gray, eds., *Cycles of Contingency: Developmental Systems and Evolution*. MIT Press, Cambridge, pp. 333–349.

Szathmáry, E. (2006). The origin of replicators and reproducers. *Philosophical Transactions of the Royal Society of London B*, **361**, 1761–76.

Tal, O., Kisdi, E., and Jablonka, E. (2010). Epigenetic contribution to covariance between relatives. *Genetics*, **184**, 1037–50.

Waters, C.K. (2008a). Beyond theoretical reduction and layer-cake antireduction: how DNA retooled genetics and transformed biological practice. In M. Ruse, ed., *The Oxford Handbook of Philosophy of Biology*. Oxford University Press, New York, pp. 238–262.

Waters, C.K. (2008b). How practical know-how contextualizes theoretical knowledge: exporting causal knowledge from laboratory to nature. *Philosophy of Science*, **75**, 707–19.

West-Eberhard, M.J. (2003). *Developmental Plasticity and Evolution*. Oxford University Press, New York.

Wimsatt, W.C. (1987). False models as means to truer theories. In M. Nitecki and A. Hoffman, eds., *Neutral Models in Biology*. Oxford University Press, London, pp. 23–55.

Wimsatt, W.C. (1992). Golden generalities and co-opted anomalies: Haldane vs. Muller and the drosophila group on the theory and practice of linkage mapping. In S. Sarkar, ed., *The Founders of Evolutionary Genetics*. Martinus Nijhoff, Dordrecht, pp. 107–166.

Wimsatt, W.C. (2001). Generative entrenchment and the developmental systems approach to evolutionary processes. In S. Oyama, P. Griffiths, and R. Gray, eds., *Cycles of Contingency: Developmental Systems and Evolution*. MIT Press, Cambridge, pp. 219–237.

Wimsatt, W.C. (2007). *Re-engineering Philosophy for Limited Beings: Piecewise Approximations to Reality*. Harvard University Press, Cambridge.

Wimsatt, W.C., and Griesemer, J. (2007). Reproducing entrenchments to scaffold culture: the central role of development in cultural evolution. In R. Sansom and R. Brandon, eds., *Integrating Evolution and Development: From Theory to Practice*. MIT Press, Cambridge, pp. 227–323.

Wiser, M.F. (1999). Cellular and molecular biology of *Plasmodium*. http://www.tulane.edu/~wiser/malaria/cmb.html#invade. Last modified 26 February 2013. Accessed 30 May 2013.

Wiser, M.F. (2000). *Plasmodium* life cycle. http://www.tulane.edu/~wiser/protozoology/notes/mal_lc.htmlhttp://www.tulane.edu/~wiser/protozoology/notes/mal_lc.html. Last update on 29 January 2009. Accessed 30 May 2013.

CHAPTER 13

Comparison of animal and plant development: a right track to establish a theory of development?

Michel Vervoort

A theory of development can be established through comparisons of convergent developmental processes

Which theory of development?

What is a theory of development? This is a quite difficult question for several reasons, including the fact that there is no real consensual view about either the definition of development or what should be a theory in biology. In this chapter, development will be used in its most common acceptation, i.e. as the process by which a single cell (in the diplobionts such as animals, a zygote resulting from the fusion of specialized cells, the gametes) will produce a new multicellular organism with a more or less vast array of cell types and capable of producing the next generation of multicellular offspring. In this view, development is strongly linked to multicellularity and therefore occurs in multicellular organisms such as animals, land plants, macroscopic mushrooms, and various types of algae. Development comprises a set of operations that allow the transition from single cells to complex multicellular assemblies. This includes coordinated cell proliferation, setting up of polarities (such as anterior–posterior polarity in animals), cell differentiation, establishment of shapes (morphogenesis) and patterns of cell differentiation (e.g. Bonner, 2000). This chapter has been written with two main aims. The first aim is to suggest that there could be principles that underly these developmental operations that would be valid for all organisms displaying a development. These principles would therefore be independent of the evolutionary position and history of the different organisms. In other words, this means that there could be ways to proceed that should be followed during development, irrespective of whether the developing organism is an animal, a plant, or any other multicellular organism. Weisblat (1998) described development as a 'quasi-historical' process, i.e. one exhibiting both non-historical (minimally contingent, reproducible) and euhistorical (highly contingent, non-reproducible) properties. What is proposed here is to identify principles that together would constitute a 'non-historical theory of development' that focusses on what unifies all types of development. Of course, there should also be other theories of development, in particular theories capable of accounting for the diversity of development that is observed and of the importance of evolution in shaping the development of present-day organisms. The second aim of this chapter is to suggest an approach to define these principles. This approach will be detailed in the next section.

How to establish this theory?

It is widely believed that development relies on general principles and that a main aim of developmental biology is to define these principles—many influential textbooks in the field (Gilbert, 2010; Martinez-Arias & Stewart, 2002; Wolpert & Tickle, 2011) include 'principles' in their main title or in the title of some of their chapters. The idea that there are general principles in development is also at the

Towards a Theory of Development. Edited by Alessandro Minelli and Thomas Pradeu

heart of the widespread use of model organisms in developmental biology: to understand development in animals, for example, it is widely assumed that it is not necessary to study this process in many different species but that the analyses can be focussed on a few model organisms such as *Drosophila melanogaster*, *Caenorhabditis elegans*, and *Mus musculus* to be extrapolated to other species and even to other phyla (Bolker, 1995). As a consequence, most of our knowledge about animal development, and most of what is written in animal developmental biology textbooks, comes from studies on less than ten different species. Emphasis is often put on the existence of seemingly similar processes or mechanisms in different species and/or at different time-points in the development of a given species: one can reasonably argue that 'principles of development' will be related to observations that have been made in more than one species and ideally in many or all of the species studied. On the contrary, mechanisms or processes that are specific to a single species and/or a single developmental event are unlikely to be considered as general developmental principles.

Let's take an example: careful studies of cell lineages in the nematode *Caenorhabditis elegans*, combined with detailed genetic analyses, have shown that programmed cell death occurs during normal development and is based on a well-defined cell-death machinery (Jacobson et al., 1997). The proposal that 'it has become clear that "programmed cell death" . . . is an integral part of development and homeostasis in all multicellular animals' (Martinez-Arias & Stewart, 2002: 216) crucially depends on the subsequent observations that programmed cell death also occurs during several developmental events in several model species and involves a core module of homologous proteins in these diverse animals (Jacobson et al., 1997). Arguably, the way to consider programmed cell death as a developmental principle—'it appears that during development, more cells are produced than those that contribute to the final organism' (Martinez-Arias & Stewart, 2002: 216)—would have been different if shown to be specific to the development of *Caenorhabditis* rather than a widespread phenomenon found in many different animals. More generally, if considering a theory as 'a well-substantiated explanation of some aspect of the natural world, based on a body of facts that have been repeatedly confirmed through observation and experiment' (as defined on the website of the American Association for the Advancement of Science, <http://www.aaas.org>), it seems clear that building a theory of development requires comparative analyses as a way to provide the repeated observations that confirm the likelihood and the explanatory nature of the proposed theory. Back to our example, one can say that the existence of programmed cell death would be included in a theory of development because it has been repeatedly observed during the development of many different animals.

This reasoning however faces one major obstacle, which is homology. In its evolutionary sense, homology denotes a similarity that is attributable to a common origin. Homologous structures or processes found in different species derive from a structure or a process that was already present in the last common ancestor of the considered species. Homology is an hypothesis that is proposed because of similarities that are observed between the structures or processes under study. The plausibility of homology vs homoplasy (similarities that are not due to common ancestry, the most common homoplasy being evolutionary convergence) is usually assessed by studying the extent of the similarities and by making careful analysis of the evolution of the structures or processes, in particular in light of the parsimony principle (which leads to favouring the evolutionary scenario requiring the least number of evolutionary changes). In case of homologous structures or processes, similarities are due to inheritance and conservation; e.g. two orthologous organs found in two different species look similar because they have inherited without major evolutionary changes some of their features from the corresponding organ that was present in their last common ancestor. If applied to the problem of building a theory of development, the repeated observation in different species of similar mechanisms or processes may therefore not point out the type of principles of development that I propose to define but simply reflect the inheritance of developmental processes or mechanisms by the present-day species from their last common ancestor. Let's take an example: large bodies of work have shown that in many species belonging to the bilaterians (i.e. most of the extant animals including all the well-studied model species of animal developmental biology), during anterior–posterior axis formation, a set of genes, known as the *Hox* genes, are expressed in well-defined spatial domains and are required to define cell identities along this axis (Carroll et al. 2004; Gehring et al., 2009). One could propose that this role of *Hox* genes might represent a general principle of development in bilaterian animals due to its repeated observation in many animals and the fact that we can predict the existence and expression of *Hox* genes in many other species in which these genes have still not been studied. Strong evidence indicates however that the involvement of *Hox* genes in axis formation in many different species is a homologous feature, i.e. a feature already present in the last common ancestor of all bilaterian animals (a hypothetical animal known as *Urbilateria*) that has been strongly conserved in most or all bilaterian lineages that derive from this remote ancestor. The reasons why and how this '*Hox* system' was established in *Urbilateria* is beyond the scope of possible investigations. The conclusions that we can infer from the use of *Hox* genes in axis formation in different animals is that key developmental processes can be conserved over long evolutionary periods and that *Urbilateria* already possessed an anterior–posterior axis patterned by *Hox* genes (Carroll et al., 2004). These are important ideas for evolutionary biology, but they cannot help build the non-historical theory of development at which I aim.

In summary, comparisons between the development of different species is required to establish principles of development, but this method faces the problem of the possible homology of the compared processes or mechanisms. I argue that only the comparison of convergent developmental processes can be useful to establish a non-historical theory of development, as homologous processes will only tell us about their ancestry and conservation during evolution, and nothing about the underlying principles and logic. In contrast, similarities in convergent developmental processes, as the consequence of repeated occurrence and not of common ancestry, may reveal inescapable steps or ways to proceed during development, thereby allowing the establishment of general principles that would be valid for all organisms displaying development. A parallel can be drawn with the recent attempts to build a theory of morphological evolution (Carroll, 2008; Stern & Orgogozo, 2009). One aspect of this theory—the '*cis*-regulatory hypothesis', which suggests that most mutations producing morphological variations are expected to reside in the *cis*-regulatory regions of developmental genes (rather than in their coding regions)—is mainly based on the observation of many different mutations and phenotypic changes that have independently occurred in many different lineages, i.e. homoplasies. In addition, these observations have pointed out the pervasiveness of repeated or parallel evolution, i.e. 'the independent evolution of similar phenotypic changes in different species due to changes in homologous genes' (Stern & Orgogozo, 2009: 747), a specific case of convergent evolution. I similarly propose that the study of convergent developmental processes may pave the way towards a theory of development. The question now is how we can define whether similarities in developmental processes are due to homology or convergence. A related question is whether homology of developmental processes is a frequent process or not, in particular when comparing distantly related species. These questions are addressed in the next section, taking as an example the comparative study of animal development.

Homology vs convergence of developmental processes

In the last 20 years, evolutionary developmental biology (evo-devo) has challenged the traditional view of animal evolution, mainly through pointing out unsuspected homologies between key developmental processes at the scale of bilaterians or even of the whole animal kingdom (De Robertis, 2008). The following cases will exemplify the pervasive nature of evolutionary conservation of developmental processes in animals.

The first example is the formation of the dorsal–ventral axis in bilaterians. Some tissues and organs develop at specific locations along this axis—the most obvious example being the central nervous system (CNS) which, when present, forms at the ventral side in species, such as *Drosophila*, which belong to the protostome clade, and at the dorsal side in chordate species, such as *Xenopus laevis*, which belong to the deuterostome clade (Arendt & Nübler-Jung, 1994; De Robertis, 2008). Molecular and genetic studies in *Drosophila* and *Xenopus* have shown that the Bone morphogenetic protein 2/4 (BMP2/4)-Decapentaplegic (Dpp) signalling pathway has similar key roles in dorsal–ventral axis formation in these two species, highlighting unsuspected similarities in the determination of this axis in insects and vertebrates (De Robertis, 2008; De Robertis & Sasai, 1996). While initially highly debated in terms of interpretation as homology or convergence (e.g. De Robertis & Sasai, 1996; Gerhart, 2000), these observations, combined with similar ones obtained on other bilaterians with key phylogenetic positions (Denes et al., 2007; Lowe et al., 2006), have led to the consensus view that *Urbilateria* had a dorsal–ventral axis patterned by the BMP2/4–Dpp pathway and that this patterning system has been widely conserved in bilaterians, although an inversion of this axis occurred at the root of the chordate lineage (De Robertis, 2008).

A second striking example concerns nervous system development. While obviously very different, the CNS of vertebrates and some protostomes are patterned through very similar mechanisms that involve homologous proteins, leading to a well-supported hypothesis of the ancestrality in bilaterians of a complex CNS and its patterning system, as well as to the hypothesis that similarities in CNS patterning in extant bilaterians is mainly due to evolutionary conservation (Arendt et al., 2008; Denes et al., 2007; De Robertis, 2008).

The third example also relates to nervous system formation. It has been shown that the genetic core module, comprising genes which encode specific transcription factors and cell signalling molecules, that controls neural progenitor formation appears to be conserved at the scale of the animals, as homologous molecules are similarly involved in this process in both bilaterians and non-bilaterians, including sponges, which are nerveless but in which the genetic module seems to be involved in the formation of some sensory-like cells (Bertrand et al., 2002; Richards et al., 2008; Simionato et al., 2008; Vervoort & Ledent, 2001). This module was therefore likely established at the dawn of animal evolution and subsequently maintained, although with modifications and further complexifications, in the different animal lineages.

Of course, the message here is not to claim that every similarity observed in the development of different animal species is due to homologies. What I want to stress is that both homologies and homoplasies are frequent and that we therefore have to be sure whether we are dealing with one or the other when comparing developmental processes. Unfortunately, this is not an easy task. When comparing the development of animal (or plant) species, even distantly related ones, we can never rule out that some, most, or all the similarities we detect are due to common inheritance and therefore that these comparisons are not suited to establish widely applicable principles of development. To circumvent this problem, we have to focus on processes for which we have compelling evidence for the absence of homologies in the underlying mechanisms. A way to do so is to compare the development of multicellular organisms that belong to distantly related eukaryote lineages, in particular plants and animals. This comparison might be the most meaningful one when aiming at the establishment of a non-historical theory of development. A similar suggestion has previously been advocated by E. Meyerowitz (Meyerowitz, 1999, 2002). This idea is further developed in the next sections, by first summarizing the evidence of the convergent nature of animal and plant development, then discussing the possible obstacles and pitfalls of this approach, and finally showing a case study of such comparisons.

Comparisons between animal and plant development: promises and pitfalls

Development of animals and plants largely relies on convergent processes—Phylogenetic evidence: the last common ancestor of plants and animals was a unicellular organism

In the sense the term 'development' is used in this chapter (and defined at the beginning of this text), developmental processes are closely related to the existence of multicellularity. I will review here the evidence in favour of independent occurrences of multicellularity in animal and plants and therefore in favour of multiple 'inventions' of developmental processes. Until now, I have used the terms 'animals' and 'plants' in their common sense. However, at this stage of the chapter, a refreshments of these terms in a phylogenetic framework is required (Figure 13.1). 'Animals' thus refers to a well-supported monophyletic group (i.e. a group that contains an ancestral species and all its descendants and that is characterized by shared derived characteristics) known as Metazoa (e.g. Philippe et al., 2009; Ruiz-Trillo et al., 2008; Torruella et al., 2012).

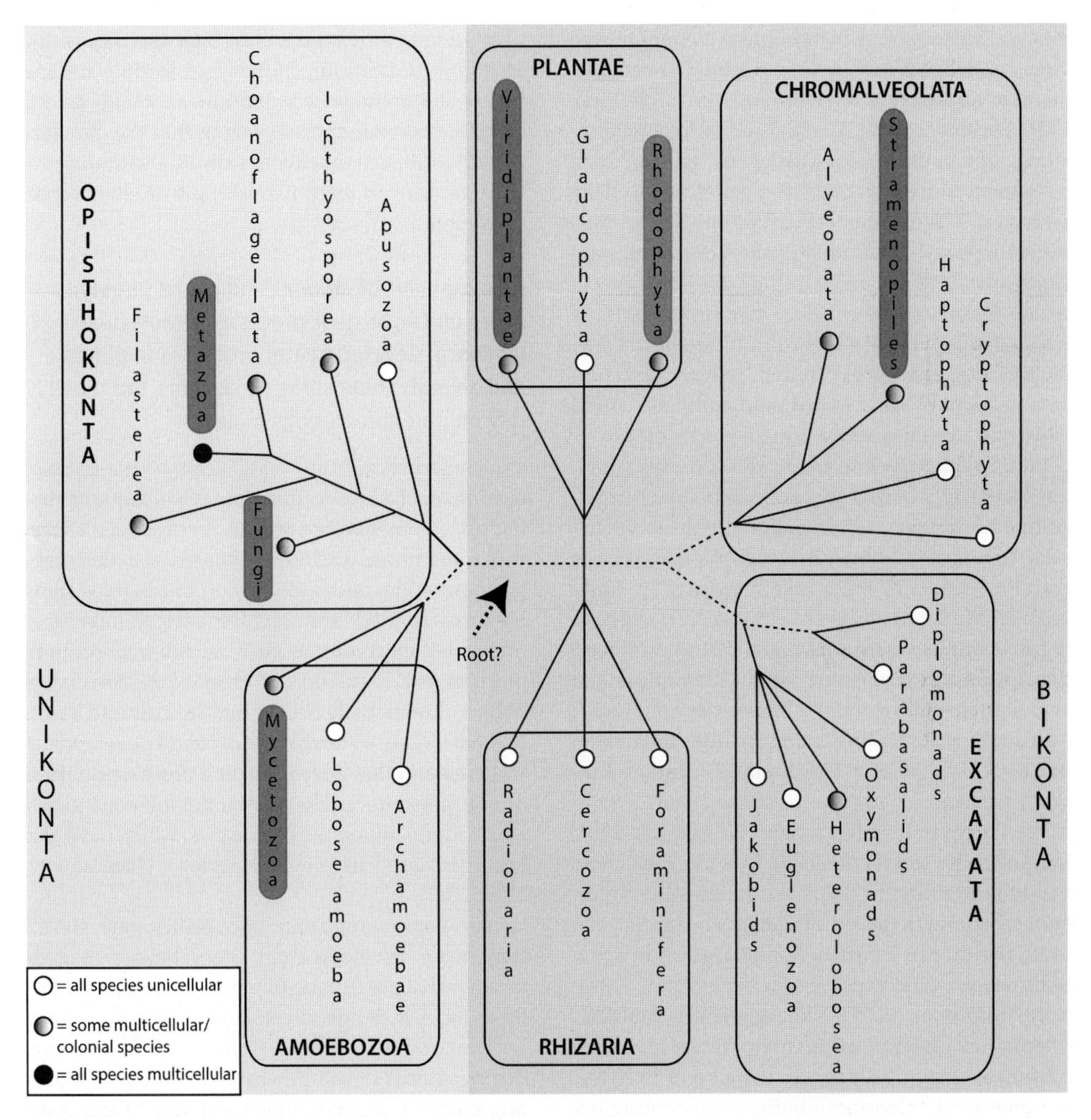

Figure 13.1 Consensus phylogeny of eukaryotes. The six proposed super-groups of eukaryotes, including their major constituting clades, are shown. Uncertain relationships are indicated by dotted lines. The arrow and the question mark indicate the probable location of the root of the eukaryotic tree. Based on this rooting, eukaryotes can be subdivided into two large groups, Unikonta and Bikonta, as shown. Groups that include many multicellular and/or colonial species are indicated by grey boxes—these are Metazoa (animals), Fungi (includes macroscopic mushrooms), Viridiplantae (includes land plants and green algae), Rhodophyta (red algae), Stramenopiles (includes brown algae), and Mycetozoa (includes slime molds). This figure has been drawn based on data reported in King (2004), Brinkmann and Philippe (2007), and Parfrey and Lahr (2013).

'Plants' is a more ambiguous term—it is used here as synonym of 'land plants' that corresponds to the monophyletic group of Embryophyta, which includes mosses, ferns, gymnosperms, and angiosperms (flowering plants) (e.g. Laurin-Lemay et al., 2012) and to which belong the most-studied plant developmental model species such as *Arabidopsis thaliana*. Metazoa and Embryophyta belong to two larger monophyletic groups, Opisthokonta and Plantae, respectively (e.g. Brinkmann & Philippe, 2007; Derelle & Lang, 2012). The phylogenetic relationships of the major eukaryotic groups, such as Plantae and Opisthokonta, are still not well resolved and consensual (Brinkmann & Philippe, 2007)—in some recent phylogenies, eukaryotes are subdivided into two large groups, the unikonts, to which Opisthokonta belongs, and the bikonts, to which Plantae belongs (Figure 13.1; Derelle & Lang, 2012; Hampl et al., 2009). In this view, the last common ancestor of 'land plants' and 'animals' would be the last common ancestor of all eukaryotes.

Within Opisthokonta, which also includes Fungi, the closest relatives of Metazoa are Choanoflagellata and Filasterea, which mainly include unicellular organisms, although some colonial species are also present in these taxa (Figure 13.1; Ruiz-Trillo et al., 2008; Torruella et al., 2012), suggesting that multicellularity evolved in the metazoan lineage after its divergence from Choanoflagellata and Filasterea (King, 2004; Parfrey & Lahr, 2013; Ruiz-Trillo et al., 2007). Plantae includes three main lineages, one of which (Viridiplantae) contains Embryophyta and several groups of organisms known as 'green algae' (e.g. Brinkmann & Philippe, 2007; Laurin-Lemay et al., 2012). It is widely believed that multicellularity evolved independently at least two times in the Viridiplantae lineage: once in some ancestor of Embryophyta and once in some ancestor of multicellular 'green algae' (King, 2004; Ruiz-Trillo et al., 2007). More generally, multicellularity has probably arisen many times from unicellular ancestors in various eukaryotic lineages, as some kind of multicellularity is also observed in species belonging to Mycetozoa ('slime molds'), Fungi (macroscopic mushrooms), Rhodophyta ('red algae'), and Stramenopiles ('brown algae'; Figure 13.1; King, 2004; Parfrey & Lahr, 2013; Ruiz-Trillo et al., 2007). In addition, within some of these lineages, such as Viridiplantae and Rhodophyta, multicellularity probably evolved independently several times (Parfrey & Lahr, 2013), suggesting that multicellularity is quite 'easy' to evolve and/or that its repeated evolution involved strong positive selection pressure (Grosberg & Strathmann, 2007; Ratcliff et al., 2012).

In summary, it seems clear that the acquisition of multicellularity in the lineages leading to land plants and animals occurred independently and is therefore convergent, suggesting that the developmental processes that allow to build the multicellular organisms in these two lineages are themselves convergent.

Development of animals and plants largely relies on convergent processes—Molecular evidence: developmental processes in plants and animals show some similarities, but mostly rely on non-homologous molecules

The construction from a single cell, be it a plant or an animal, of a multicellular organism requires similar developmental operations. As described above, these operations include regulated cell divisions, axis formation, and specification of cell fates along the axes formed and involve basic cellular and molecular mechanisms such as cell–cell communication and regulation of gene expression. Large bodies of work have shown that animals and plants possess largely different repertoires of developmental genes and that developmental processes in these organisms mainly rely on non-homologous genes. I will briefly illustrate both aspects, which have also been discussed in previous reviews (Meyerowitz, 1999, 2002).

First, comparative genomic studies have shown that many developmental genes belong to gene families that are unique to plants or animals or that have independently diversified in these lineages. Let's take some examples. Cell–cell communication during animal development relies on a few signalling pathways such as the Notch, BMP, Hedgehog, and Wnt pathways (Barolo & Posakony, 2002). Most of these pathways were established during early metazoan evolution (e.g. Gazave et al., 2009; Richards & Degnan, 2009; Srivastava et al., 2010), meaning that homologous pathways are not present in plants and can not therefore act during their development. Conversely, some of the most important cell signalling pathways that have been shown to act during plant development have no orthologues in animals—this is for example the case for the pathway in *Arabidopsis* that uses CLAVATA1 as receptor and controls meristematic growth (Meyerowitz, 2002). In other cases, plant and animal developmental genes belong to gene families that have undergone lineage-specific expansion and diversification. This can be well exemplified using the homeobox and basic helix-loop-helix (bHLH) gene families. In both cases, these genes encode transcription factors that have key functions in both plant and animals, including roles during their development. Genomic analyses have shown that homeobox genes are present in all main eukaryotic lineages and that the presence of at least two different types of homeobox genes (TALE and non-TALE) is ancestral in eukaryotes (Derelle et al., 2007). In both plant and animal lineages, subsequent diversifications occurred independently, leading to plant- and animal-specific subfamilies (e.g. Degnan et al., 2009; Derelle et al., 2007; Hay & Tsiantis, 2010; van der Graaff et al., 2009): many homeobox genes involved in animal development, such as the *Hox* and *NK* genes, belong to the animal-specific Antennapedia class, whereas the *KNOX* and *WOX* subfamilies, which contain members that act during various aspects of plant development, are plant-specific subfamilies. Similarly, bHLH genes independently diversified in animal and plants from a small set of ancestral genes, leading to plant- and animal-specific subfamilies to which belong the bHLH genes involved in development in the two lineages (Carretero-Paulet et al., 2010; Degnan et al., 2009; Feller et al., 2011; Ledent et al., 2002). More generally, genomic analyses have shown that most transcription factors involved in animal development evolved either in the holozoan lineage or in the metazoan lineage (Degnan et al., 2009), i.e. after the divergence of Opisthokonta and Plantae.

Second, when comparing developmental processes that show similarities between plants and animals, it has been observed, in most cases, that the molecules involved in these processes are not homologous. One of the first and most compelling examples, which will be discussed in more detail in the third part of this chapter, is that of plant and animal homeotic genes that specify body parts, such as segments in animals or flower structures in plants: whereas animal homeotic genes are homeobox genes, those of plants belong to a completely different gene family known as the MADS-box family (Meyerowitz, 2002). Small non-coding RNAs, such as microRNAs (miRNA), have been shown to have multiple roles during development, including the control of developmental timing in both plants and animals (e.g. McKim & Hay, 2010; Moss, 2007). There is, however, compelling evidence that miRNA evolved in animals and plants independently and that therefore the miRNAs involved in their developmental processes are non-homologous molecules (Shabalina & Koonin, 2008). Another striking example is the formation of spaced epidermal structures, such as trichomes in plants and bristles in *Drosophila*, that involve a process, known as lateral inhibition, by which a cell that adopts a defined fate prevents neighbouring cells from adopting the same fate (Balkunde et al., 2010; Ghysen et al., 1993). While lateral inhibition in *Drosophila* (and other animals) relies on the Notch signalling pathway and therefore involves ligand/receptor interactions at the plasma membrane and signal transduction within the cell, in *Arabidopsis*, lateral inhibition is based on the movement of MYB transcription factors from one cell to the others. Other clear examples include tissue growth and axis formation that mainly rely on non-homologous cell signalling pathways and transcription factors belonging to animal- or plant-specific subfamilies, as described above. We can therefore reasonably argue that plant and animal developmental processes mainly rely on non-homologous molecules, strongly strengthening the convergent nature of development in these two distantly related eukaryotic groups.

Possible pitfalls and obstacles in the use of the comparison of animal and plant development to build a theory of development

We can recognize five possible pitfalls in and obstacles to establishing a theory of development based on comparisons of plant vs animal development. The first one is the paucity of such comparisons in the existing literature—searches performed on the

PubMed database with keywords such as 'comparison of plant and animal development' only retrieved very few significant results. This is arguably due to the fact that most developmental biologists working on animals do not feel competent to address and discuss plant developmental data, and vice versa. As an example, in his seminal book *Developmental Biology* (Gilbert, 2010), Scott Gilbert described a vast array of developmental processes in various animals, with chapters ranging from very basic cellular and molecular processes to medical or evolutionary considerations—however, the single chapter on plant development (first introduced in the 8th edition of the book) was written by another author (Susan Singer). In this chapter on plant development in Gilbert's book, very few parallels are drawn with the 'animal chapters', besides an introductory sentence stating that 'the two kingdoms have many commonalities (and the land plants are sometimes referred to as "embryophytes", calling attention to the significance of the embryo in their life histories)' (Gilbert, 2010: 627). I would argue that developmental biologists, be they specialists of animal or plant development, can be competent to understand, discuss, and compare the development of both types of organisms. However, the paucity of comparisons published so far is an obstacle in the use of such comparisons with the general aim to establish principles of development.

A second and very serious pitfall is that the comparisons may lead to useless truisms rather than useful principles. Having chosen very distant organisms, one can expect their developmental processes to be highly divergent and therefore their similarities to be limited to superficial aspects. 'Principles' to be drawn from such similarities might therefore be of limited interest. As an example, it is obvious that the construction of a multicellular organism requires a tight control of cell divisions in both space and time—it is therefore expected (and observed) that such control will take place during development of both plants and animals. If a more detailed comparison does not point out any significant similarities in the ways this control is achieved in animals and plants, the only 'principle' that can be established would be 'the construction of a multicellular organism requires a tight control of cell divisions in both space and time'—one can reasonably question whether we should really consider this as a 'principle' and whether a theory of development only based on such 'principles' might really be useful. More precise and significant conclusions should be aimed at. Whether comparisons of plant vs animal development may lead to such conclusions or not cannot be determined a priori—in the third part of this chapter, I will address this issue through a case study.

A third possible obstacle is the representativeness of the species from which the data to be compared have been acquired. Indeed, what can be compared is not the development of plants and animals as a whole, but data obtained from a few plant and animal model species. To what extent that which has been observed in a small sample of species can be generalized to the whole animal or plant clade is an open question that is related to a more general debate about the use and the choice of model organisms in biology (e.g. Jenner, 2006). I will not enter this debate in this text, but I want to stress that it is an important issue for the inferences about principles of development that can be drawn from animal–plant comparisons. The developmental processes to be compared, while having to be convergent between animals and plants, should nevertheless be conserved, as much as possible, within each of these clades.

A fourth possible problem is that the observed similarities between plant and animal development, although the processes have been independently acquired, might nevertheless be related to homologies. Plants and animals do of course have a common ancestor, a unicellular organism for which it is difficult to envision what its characteristics and what the genetic networks and cell routines already functionning to control its behaviour, divisions, and so forth were. We can therefore imagine that some of these networks and routines may have been independently recruited when multicellularity independently evolved in the plant and animal lineages. In this view, similarities between animal and plant development might be the consequence of the use of ancestral cellular routines and genetic networks that have been independently co-opted in the two lineages. This is related to what is often termed 'deep homology' (Shubin et al., 2009). One classical example of deep homology concerns the paired appendages that are observed in arthropods and vertebrates—these structures are not homologous and evolved in their respective lineage independently (Shubin et al., 2009). Nevertheless, striking similarities are observed in the genetic network that controls the formation in these limbs in arthropods and vertebrates, including the use of several homologous molecules. The 'deep homology' interpretation of these observations, is that this genetic network (or at least a part of it) was already present in *Urbilateria*, the last common ancestor of arthropods and vertebrates, used to promote the formation of an unknown type of outgrowths of the body (it could be a limb or not), and that this network was independently recruited at the time arthropod and vertebrate limb evolved. As in the case of 'classical' homology, similarities would therefore not point out the logic or principles of the development of limbs and only highlight common ancestry. A solution to circumvent this problem, in the comparison between plant and animal development, is to focus mainly (or only) on those developmental events that rely on non-homologous molecules in plants and animals. Fortunately, as briefly discussed in the previous section, this seems to be the most common case.

A fifth putative problem is the issue of whether we could, in fact, define a theory that would lead to precise predictions and would be falsifiable. Predictability is not a real problem, as the principles established through the comparison of animal and plant development can be tested by looking whether these principles apply to the development of unrelated multicellular organisms, such as fungi and brown algae (Figure 13.1). Falsifiability is a more complex problem—without entering the debate of whether falsifiability should be a key property of a theory or not (Kitcher, 1982; Popper, 1963), we are facing the problem that the development of present-day organisms is not solely dependent on general non-historical principles but also on the evolutionary history of these organisms. It is not clear to which extent the non-historical principles may be modified or obscured by euhistorical processes. One can argue that any principles about any biological processes would have exceptions due to the particular evolutionary history of the organisms in which the processes have been studied. The important question is therefore how many exceptions would be acceptable without falsifying the theory. Unfortunately, there is no obvious, general, a priori answer to this question, and it has to be addressed on a case by case basis.

Comparisons between animal and plant development: a case study—the specification of body parts and tissue identities

One fundamental developmental operation is to specify the identity of cells, tissues, and organs in relationship to their positions within the overall architecture of the organism. Striking examples of such an operation are the specification of segment identity along the anterior–posterior axis in bilaterian animals and the specification of whorls of parts (sepals, petals, and reproductive organs) in the flowers of angiosperms. These processes are controlled by sets of genes, *Hox* genes in animals and floral identity genes in plants, that have been discovered because loss-of-function mutations of these genes produce homeotic phenotypes, i.e. the transformation of the identity of particular segments or whorls into that of other ones (Causier et al., 2010; Garcia-Fernàndez, 2005). As mentioned above, *Hox* and floral identity genes encode unrelated transcription factors and are therefore non-homologous genes (Meyerowitz, 2002). Nevertheless, *Hox* genes, initially discovered in *Drosophila*, are strongly conserved in bilaterian animals, and floral identity genes, first studied in *Arabidopsis* and *Antirrhinum majus*, are also conserved in various lineages of flowering plants (Causier et al., 2010; Garcia-Fernàndez, 2005). While it is beyond the scope of this chapter to extensively review the functions of these genes, I will here only point out some key shared features of these genes that may help to define some principles that underly the specification of body-part identities.

An important feature of the *Hox* and floral identity genes is that these genes are 'selector genes' (García-Bellido, 1975, 1981), i.e. genes that cell-autonomously control which part of the genome will be active in the cells in which these genes are expressed. Selector genes are thought to have very precise expression patterns that are controlled by 'activator genes' (which may include genes encoding members of signalling pathways) and to act through the control of 'realizator genes' whose expression allows the acquisition by the cells of specific properties. While initially proposed for *Drosophila Hox* genes, this applies to animal *Hox* genes in general, as well as to plant floral identity genes (Causier et al., 2010; Garcia-Fernàndez, 2005). In both cases, the activity of activator genes, such as 'gap' and 'pair-rule' genes in *Drosophila* or *wuschel*, *leafy*, and *ufo* in *Arabidopsis*, leads to the restricted transcription and activity of the *Hox* and floral identity genes in defined regions of the animal body axis and floral meristem, respectively. In turn, the proteins encoded by the selector genes regulate the transcription of a very large set of target genes (these proteins have been shown to bind to hundreds or thousands different genomic sites), influencing a wide array of developmental and cellular processes required for the formation of the specified structures (Hueber & Lohmann, 2008; Sablowski, 2010).

Hox and floral identity genes thus behave as 'input–output genes' (Davidson et al., 2002; Stern & Orgogozo, 2009) which integrate spatio-temporal information (input) and direct differentiation through the coordinated control of many target genes (output; Figure 13.2a). In theory, other ways to control the specification of the segment or flower whorl identities are conceivable, including the parallel regulation of various aspects of identity specification or the use of several 'relay' transcription factors in this process (Figure 13.2b, c). The observation of a similar 'input–output gene' architecture in plant and animal body-part specification may therefore point to an underlying logic in the way this developmental step is controlled. Interestingly, a similar situation is observed for several other selector genes, in animals at least, including the *shavenbaby* and *scute* genes in *Drosophila*, as well as the *ASCl1* gene in mouse (Castro et al., 2011; Stern & Orgogozo, 2009). These genes are cell-type selector genes (Carroll et al., 2005) that control the differentiation of a specific type of epidermal cells (*shavenbaby*) or of neurals cells (*scute* and *ASCl1*). Like *Hox* and floral identity genes, the cell-type selector genes act through the regulation of many targets genes involved in various aspects of the differentiation of the cells (Figure 13.2). I would therefore suggest that a principle of cell, tissue, organ, or body-part specification is to rely on the function of 'input–output genes' that make the link between various spatio-temporal information and the differentiation of specific structures. A testable prediction of this

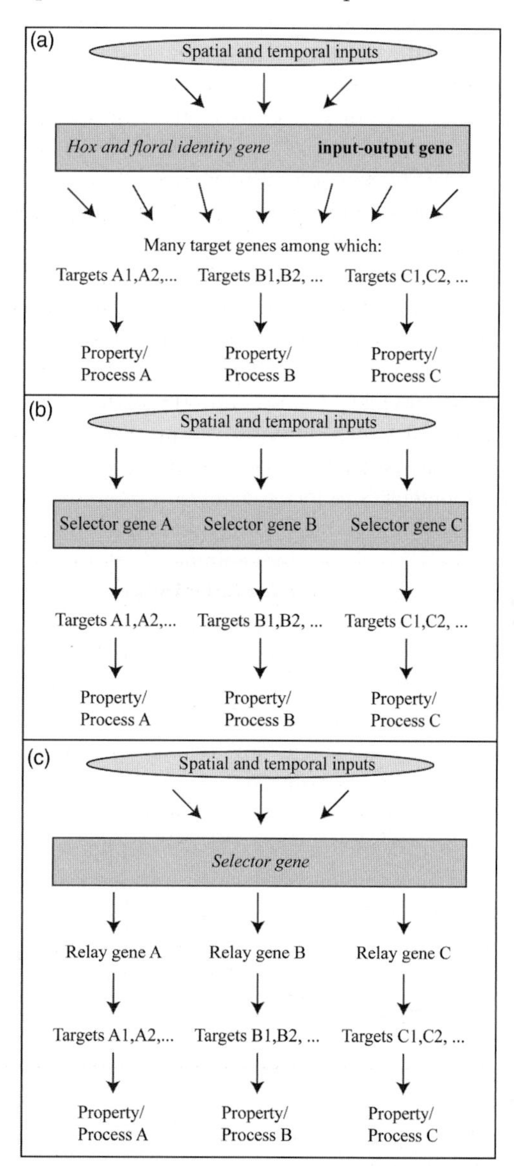

Figure 13.2 Possible ways to control the identity of body and flower structures. (a) *Hox* and floral identity genes are 'input–output genes'. See text for details. (b) and (c) Two other theoretically possible ways to control specification, either through the use of several selector genes acting in parallel to control various aspects of body/flower part specification (b) or through the use of several relay genes, each controlling a part of the specification process (c).

suggestion is that 'input–output genes' should be found in a large number of identity specification events during the development of animals, plants, and other eukaryotic multicellular organisms.

A second interesting property that is shared by *Hox* and floral identity genes is the complexity of the regulation of their expression (Figure 13.3). In addition to the initial regulation of their transcription by 'activator genes', there are four additional layers of regulation of *Hox* and floral identity genes. A first level is the self-maintenance of transcription of *Hox* and floral identity genes by binding of the encoded proteins to their own regulatory regions (Figure 13.3a; Hueber & Lohmann, 2008; Liu & Mara, 2010). This autoregulative capability, which is also found for many other selector genes, allows the stable expression of the genes irrespective of the presence and activity of the proteins that initially lead to the initiation of their transcription. Cross-inhibition represents a second important layer of regulation (Figure 13.3b): some Hox and floral identity proteins interfere with the expression and/or activity of other *Hox* and floral identity genes (Causier et al., 2010; Duboule & Morata, 1994; Liu & Mara, 2010). The transcriptional repression of *APETALA1* by AGAMOUS (AG), and that of *AG* by APETALA2 constitutes an important mechanism for the development of the flowers in *Arabidopsis*. In animals, the posterior *Hox* genes suppress the activity of the more anterior ones, a process known as 'posterior prevalence' or 'posterior dominance' and which is required for the specification of the posteriormost segments—in *Drosophila*, for example, several *Hox* genes are expressed in the posterior segments, but only Abdominal-B (Abd-B) is active in these segments, allowing the acquisition of their specific properties. A third level of regulation involves miRNAs that negatively regulate some of the *Hox* and floral identity genes at the post-transcriptional level (Figure 13.3c; Liu & Mara, 2010; Mansfield & McGlinn, 2012; Nag & Jack, 2010; Yekta et al., 2008). In plants, *miR172* (in *Arabidopsis*)

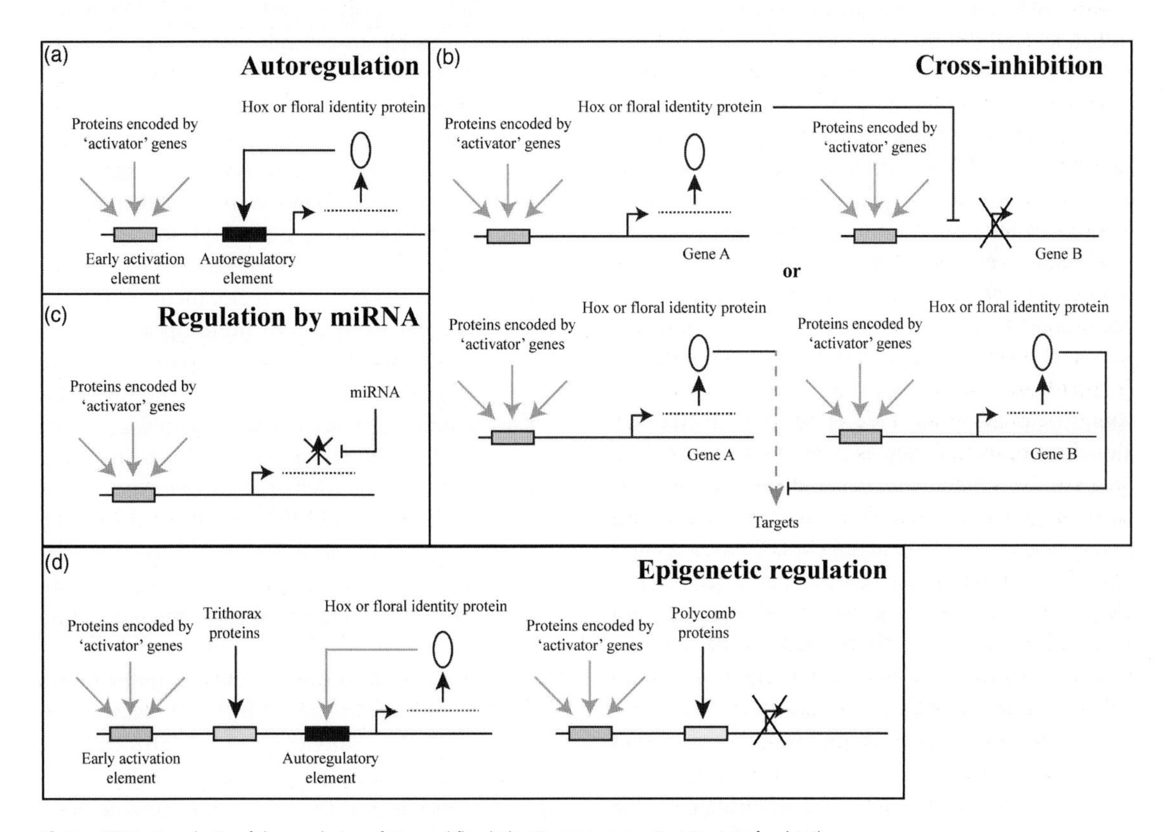

Figure 13.3 Complexity of the regulation of *Hox* and floral identity gene expression. See text for details.

and *miR169* (in *Antirrhinum majus* and *Petunia hybrida*) are important regulators of floral identity genes and are required for proper flower development. In animals, several miRNAs such as *miR-10*, *miR-933*, and *miR-196*, which are embedded within the *Hox* clusters, target some of the *Hox* genes and have been shown to participate in posterior prevalence. These miRNAs more generally contribute to establishing precise *Hox* expression domains and probably help ensuring the clearance of unwanted transcripts. Finally, the on-or-off state of the transcription of the *Hox* and floral identity genes is maintained over very long developmental time by stable modifications of their chromatin state (epigenetic regulation; Figure 13.3d) through binding of the Polycomb group and Trithorax group proteins (Liu & Mara, 2010; Schuettengruber et al., 2007; Soshnikova & Duboule, 2009). This leads to the continuous presence (or absence) of the Hox and floral identity proteins during the whole development of the segments and the whorls, respectively, allowing these proteins in some segments or whorls to control all the steps of these processes.

This very sophisticated regulation of the activity of *Hox* and floral identity genes illustrates two important features of their role during body-part specification. First is the fact that in the case of selector genes, it is equally important that these genes are expressed and active in some cells and that they are not expressed and/or active in other ones. This is of course the basis of the existence of different body parts, such as different segments or different whorls of flower structures. It is likely that the proteins encoded by the selector genes can act even at low concentrations; therefore, very elaborated regulations are needed to ensure the absence of activity of these genes in some of the cells. Several layers of negative regulation are thus required; and the comparison between plant and animal homeotic genes suggests that cross-inhibition, i.e. the activity of some selector genes interfering with that of (an)other selector gene(s), and regulation by miRNAs seem to be efficient ways to achieve the required tight regulation of the expression of selector genes.

A second key feature is the need for sustained expression of the selector genes during the whole body-part specification process. This is related to the 'input–output gene' mode of action of these selector genes which control, in a more or less direct fashion, most or all aspects of the process in which they are involved and are therefore required at most or all steps. Based on the comparison between animal and plant homeotic genes, the combination of autoregulation and epigenetic maintenance seems to be the most efficient way to achieve the sustained expression of selector genes involved in body plan. One may therefore expect to find similar mechanisms (cross-inhibition, regulation by miRNAs, autoregulation, and epigenetic regulation) during body-part specification in other multicellular organisms.

Conclusion

In this chapter, I developed the idea that a theory of development can be built through the comparison between developmental processes in animals and plants. More specifically, this comparison would lead to define common non-historical properties of development in these two lineages, as the convergent nature of their developmental processes makes it possible to exclude euhistorical properties due to homology and common ancestry. This would hopefully lay the groundwork for the identification of general principles of development that would be valid for all multicellular organisms. Only few detailed comparisons of animal and plant development have been published so far—much more comparative research, conducted by specialists of various aspects of development, would be required to reach the aim of building a theory of development based on these comparisons. More data on developmental processes in multicellular species belonging to other eukaryote lineages, such as fungi and brown algae, would also be required in order to test the predictive value of the principles established in plants and animals. It would also be important to obtain more data about the development of organisms that alternate between unicellular and multicellular states either by cell agregation (as observed in *Dictyostelium*; Urushihara, 2008), or by cell division (for example the choanoflagellate *Salpingoeca rosetta* and the ichthyosporean *Creolimax fragrantissima*; Fairclough et al., 2010; Suga & Ruiz-Trillo, 2013) to see whether the defined principles are applicable to these organisms as well. Obviously, understanding the euhistorical aspect of development is also very important—the theory of development I propose to establish should be complemented by a theory of developmental evolution, as advocated elsewhere in this book by Armin P. Moczek, in order to take into account the importance of evolution in understanding how development is organized.

Acknowledgments

I thank Alessandro Minelli and Thomas Pradeu for the opportunity to contribute this chapter and for their helpful comments on the manuscript. I am grateful to Thierry Hoquet, Lucie Laplane, Michel Morange, Virginie Orgogozo, Thomas Pradeu, and Karine Prévot for helpful and stimulating discussions. Work in my lab is funded by the CNRS, the Institut Universitaire de France, the Agence National de la Recherche (ANR grant BLAN-0294), and the Who am I? laboratory of excellence (No. ANR-11-LABX-0071) funded by the French Gouvernement through its 'Investments for the Future' program operated by ANR under grant No.ANR-11-IDEX-0005–01.

References

Arendt, D., Denes, A.S., Jékely, G., et al. (2008). The evolution of nervous system centralization. *Philosophical Transactions of the Royal Society of London B*, **363**, 1523–8.

Arendt, D., and Nübler-Jung, K. (1994). Inversion of dorsoventral axis? *Nature*, **371**, 26.

Balkunde, R., Pesch, M., and Hülskamp, M. (2010). Trichome patterning in *Arabidopsis thaliana* from genetic to molecular models. *Current Topics in Developmental Biology*, **91**, 299–321.

Barolo, S., and Posakony, J.W. (2002). Three habits of highly effective signaling pathways: principles of transcriptional control by developmental cell signaling. *Genes and Development*, **16**, 1167–81.

Bertrand, N., Castro, D.S., and Guillemot F. (2002). Proneural genes and the specification of neural cell types. *Nature Reviews Neurosciences*, **3**, 517–30.

Bolker, J.A. (1995). The choice and consequences of model systems in developmental biology. *Bioessays*, **17**, 451–5.

Bonner, J.T. (2000). *First Signals*. Princeton University Press, Princeton.

Brinkmann, H., and Philippe, H. (2007). The diversity of eukaryotes and the root of the eukaryotic tree. *Advances in Experimental Medicine and Biology*, **607**, 20–37.

Carretero-Paulet, L., Galstyan, A., Roig-Villanova, I., et al. (2010). Genome-wide classification and evolutionary analysis of the bHLH family of transcription factors in *Arabidopsis*, poplar, rice, moss, and algae. *Plant Physiology*, **153**, 1398–412.

Carroll, S.B. (2008). Evo-devo and an expanding evolutionary synthesis: a genetic theory of morphological evolution. *Cell*, **134**, 25–36.

Carroll, S.B., Grenier, J., and Weatherbee, S. (2004). *From DNA to Diversity: Molecular Genetics and the Evolution of Animal Design*. Blackwell, Malden.

Castro, D.S., Martynoga, B., Parras, C., et al. (2011). A novel function of the proneural factor Ascl1 in progenitor proliferation identified by genome-wide characterization of its targets. *Genes and Development*, **25**, 930–45.

Causier, B., Schwarz-Sommer, Z., and Davies, B. (2010). Floral organ identity: 20 years of ABCs. *Seminars in Cell and Developmental Biology*, **21**, 73–9.

Davidson, E.H., Rast, J.P., Oliveri, P., et al. (2002). A genomic regulatory network for development. *Science*, **295**, 1669–1678.

De Robertis, E.M. (2008). Evo-devo: variations on ancestral themes. *Cell*, **132**, 185–95.

De Robertis, E.M., and Sasai, Y. (1996). A common plan for dorsoventral patterning in Bilateria. *Nature*, **380**, 37–40.

Degnan, B.M., Vervoort, M., Larroux, C., et al. (2009). Early evolution of metazoan transcription factors. *Current Opinion in Genetics and Development*, **19**, 591–9.

Denes, A.S., Jékely, G., Steinmetz, P.R., et al. (2007). Molecular architecture of annelid nerve cord supports common origin of nervous system centralization in Bilateria. *Cell*, **129**, 277–88.

Derelle, R., and Lang, B.F. (2012). Rooting the eukaryotic tree with mitochondrial and bacterial proteins. *Molecular Biology and Evolution*, **29**, 1277–89.

Derelle, R., Lopez, P., Le Guyader, H., et al. (2007). Homeodomain proteins belong to the ancestral molecular toolkit of eukaryotes. *Evolution and Development*, **9**, 212–9.

Duboule, D., and Morata, G. (1994). Colinearity and functional hierarchy among genes of the homeotic complexes. *Trends in Genetics*, **10**, 358–64.

Fairclough, S.R., Dayel, M.J., and King, N. (2010). Multicellular development in a choanoflagellate. *Current Biology*, 20, R875–R876.

Feller, A., Machemer, K., Braun, E.L., et al. (2011). Evolutionary and comparative analysis of MYB and bHLH plant transcription factors. *Plant Journal*, **66**, 94–116.

García-Bellido, A. (1975). Genetic control of wing disc development in *Drosophila*. In S. Brenner, ed., *Cell Patterning* (*Ciba Foundation Symposium* 29). Elsevier, Amsterdam, pp. 161–182.

García-Bellido, A. (1981). The Bithorax-syntagma. In S. Lakovaara, ed., *Advances in Genetics, Development and Evolution in* Drosophila. Plenum, New York, pp. 135–148.

Garcia-Fernàndez, J. (2005). Hox, ParaHox, ProtoHox: facts and guesses. *Heredity*, **94**, 145–52.

Gazave, E., Lapébie, P., Richards, G.S., et al. (2009). Origin and evolution of the Notch signalling pathway: an overview from eukaryotic genomes. *BMC Evolutionary Biology*, **13**, 249.

Gehring, W.J., Kloter, U., and Suga, H. (2009). Evolution of the Hox gene complex from an evolutionary ground state. *Current Topics in Developmental Biology*, **88**, 35–61.

Gerhart, J. (2000). Inversion of the chordate body axis: are there alternatives? *Proceedings of the National Academy of Sciences USA*, **97**, 4445–8.

Ghysen, A., Dambly-Chaudière, C., Jan, L.Y., et al. (1993). Cell interactions and gene interactions in peripheral neurogenesis. *Genes and Development*, **7**, 723–33.

Gilbert, S.F. (2010). *Developmental Biology*. 9th ed. Sinauer Associates, Sunderland.

Grosberg, R.K., and Strathmann, R.R. (2007). The evolution of multicellularity: a minor major transition? *Annual Review of Ecology, Evolution and Systematics*, **38**, 621–54.

Hampl, V., Hug, L., Leigh, J.W., et al. (2009). Phylogenomic analyses support the monophyly of Excavata and resolve relationships among eukaryotic 'supergroups'. *Proceedings of the National Academy of Sciences USA*, **106**, 3859–64.

Hay, A., and Tsiantis, M. (2010). KNOX genes: versatile regulators of plant development and diversity. *Development*, **137**, 3153–65.

Hueber, S.D., and Lohmann, I. (2008). Shaping segments: Hox gene function in the genomic age. *Bioessays*, **30**, 965–79.

Jacobson, M.D., Weil, M., and Raff, M.C. (1997). Programmed cell death in animal development. *Cell*, **88**, 347–54.

Jenner, R.A. (2006). Unburdening evo-devo: ancestral attractions, model organisms, and basal baloney. *Development, Genes and Evolution*, **216**, 385–94.

King, N. (2004). The unicellular ancestry of animal development. *Developmental Cell*, **7**, 313–25.

Kitcher, P. (1982). *Abusing Science: The Case Against Creationism*. The MIT Press, Cambridge.

Laurin-Lemay, S., Brinkmann, H., and Philippe, H. (2012). Origin of land plants revisited in the light of sequence contamination and missing data. *Current Biology*, **22**, R593–R594.

Ledent, V., Paquet, O., and Vervoort, M. (2002). Phylogenetic analysis of the human basic helix-loop-helix proteins. *Genome Biology*, **3**, RESEARCH0030.

Liu, Z., and Mara, C. (2010). Regulatory mechanisms for floral homeotic gene expression. *Seminars in Cell and Developmental Biology*, **21**, 80–6.

Lowe, C.J., Terasaki, M., Wu, M., et al. (2006). Dorsoventral patterning in hemichordates: insights into early chordate evolution. *PLoS Biology*, **4**, e291.

Mansfield, J.H., and McGlinn, E. (2012). Evolution, expression, and developmental function of Hox-embedded miRNAs. *Current Topics in Developmental Biology*, **99**, 31–57.

Martinez-Arias, A., and Stewart, A. (2002). *Molecular Principles of Animal Development*. Oxford University Press, Oxford.

McKim, S., and Hay, A. (2010). Patterning and evolution of floral structures—marking time. *Current Opinion in Genetics and Development*, **20**, 448–53.

Meyerowitz, E.M. (1999). Plants, animals and the logic of development. *Trends in Cell Biology*, **9**, M65–M68.

Meyerowitz, E.M. (2002). Plants compared to animals: the broadest comparative study of development. *Science*, **295**, 1482–5.

Moss, E.G. (2007). Heterochronic genes and the nature of developmental time. *Current Biology*, **17**, R425–R434.

Nag, A., and Jack, T. (2010). Sculpting the flower; the role of microRNAs in flower development. *Current Topics in Developmental Biology*, **91**, 349–78.

Parfrey, L.W., and Lahr, D.J. (2013). Multicellularity arose several times in the evolution of eukaryotes. *BioEssays*, **35**, 339–47.

Philippe, H., Derelle, R., Lopez, P., et al. (2009). Phylogenomics revives traditional views on deep animal relationships. *Current Biology*, **19**, 706–12.

Popper, K. (1963). *Conjectures and Refutations*, Routledge and Kegan Paul, London.

Ratcliff, W.C., Denison, R.F., Borrello, M., et al. (2012). Experimental evolution of multicellularity. *Proceedings of the National Academy of Sciences USA*, **109**, 1595–1600.

Richards, G.S., and Degnan, B.M. (2009). The dawn of developmental signaling in the Metazoa. *Cold Spring Harbor Symposia on Quantitative Biology*, **74**, 81–90.

Richards, G.S., Simionato, E., Perron, M., et al. (2008). Sponge genes provide new insight into the evolutionary origin of the neurogenic circuit. *Current Biology*, **18**, 1156–61.

Ruiz-Trillo, I., Burger, G., Holland, P.W., et al. (2007). The origins of multicellularity: a multi-taxon genome initiative. *Trends in Genetics*, **23**, 113–18.

Ruiz-Trillo, I., Roger, A.J., Burger, G., et al. (2008). A phylogenomic investigation into the origin of Metazoa. *Molecular Biology and Evolution*, **25**, 664–72.

Sablowski, R. (2010). Genes and functions controlled by floral organ identity genes. *Seminars in Cell and Developmental Biology*, **21**, 94–9.

Schuettengruber, B., Chourrout, D., Vervoort, M., et al. (2007). Genome regulation by polycomb and trithorax proteins. *Cell*, **128**, 735–45.

Shabalina, S.A., and Koonin, E.V. (2008). Origins and evolution of eukaryotic RNA interference. *Trends in Ecology and Evolution*, **23**, 578–87.

Shubin, N., Tabin, C., and Carroll, S.B. (2009). Deep homology and the origins of evolutionary novelty. *Nature*, **457**, 818–23.

Simionato, E., Kerner, P., Dray, N., et al. (2008). *atonal*- and *achaete-scute*-related genes in the annelid *Platynereis dumerilii*: insights into the evolution of neural basic-Helix-Loop-Helix genes. *BMC Evolutionary Biology*, **8**, 170.

Soshnikova, N., and Duboule, D. (2009). Epigenetic regulation of vertebrate Hox genes: a dynamic equilibrium. *Epigenetics*, **4**, 537–40.

Srivastava, M., Simakov, O., Chapman, J., et al. (2010). The *Amphimedon queenslandica* genome and the evolution of animal complexity. *Nature*, **466**, 720–26.

Stern, D.L. and Orgogozo, V. (2009). Is genetic evolution predictable? *Science*, **323**, 746–51.

Suga, H., and Ruiz-Trillo, I. (2013). Development of ichthyosporeans sheds light on the origin of metazoan multicellularity. *Developmental Biology*, **377**, 284–92.

Torruella, G., Derelle, R., Paps, J., et al. (2012). Phylogenetic relationships within the Opisthokonta based on phylogenomic analyses of conserved single-copy protein domains. *Molecular Biology and Evolution*, **29**, 531–44.

Urushihara, H.(2008). Developmental biology of the social amoeba: history, current knowledge and prospects. *Development Growth and Differentiation*, **50**, S277–S281.

van der Graaff, E., Laux, T., and Rensing, S.A. (2009). The WUS homeobox-containing (WOX) protein family. *Genome Biology*, **10**, 248.

Vervoort, M., and Ledent, V. (2001). The evolution of the neural basic Helix-Loop-Helix proteins. *Scientific World Journal*, **1**, 396–426.

Weisblat, D.A. (1998). Embryonic development as a quasi-historical process. *International Journal of Developmental Biology*, **42**, 475–8.

Wolpert, L., and Tickle, C. (2011). *Principles of Development*. 4th ed. Oxford University Press, Oxford.

Yekta, S., Tabin, C.J., and Bartel, D.P. (2008). MicroRNAs in the Hox network: an apparent link to posterior prevalence. *Nature Reviews Genetics*, **9**, 789–96.

CHAPTER 14

Towards a theory of development through a theory of developmental evolution

Armin P. Moczek

Overview

This chapter explores the relationship between a theory of development and a theory of developmental evolution, in three parts. The first part reviews points of tension between different perspectives on the importance of understanding development in order to understand organismal form, function, and evolution, and highlights persistent empirical roadblocks within sub-disciplines that could potentially be resolved through contributions from companion perspectives. In the second part I ask if a theory of development could be formulated that could serve as a conceptual mediator to revise existing disconnects. I posit that to achieve this goal a theory of development should be nested within a theory of developmental evolution. Specifically, I propose a two-step approach to construct a scaffold along which a theory of development could be built. Step 1 would accumulate the knowledge base of development, focussed on identifying and linking developmental products and processes. Step 2 would then organize this information using a three-layered approach, focussed on the development of homologues (layer 1), a nested hierarchy of homologues (layer 2), and a description of patterns and causes of variation within homologues (layer 3). I argue that by nesting a theory of development within a theory of developmental evolution, we will be able to go beyond understanding the nature of development and towards a historic and phylogenetic understanding of this nature. The strategy outlined here would allow a conceptualization of development that (i) is biologically realistic, (ii) refineable alongside a growing knowledge base, (iii) flexible to incorporate both homology (descent) and variation (modification), and (iv) capable of bridging to relevant conceptual frameworks that are developing in adjacent biological fields. Examples of such frameworks, the bridges they provide, and future challenges are discussed in the third and last part.

Stepping back: what needs to be explained?

Before I embark on the main objectives of this essay, at least two terms require clarification. The first is development. For the purposes of this essay, I would like to define development as the sum of all processes and interacting components that are required to allow organismal form and function, on all levels of biological organization, to come into being. Components are defined to specifically include all the products of developmental processes (from nucleic acids and morphogen gradients to tissues and organ systems) that influence subsequent developmental outcomes, whereas form and function are defined broadly, including morphology, physiology, behaviour, and the complex phenotypes that emerge through their interactions.

If this is how I conceptualize development, what then would I consider a theory of development? I would consider a theory of development any conceptual framework that is applicable to a wide range

Towards a Theory of Development. Edited by Alessandro Minelli and Thomas Pradeu

of organismal diversity and across levels of biological organization and which would allow us to identify, understand, analyse, and derive predictions about the nature of development. In other words, a theory of development should allow us to generalize features of organismal development (what is development like?), understand the forces that shape development (why is it the way it is?), and provide starting points for further expanding and applying this framework to new organisms and traits (given what we know about development, what hypotheses can we generate to explain the developmental origins of previously unexplained organismal diversity?).

In this essay I will focus in many ways on the relationship between development and evolutionary biology. Specifically, I will argue that, to maximize the usefulness of a theory of development, it should be nested within a theory of developmental evolution, and that doing so will not diminish our understanding of the principles of phenotype construction but will instead ensure that one of the dominating forces that has shaped the way present-day organisms build themselves, decent with modification, is adequately incorporated as we further analyse and begin to derive predictions regarding the nature of development in as-of-yet unexplored phenotype space. I argue in favour of such an approach, despite the historically rather heterogeneous appreciation that evolutionary biology has harbored for development (Amundson, 2005; Gilbert & Epel, 2009). What do I mean by that?

Population geneticists, for instance, frequently define evolution as a change in the genetic composition of a population, enabling a corresponding change in the population-wide distribution of a given phenotype of interest. Describing such changes and identifying the relative contributions of different evolutionary mechanisms (selection, drift, recombination, mutation) to the process represent key objectives. Such efforts depend on a close association between the occurrence of specific phenotypic variants in a population and some marker linking them to the assumed genetic basis of different variants, such as alleles, quantitative trait loci, single nucleotide polymorphisms, etc. As long as such associations exist, traits and variants, and their genetic proxies, can be counted in populations, their changes in frequency can be traced across generations, and the nature of change can be compared to model predictions. Importantly, as long as such associations exist, it is secondary exactly how a given trait or trait variant of interest comes into being through ontogeny. Instead, the evolutionary process is studied through the methodologies of *transmission genetics*, and heredity can be studied without reference to ontogeny. Consequently, traditional population genetics does not have to draw from developmental biology in order to answer the questions it poses, and a theory of development may seem of little value.

Embryologists, comparative morphologists, and practitioners of evo-devo are also interested in phenotypic changes over time, but the starting point of their investigation is different (Gilbert & Epel, 2009; Minelli, 2003). Traits are generally viewed as the products of development. To change these products over evolutionary time, aspects of their development must change. To understand phenotypic evolution thus requires an understanding of the evolution of developmental properties, which in turn requires an understanding of development. To the extent to which such an investigation is carried out on the level of genes and pathways, *comparative developmental genetics* takes the place of transmission genetics as an important means of inquiry into the evolutionary process. Rather than divorced from it, heredity emerges through ontogeny, and a comprehensive theory of development may constitute a key resource to advance the field.

Given the uneven relationship between developmental and evolutionary biology, why then do I advocate nesting a theory of development within a theory of developmental evolution? The most important reason is that, given the role of phylogeny in shaping organismal development, this simply makes a lot of sense. But an important second reason is that doing so would enable both developmental and evolutionary biologists to go beyond where each discipline has been able to go thus far, and to approach areas in which fundamental problems have remained stubbornly resistant to resolution because appropriate empirical and conceptual tools are missing within disciplines and points of exchange across disciplines have not yet been cultivated. One of the most productive contributions of a theory of development that is explicitly integrated into a theory of developmental evolution may thus lay in providing opportunities for such cross-fertilization.

For example, the origin of phyla and *Bauplan* diversity is a fundamental question in evolutionary biology (Raff, 1996), but one that population-genetic approaches are unable to address because they lack the necessary phenotypic variation accessible via the methods of transmission genetics (Amundson, 2005). *Baupläne*, while highly diverse among phyla, are nearly invariant within them; thus, no variation exists that could be marked, followed across generations, or artificially selected upon. The same problem arises for any highly canalized, phenotypically invariant trait. Not that these traits do not (or did not) evolve in natural populations or that they are somehow less relevant—on the contrary. But unlike bristle patterns or eye color, their respective patterns of variation simply render them difficult to study through traditional population-genetic means.

A similar problem arises in the study of novelty and innovation in evolution, one of the oldest, most fundamental, and still largely unresolved questions in evolutionary biology (Moczek, 2008). Population-genetic approaches allow an investigation into how and why the composition of trait variations within a population changes over time, but provide no satisfying understanding of how novel traits come into being in the first place beyond postulating the occurrence of key mutations that must somehow have enabled a particular phenotypic transformation. In contrast, comparative developmental and developmental-genetic approaches can overcome these limitations, because in this case, evolutionary insights derive from the comparison of phenotype construction during development across taxa, rather than following the differential spread of variants across generations. The resulting efforts have permanently enriched evolutionary biology by contributing fundamental new concepts such as co-option, deep and partial homology, and developmental systems drift (Abouheif, 1997; Shubin et al., 2009; True & Haag, 2001).

Inversely, developmental biologists and evo-devo practitioners have encountered their own persistent empirical roadblocks. For example, comparative embryologists have known for a long time that life stages differ dramatically in the degree to which they have diversified across taxa, giving rise to the concepts of the hourglass of development and the phylotypic stage (Raff, 1996). Several hypotheses have been advanced to explain these patterns, arguably without resolving the issue in any satisfying manner. Recent genomic and population-genetic approaches have now provided important new considerations, suggesting for instance that relaxed selection on strictly maternally acting genes (such as those guiding much of early embryonic development in many organisms) may be sufficient to explain their elevated accumulation of sequence variation within, and differential divergence among, species (Cruickshank & Wade, 2008; Demuth & Wade, 2007). While these results do not yet resolve the issue of the developmental hourglass, they nevertheless highlight how population-genetic approaches can contribute relevant and novel insights beyond where comparative developmental approaches have been able to go on their own.

Thus, integrating a theory of development into a larger theory of developmental evolution may benefit all involved, regardless of their initial perspectives on exactly what matters in development or evolution and what deserves explanation. The remainder of this chapter seeks to explore how a theory of development, and what kind of a theory, may be most conducive towards that goal.

How to build a productive theory of development—a two-step proposal

A productive theory of development in most general terms should provide a meaningful framework for understanding the nature of development. But to do so, as a first step, we must agree what we mean by the nature of development and what it is that is worth explaining and generalizing about development. What should be the foci around which a theory of development should be structured to be most useful to developmental biology itself and to other disciplines with which it interacts?

I posit that, to be most useful, a theory of development should have three foci. First, it should focus on the outputs of development, across all levels of biological organization, from bona fide traits (e.g. cells, tissues, organs) to more transient products (e.g. expression domains, gradients, thresholds). Second, it should focus on the developmental processes that generate these products, across the domains of molecular and developmental biology as well as physiology. Third, and most importantly, it should focus on linking products to processes, recognizing that this relationship is not linear: developmental processes generate many products, products require many processes, and frequently the product of one developmental process itself constitutes a critical component of another process generating yet another developmental output. Clearly, this is not a simple and straightforward starting point for a theory. But it has several key advantages that may make it worth the effort. By identifying, and linking, developmental products and processes, this strategy allows us to begin to organize the complexities of organismal development, to make room to accommodate the self-constructing nature of ontogeny (more on this in the next section), and to facilitate an understanding of development across levels of biological organization.

Next, to be meaningful, a theory of development must provide opportunity to conceptualize the diversity of development across the enormity of organismal diversity, to discover general rules and principles, should they exist, and to connect to relevant conceptual frameworks that are developing outside its area of focus. To do so I propose as a second step to organize and expand the knowledge base accumulated during step 1 using a three-layered framework to begin constructing a theory of development, with the goal to simultaneously incorporate homology and diversification, developmental descent and developmental variation.

The first, foundational level of a theory of development would catalogue the developmental means by which homologous traits come into being. The use of homologous traits allows us to draw from all of organismal diversity while simultaneously reducing this diversity to a more manageable level, namely that of homologs. As such it establishes homology as a structuring principle of a theory of development (for a contrasting view, see Vervoort, this volume). The second level then establishes a nested hierarchy of homologues and their developmental basis, from genes, to pathways, to networks, and to cell and tissue types, organs, and bodies. This effort facilitates immediate recognition of where developmental processes are reused to generate diverse products, how similar products are made in diverse ways, and on what level of biological organization any of this is occurring. Collectively, this permits co-option and convergence, as well as partial and deep homology to emerge naturally from within the approach (Shubin et al., 2009). The third and last level then focusses on the patterns and causes of variation inherent in the development of each homologue on a microevolutionary level, i.e. among populations and closely related species.

Combined, steps 1 and 2 thus catalogue the genesis of form and function across the diversity of developmental products and processes and their various homologous manifestations. Steps 2 and 3 then expand this framework for understanding the nature of development into a framework for understanding the nature of developmental evolution by (i) providing an deeper appreciation of how development has (or has not) diversified at different levels of biological organization and in different lineages; (ii) identifying patterns and mechanistic causes of variation available in natural populations; and (iii) doing all of this in a framework of homology and descent with modification. Thus, by nesting a theory of development within a theory of developmental evolution, we can go beyond understanding the nature of development and towards a historic and phylogenetic understanding of this nature; moreover, we may be able to connect such a theory to existing, independent frameworks in ecological and evolutionary genetics.

How to build a productive theory of development—an example

Let's now try to implement the strategy sketched out above in the concrete example of appendage development in insects. Specifically, as posited above, a theory of development should begin by focussing on identifying and linking products and processes. With respect to insect appendage formation, we can identify many concrete products, such as segments, joints, spines, bristles, cuticle, tissue types, attachment sites, etc., but also proximal–distal and anterior–posterior axes or sizes of parts, in absolute terms as well as relative to other traits. And we can identify a diversity of processes that underlie their production, from gene expression, paracrine signaling, and pattern formation to the behaviour of cells and the interactions among tissues (e.g. Angelini & Kaufman, 2005; Kojima, 2004; Snodgrass, 1935). Accumulating this information across as much of insect diversity as possible, we are essentially building an ever-growing knowledge base of everything it takes to build an insect appendage, no matter what the appendage, or the insect, at least for starters.

With this raw material as a starting point, we can begin to organize and expand this information using the three-layered approach proposed above. On the foundational level our understanding of insect appendage development would be organized according to the developmental means by which homologous appendages (and their component parts) come into being, as well as the processes that underlie their formation. For example, we would note that all insect appendages are of epidermal origin, that most emerge late in larval development as epidermal outbuddings while others derive from early developing imaginal discs, that distal identity in all but the mandible requires the correct expression of the transcription factor *Distal-less*, that programmed cell death plays a key role in delineating the exact final shape of at least some appendages, etc. On the second level, homologous relationships would be refined further by nesting them within each other: for example, *hedgehog* expression and function would be nested within the *hedgehog* pathway, which in turn would nest within anterior–posterior axis formation and growth regulation, etc., which would be nested within the particular appendage types and regions to which we know this applies (Angelini & Kaufman, 2005; Kojima, 2004). The third and final layer would then add information regarding variation in product and process homologues present in natural populations or among closely related species. In the context of insect appendage development, we would note, for example, that all appendages of adult fruit flies form from imaginal discs specified during late embryonic development, which then grow during most of larval development as two-dimensional invaginations into the larval body, and of which at least the legs functionally require the expression of the morphogen Decapentaplegic for initial specification (Kojima, 2004; Ober & Jockusch, 2006). We would further note that a subset of this machinery is also involved in the making of genital claspers in sepsid flies, highly modified serial homologues of traditional appendages, even though they derive from histoblasts rather than imaginal disks (Bowsher & Nijhout, 2007, 2009). This would be in contrast to the homologous appendages of adult beetles or butterflies which either derive from early developing imaginal disks, or more frequently, from late-forming, three-dimensional evaginations which—at least in part—do not require Decapentaplegic for normal development (Švácha, 1992;). This view could then be expanded to the horns of beetles, which, unlike sepsid claspers, lack even remote homology to traditional appendages, but like sepsid claspers, exhibit in their development a certain degree of partial, and deep, homology to that of legs and antennae (Moczek, 2009). Current understanding of beetle horn development then also allows us to contrast the degree of variability that exists in their development on a microevolutionary level as a function of sex, population, or species, providing starting points to link such variation to the evolutionary processes that might shape it in natural populations (Kijimoto et al., 2012).

Combined, this approach offers several key opportunities: first, there is no predefined starting point or directionality. Instead, investigators can begin *anywhere*, with *any kind of trait on any taxonomic level*, to contribute specific observations towards an eventual, general understanding of the nature of development. For our example above, any insect, any appendage, any aspect of appendage development, and any kind of comparison is relevant. Second, using homology as an organizing principle (levels 1 and 2) immediately identifies the presence or absence of correspondence between homology across products and across processes and whether this correspondence changes as a function of the level of biological organization or phylogenetic distance that is considered. On one side, whatever patterns emerge can then be compared to other efforts elsewhere in phenotype space (e.g. appendages, anterior–posterior axes, or growth control), providing deep resolution to identify general principles by which development enables its products, and vice versa. On the other, such patterns of variation can be compared to those present among populations and closely related species, providing insights into where and how micro- and macroevolution of development might intersect.

Strengthening the model and building bridges

I will end this chapter by highlighting several problem areas associated with the framework sketched out above. If overcome, however, these problem areas transform into key opportunities to build bridges between a theory of development and recent expansions of adjacent areas of biology.

The nature of nurture

Our understanding of the developmental basis of traits is rapidly advancing, including homologous and partially homologous traits across an ever-growing diversity of organisms. At the same time developmental genetics and evo-devo in general remain steeped in rather traditional perspectives on the causes of traits. In many ways, we continue to assume that traits and organisms essentially pre-exist their development and are programmed somehow in the genome, ready to unfold if the right opportunities present themselves. Moreover, we remain convinced that organismal development can be partitioned into genetic and environmental contributions and their respective interactions. But the metaphors of genes and genomes as blueprints of development, and the separability of genes and environment as contributors to trait formation, have outlived their usefulness: instead it is becoming increasingly clear that while genes and genomes matter enormously in development, they neither suffice to make traits nor organisms. Similarly, while both genes and environmental conditions interact in their contributions to trait formation, it has become clear that describing their relationships as merely interactive is insufficient. Yes, both contributors do interact, but more often than not they are both *cause and effect* of each other: genes and their products help generate environmental conditions within which the next round of gene expression can contribute to shaping subsequent developmental outcomes. Thus, traits and organisms need to be viewed as the products of developmental *systems* to which genes contribute important interactants. Or put another way: the development of a trait of interest begins with a gene *only* if this is where our investigation starts (Moczek, 2012; Oyama, 1985).

As ongoing efforts in developmental biology provide much of the knowledge base that would allow us to form a theory of development, the way we think of the genesis of traits must therefore become more biologically realistic. In particular, we need to arrive at a deeper appreciation of the contingent nature of developmental processes, and the interdependencies of genetic, developmental, and environmental contributions (Gilbert, 2002; Gilbert & Epel, 2009; Keller, 2010; Moczek, 2012; Oyama, 1985). Several interrelated conceptual frameworks already exist that could facilitate such a process.

For example, the *theory of facilitated variation* as formulated by Kirschner and Gerhart (2005) and Gerhart and Kirschner (2007, 2010) proposes that the combination of exploratory behaviour and weak linkage (between inputs and outputs) inherent in core developmental processes enable developmental systems to be adaptably responsive to conditions. Developmental processes therefore facilitate ontogenetic changes, because they enable adjustments to developmental context, and facilitate evolutionary change, because they enable random genetic variation to give rise to non-random and functionally integrated phenotypic variants. The theory of facilitated variation makes several important contributions towards a theory of development. For instance, it emphasizes that traits and trait variation do not pre-exist in genes and genetic variation but instead emerge through development. Genes and genetic variation are key contributors but by themselves do not suffice to understand the genesis of traits. The theory of facilitated variation thus provides important opportunities to fill (or replace) an abstract, assumed genotype–phenotype map with biological reality. Similarly, this theory provides a framework for understanding the mechanisms by which random and modest genetic changes can elicit substantial and well-integrated phenotypic changes, guided by the facilitating nature of development (Moczek, 2012).

The *theory of evolution by genetic accommodation*, developed by West-Eberhard (2003, 2005a, b), similarly explores the interplay between environmental conditions and developmental processes in the expression of phenotypic variation. In particular, it emphasizes that environmental changes can elicit, through the condition-sensitivity inherent in developmental processes, phenotypic transformations that can subsequently be stabilized genetically through selection operating on genetic variation present, or newly arising, in a population. As such, genetic accommodation theory critically extends the roles of development and environment in the evolutionary process by emphasizing that the interactions among them determine which genetic variants will be phenotypically expressed and thus selectable and which will remain cryptic (reviewed by Moczek et al., 2011; Pfennig et al., 2010).

Lastly, *niche construction theory* (Lewontin, 1983; Odling-Smee, 2010; Odling-Smee et al., 2003) focusses on the interplay between organisms and their niche, which we generally tend to view as existing separate from each other, i.e. without the organism, the niche should still be there. Niche construction theory challenges this dichotomy and argues instead that organisms actively construct their niches, which in turn affect their development and fitness, with effects often extending across generations (Lewontin, 1983; Odling-Smee et al., 2003). Such niche construction is perhaps most obvious in the manufacturing of shells, cocoons, casings, and nests but also in the alteration of soil properties by fungi or earthworms, the alteration of fire regimes by plant communities, and in fact any kind of parental care. Niche construction theory thus makes room to understand the immediate developmental environment experienced by individuals not as separate from them but instead as being constructed, shaped, and modified by their actions as well as those of their ancestors. Because the environment is now in part generated by the organism itself, it too has a heritable component and can evolve. Most importantly, this allows environmental factors to be incorporated into population-genetic models and predictions even if the environment has no genes that can be passed on to the next generation. Instead what is passed on is the selective environment as generated by individuals and as experienced by descendant generations. More generally, much like the theory of facilitated variation and genetic accommodation, niche construction emphasizes how the contingent nature of development (and physiology and behaviour) facilitates the production of adaptive phenotypes by improving the match between phenotypes and the selective contexts within which they function.

Collectively, the three theories summarized above provide a rich, interlocking construct within which to begin frame a more realistic understanding of the genesis of form and function in development and evolution beyond the persisting—yet useless—'blueprint' and 'program metaphors' for development. We should not let this opportunity pass us by.

The Lego fallacy

Above, I proposed to organize developmental products and processes by utilizing a nested hierarchy of homologies. Implicit in this approach is an assumption of modularity: distinct sets of developmental processes work together to generate a developmental product, which differs distinctly from other such products. Processes and products can then be compared, homologized, and related to each other. Furthermore, we should be able to divide both processes and their products into their component parts, allowing further comparison. One major advantage of this approach is that it allows partial and deep homology as well as developmental systems drift to emerge from within the framework as we document the differential reuse of component parts and processes. But such an approach can quickly exhausts its usefulness if we do not recognize the limits of modularity in development. Organisms and their traits are not like Lego bricks, with all parts separable and recombinable in the precise same way. Neither are the developmental processes that produce organisms and their parts. Instead, module boundaries may be more or less definable depending on ontogenetic timing and level of biological organization. Moreover, the relationship between products and processes in development is complicated: a single developmental process generates (or interacts with a varying cast of other processes to generate) many products, a single product requires many interacting processes, and the same thing can both be product and part of a process. Furthermore, developmental products and processes influence other products and processes, reciprocally inducing, shaping, and modifying phenotypic outcomes and properties. Where one trait ends and another begins is often remarkably difficult to assess. Thus, as we organize organismal development into nested hierarchies of homologues, we must be mindful that modularity and homology are matters of degree and that we may learn much from shifting, loosening, or otherwise adjusting how we subdivide the developing organism into parts and processes (for detailed discussion of these and related topics see Minelli (1997) and Moczek (2008)).

Conclusion

Nesting a theory of development within a theory of developmental evolution offers the opportunity to acquire an understanding of the nature of development alongside a historic and phylogenetic understanding of this nature. Here I have proposed a framework by which the growing richness of our understanding of organismal development could be organized to structure the formulation of a theory of developmental evolution in a way that is biologically realistic and meaningful, able to incorporate both homology and variation, and capable of linking to important conceptual developments in adjacent biological fields. Further refinement and application of such a framework may hold the key for diverse biological disciplines to grow together and to facilitate the resolution of long-standing, fundamental challenges in a productive manner.

Acknowledgments

I thank the editors for the opportunity to contribute this chapter as well as for thoughtful comments on previous drafts.

References

Abouheif, E. (1997). Developmental genetics and homology: a hierarchical approach. *Trends in Ecology and Evolution*, **12**, 405–8.

Amundson, R. (2005). *The Changing Role of the Embryo in Evolutionary Thought: Roots of Evo-Devo*. Cambridge University Press, Cambridge.

Angelini, D.R., and Kaufman, T.C. (2005). Insect appendages and comparative ontogenetics. *Developmental Biology*, **286**, 57–77.

Bowsher, J. H., and Nijhout, H.F. (2007). Evolution of novel abdominal appendages in a sepsid fly from histoblasts, not imaginal discs. *Evolution and Development*, **9**, 347–54.

Bowsher, J. H., and Nijhout, H.F. (2009). Partial co-option of the appendage patterning pathway in the development of abdominal appendages in the sepsid *Themira biloba*. *Development, Genes, and Evolution*, **219**, 577–87.

Cruickshank, T., and Wade, M.J. (2008). Microevolutionary support for a developmental hourglass: gene expression patterns shape sequence variation and divergence in *Drosophila*. *Evolution & Development*, **10**, 583–90.

Demuth, J.P., and Wade, M.J. (2007). Maternal expression increases the rate of *bicoid* evolution by relaxing selective constraint. *Genetica*, **129**, 37–43.

Gerhart, J.C., and Kirschner, M.W. (2007). The theory of facilitated variation. *Proceedings of the National Academy of Sciences USA*, **104**, 8582–9.

Gerhart, J.C., and Kirschner, M.W. (2010). Facilitated variation. In M. Pigliucci and G.B. Müller, eds., *Evolution: The Extended Synthesis*. MIT Press, Cambridge, pp. 253–80.

Gilbert, S.F. (2002). Genetic determinism: the battle between scientific data and social image in contemporary developmental biology. In A. Grunwald, M. Gutmann and E.M. Neumann-Held, eds., *On Human Nature. Anthropological, Biological, and Philosophical Foundations*. Springer, New York, pp. 121–140.

Gilbert, S.F., and Epel, D. (2009). *Ecological Developmental Biology: Integrating Epigenetics, Medicine, and Evolution*. Sinauer Associates, Sunderland.

Keller, E.F. (2010). *The Mirage of a Space between Nature and Nurture*. Duke University Press, Durham.

Kirschner, M.W., and Gerhart, J.C. (2005). *The Plausibility of Life: Resolving Darwin's Dilemma*. Yale University Press, New Haven.

Kijimoto, T., Pespeni, M., Beckers, O., et al (2012). Beetle horns and horned beetles: emerging models in developmental evolution and ecology. *WIREs Interdisciplinary Reviews in Developmental Biology*, **2**, 415-18.

Kojima, T. (2004). The mechanism of *Drosophila* leg development along the proximodistal axis. *Development Growth and Differentiation*, **46**, 115–29.

Lewontin, R. (1983). Gene, organism, and environment. In D.S. Bendall, ed., *Evolution from Molecules to Man*. Cambridge University Press, Cambridge, pp. 273–285.

Minelli, A. (1997). Molecules, developmental modules, and phenotypes: a combinatorial approach to homology. *Molecular Phylogenetics and Evolution*, **9**, 340–347.

Minelli, A. (2003). *The Development of Animal Form. Ontogeny, Morphology and Evolution*. Cambridge University Press. Cambridge.

Moczek, A.P. (2008). On the origin of novelty in development and evolution. *BioEssays*, **5**, 432–47.

Moczek, A.P. (2009). The origin and diversification of complex traits through micro- and macro-evolution of development: Insights from horned beetles. *Current Topics in Developmental Biology*, **86**, 135–62.

Moczek, A.P. (2012). The nature of nurture and the future of evodevo: toward a comprehensive theory of developmental evolution. *Integrative and Comparative Biology*, **52**, 108–19.

Moczek, A.P., Sultan, S., Foster, S., et al (2011). The role of developmental plasticity in evolutionary innovation. *Proceedings of the Royal Society of London B*, **278**, 2705–13.

Ober, K.A., and Jockusch, E.L. (2006). The roles of *wingless* and *decapentaplegic* in axis and appendage development in the red flour beetle, *Tribolium castaneum*. *Developmental Biology*, **294**, 391–405.

Odling-Smee, J. (2010). Niche inheritance. In M. Pigliucci and G.B. Müller, eds., *Evolution. The Extended Synthesis*. MIT Press, Cambridge, pp. 175–207.

Odling-Smee, F.J., Laland, K.N., and Feldman, M.W. (2003). *Niche Construction. The Neglected Process in Evolution*. Princeton University Press, Princeton.

Oyama, S. (1985). *The Ontogeny of Information: Developmental Systems and Evolution*. Cambridge University Press, Cambridge.

Pfennig, D.W., Wund, M.A., Snell-Rood, E.C., et al (2010). Phenotypic plasticity's impacts on diversification and speciation. *Trends in Ecology and Evolution*, **25**, 459–67.

Raff, R.A. (1996). *The Shape of Life. Genes, Development and the Evolution of Animal Form*. University of Chicago Press, Chicago.

Shubin, N., Tabin, C., and Carroll, S. (2009). Deep homology and the origins of evolutionary novelty. *Nature*, **457**, 818–23.

Snodgrass, R.E. (1935). *Principles of Insect Morphology*. Comstock, Ithaca.

Švácha, P. (1992). What are and what are not imaginal disks—reevaluation of some basic concepts (Insecta, Holometabola). *Developmental Biology*, **154**, 101–17.

True, J.R., and Haag, E.S. (2001). Developmental system drift and flexibility in evolutionary trajectories. *Evolution & Development*, **3**, 109–19.

West-Eberhard, M.J. (2003). *Developmental Plasticity and Evolution*. Oxford University Press, Oxford.

West-Eberhard, M.J. (2005a). Developmental plasticity and the origin of species differences. *Proceedings of the National Academy of Sciences USA*, **102**, 6543–9.

West-Eberhard, M.J. (2005b). Phenotypic accommodation: adaptive innovation due to developmental plasticity. *Journal of Experimental Zoology B, Molecular and Developmental Evolution*, **304B**, 610–18.

CHAPTER 15

Developmental disparity

Alessandro Minelli

Life cycle, individuals, and generations

'Adult phenotypes do not evolve, life cycles do; life *is* development' (Konner, 2010: 741). Shall we subscribe to this flamboyant series of telegraphic statements?

Prima facie, Konner's sentence makes a lot of sense. It points the finger to a generalized weakness of biology, the adultocentric perspective (Minelli, 2003), according to which what really matters in the living world is the adult, the fitness of which can be calculated by determining how much its offspring contributes to the next generation: non-reproductive stages are relevant only in so far as they pave the way to the coming of the adult on the scene.

Development thus reduces to the sequence of morphological events through which the egg (or the seed) is turned into the adult, perhaps together with those alternative changes (as in regeneration) that manage to save an otherwise endangered way to becoming (or to keep being) a reproductively successful adult. Post-reproductive changes, if any, should count as non-adaptive (senescence) and thus foreign to development, as also foreign to development are other non-adaptive changes, e.g. those due to pathogens or to the outbreak of cancer.

In this picture, within the scope of developmental biology, there is hardly a place for events at the level of unicells. On the one side, single-cell stages in a life cycle dominated by multicellularity, such as gametes in animals, are often seen as at best marginal with respect to developmental biology, despite the dramatic events that culminate in the production of these very atypical cells. On the other side, the generalized polyphenism of unicells, prokaryotic and eukaryotic alike, seems too far from the events leading to the production of an adult plant or animal to deserve be considered from the perspective of developmental biology. At most, some aspects of their biology are focussed on in the search for early evolutionary steps towards multicellularity: in particular, the presence in the choanoflagellates, arguably the clade of unicells closest to the metazoans (King et al., 2008; Rokas, 2008), of genes whose products play a key role in cell-to-cell adhesion and communication in animals. This is a reasonable appraoch, to be sure, provided that it does not turn into driving evolutionary biology back into the same finalistic mood still deeply entrenched in developmental biology, that is, provided that we do not regard choanoflagellates as a kind of preparatory step towards full-fledged metazoans.

Well, back to the opening quotation: 'Adult phenotypes do not evolve, life cycles do; life *is* development.' A living organism consists of its whole life cycle, simple or complex as it may be, either limited to unicellular stages or including a dominant multicellular phase. But a life cycle is also a sequence of developmental events. A conventional (unigenerational) life cycle is thus the whole of an individual organism's development. From this perspective, the individual ontogeny, as deployed along a life cycle, seems to be the most sensible unit of development, but I will contend in the following that this is not necessarily the case.

To be sure, following Konner's (2010) sentence we have already taken distance from adultocentrism and also opened a sympathetic eye towards

Towards a Theory of Development. Edited by Alessandro Minelli and Thomas Pradeu

unicellular organisms and unicellular developmental phases in a life cycle dominated by stages with multicellular organization. However, this first step is arguably not enough, if we are trying to circumscribe a framework of developmental biology broad enough to apply, if possible, to every form of life on Earth. To get closer to this target, we shall navigate through the diversity of existing life cycles and forms of change, to test against them our traditional, or naïve, concepts of development. This is the main aim of this chapter, in the hope that this exercise will eventually bring us closer to fixing a general framework, to the service of which we shall target our efforts, to articulate a theory of development.

I will start by taking a closer view of life cycles, remarking in particular that these are often multigenerational. This may conflict with the widespread, although not universal, attitude that identifies a life cycle with the developmental sequence of an individual. I will next focus on the multigenomic nature of many (all?) systems behaving as functional individuals in development. The components of multigenomic systems are functionally integrated, but to various degrees. Therefore, their involvement in developmental events can be very different. We can expect that in many cases, one component will have a leading role in a given developmental event, while other components will only behave as facilitators, serve as scaffolds (see Griesemer, this volume), or even remain neutral. As a consequence, what may be sensibly regarded as 'the' individual involved in a developmental process is something to be determined empirically, case by case.

Second, I will open a window onto the world of unicells, to flag the diversity of processes they exhibit, processes which deserve to be taken under the umbrella of developmental biology.

Third, I will discuss the common, although not commonly disclosed, opinion that only functionally adaptive aspects of change deserve to be regarded as aspects of development. We will see how this misconception supports adultocentrism and why it should be better abandoned, especially in respect to senescence and regeneration. Indeed, I regard both senescence and regeneration as pertaining to development, although neither of these classes of phenomena belongs to the 'normal' course of events along a conventional egg-to-adult life cycle.

In the subsequent three short sections I will address three difficult borders: first, the issue of the (ir)reversibility of development; second, how fuzzy the divide between development and metabolism can be at times; and third, aspects of development other than morphological change. In the final section, I will contrast the properties of development that emerge from this survey of its disparity across different forms of life with the 'current view' of development, especially in its prevailing adultocentric variety.

Life cycle vs individual ontogeny

Strictly speaking, these two concepts—life cycle and individual's ontogeny—overlap only in the case of organisms exclusively reproducing by asexual reproduction. In this case, an individual's development covers the span between the emerging of a founder cell, or a cluster of founder cells (especially in the case of aggregative multicellularity; see 'Steps in developmental evolution: unicellular to multicellular systems'), and the deployment of a new multicellular organism engaged in its turn in reproduction.

The beginning, at least, and possibly also the end of individual development, are arguably better defined in the case of sexually reproducing organisms; but in this case, the life cycle includes an additional segment which does not easily accommodate within the conventional span of an individual's development. In the case of diplobiont organisms, as most of animals are, this additional segment is gametogenesis.

Despite the obvious, and generally dramatic, prevalence of the diploid phase extending between fertilization and death, from the perspective of development, we cannot ignore the haploid phase represented by the gametes, and especially the process through which these are produced, i.e. gametogenesis—although, one might argue that fertilization is also a developmental event. This is indeed a developmental sequence of special interest, as it includes dramatic changes of single-cell systems within an otherwise multicellular organism, and opens difficult questions, e.g. to determine if or when a germ-line cell or a mature gamete acquires full-title individuality as a physically distinct entity, especially in the case of animal eggs, where the control of maternal gene products is usually dominant beyond fertilization and well into the first phase of the embryo's development.

To treat gametogenesis as a developmental process means to close the (life) cycle by adding a phase that is missing if we only consider the zygote-to-adult sequence as properly pertaining to development. Song et al. (2006) explicitly described oogenesis as a case of single-cell development and differentiation. The process indeed is one of such complexity as to suggest adopting a customized periodization based on discrete developmental transitions, such as waves of specific gene activity and the synthesis of specialized organelles. The major developmental changes highlighted by Song et al. (2006) in the oogenesis of a sea urchin species were the transition from a mitotic stem cell to a meiotically committed egg precursor, the onset of a period of rapid nutrient incorporation and storage (the culmination of vitellogenesis), and the completion of meiotic divisions, accompanied by a global change in the mRNA composition and by a massive translocation of granules to form the egg cortex. Summing up, 'Throughout its lifecycle, the oocyte constantly undergoes changes unlike any other cell in the adult' (Song et al., 2006: 401): the egg does not simply provide the theatre of metabolic, essentially reversible changes. This means that we can well speak of development *of* the egg, irrespective of what we may mean by saying that an animal develops *from* the egg.

It is, however, among the haplo-diplobionts that the lack of correspondence between the life cycle and the individual development becomes obvious. The best textbook examples of such a cycle are offered by plants such as mosses, liverworts, ferns, and horsetails, but an alternation of haploid gametophyte and diploid sporophyte happens also in the flowering plants, although their diminutive gametophyte comprises only very few cells (or nuclei)–a male gametophyte being a pollen grain, and a female gametophyte being represented by the ovule together with the associated embryo sac. The alternation of a gametophyte with a sporophyte deserves to be described as a multigenerational life cycle (Minelli & Fusco, 2010), with a gametophyte originating in well-defined way from a spore, while the sporophyte originates from a female ovule fertilized by a spermatic nucleus. Only going through spore, production of a mature gametophyte, meiosis, fertilization, and production of a mature sporophyte (in that order, but not necessarily starting with the spore) is one cycle completed, with the return to the condition from which we moved.

Multigenerational cycles are also exhibited by several animals, which are diplobionts, including many cladocerans, monogonont rotifers, and aphids, all of which alternate bisexual with obligate parthenogenetic generations (a heterogonic cycle), and also a large number of cnidarians, if we accept the conventional description recognizing the sexually produced polyp and the asexually produced medusa as two distinct generations within a metagenetic cycle (Steenstrup, 1845).

Polyp and medusa are obviously different, and morphological differences are sometimes conspicuous also between the amphigonic and parthenogenetic generations of the just-mentioned heterogonic invertebrates. But possibly more interesting from the point of view of developmental biology are those multigenerational cycles where quite similar adults are produced by asexual reproduction (e.g. by budding) and by sexual reproduction (through a more or less conventional series of embryonic stages, which is not mirrored by the early stages in the differentiation of a bud). In the colonial sea squirts of the genus *Botryllus*, the similarity between fully grown animals produced either way is very high, in gross anatomy as well as in many structural details (Manni & Burighel, 2006).

Individuals and generations

Besides the revised perspective on the relationship between life cycle and individual development necessitated by a due consideration of multigenerational life cycles, a wide-angle perspective on the diversity of life cycles calls also into question the definition of generation.

The common-sense meaning of generation, ultimately based on human biology, is a demographic one and is based on reproduction. In multicells, the offspring of parents belonging to an arbitrarily identified F_0 generation belong to the next, F_1, generation. As in many other instances of generalized extension to different kinds of organisms of concepts originally applied to humans and other vertebrates (think of 'individual' but also 'brain', 'eye', etc.), we also meet with problems with the concept of generation. Still in the domain of demography, unicells are problematic, especially when they multiply by symmetrical cell division. The problem is only superficially solved by saying that the F_0 cell is totally replaced by two F_1 cells, that is, that no F_0 individual survives, even for a short while, alongside its offspring. The question is, whether two bacterial or protist cells deriving from a 'founder' cell represent a new generation in the same sense as the F_1 offspring of a multicellular organism (or a couple of multicells) in respect to F_0. Would it be more sensible to equate the growing clone of unicells to the growing clone of blastomeres in the embryo of the multicell? I do not think so. The demographic concept of generation is tightly related to the concept of an individual as a physically distinct object (*pace* Janzen, 1977). This is a reasonable though not universally valid and clear-cut criterion of individuality, and it applies to the unicells and to the multicells, rather than to the individual cells within the latter.

A strictly demographic concept of generation is seriously challenged by those organisms in which reproduction is decoupled from sexuality, but sexuality anyway occurs regularly, as in ciliates. In this clade of protists, the sexual process is conjugation, by which two unicells (conjugants) exchange a haploid copy of their genome but do not fuse together to form a zygote. Following nucleus exchange, two ex-conjugants split off: both of them have a new nuclear asset, but each of them retains the cytoplasm (including the sophisticated architecture of the ciliated cell cortex) of one of the conjugants. The question then arises, do the ex-conjugants belong to the same generation as the conjugants or to a distinct one? As conjugation does not increase the number of individuals, we should conclude that the ex-conjugants do not represent a new generation. However, from a genetic point of view, they differ from the conjugants in the same way as the F_1 offspring of a sexually reproducing animal differ from their F_0 parents. To clear up this difficulty, it may be sensible to split the traditional concept of generation into two separate concepts, by adopting the following definitions:

Demographic generation—a set B of individuals produced by a set A of individuals (representing a distinct demographic generation) by sexual or asexual reproduction.

Genetic generation—a set B of individuals produced by a set A of individuals (representing a distinct genetic generation) by sexual reproduction or by pure sexuality (i.e. sexuality without reproduction, as in the case of ciliates).

In the case of sexual reproduction, a demographic generation is also a genetic generation, and vice versa. In the case of asexual reproduction, it is only matter of demographic generations: an uninterrupted series of events of asexual reproduction gives rise to a series of demographic generations, within one and the same genetic generation. In the case of pure sexuality, it is only a matter of genetic generations: for example, in ciliates, ex-conjugants belong to a new genetic generation but to the same demographic generation as the conjugants.

A problem is, how much sex is required to set a divide between two genetic generations: literally, 'how much sex' (or, what kind of sexual process), not 'how much the two partners should differ in genotype'. The question arises from the fact that, in many organisms, sexuality does not occur through typical meiosis and karyogamy. Bacteria, for example, do not have meiosis and exchange genes through different mechanisms, often involving a virus as a carrier of DNA segments. This difficulty, however, should not surprise us. It is not worse than the difficulty in applying the biological species concept to organisms with uniparental reproduction. Let's anyway assume that the proposed definition of genetic generation applies quite generally, at least to eukaryotes.

My definition of genetic generation corresponds to Gorelick's (2012) generation, but I think that his suggestion to conflate into a single generation everything between meiosis and karyogamy, or between karyogamy and meiosis, gathers too different items under one term; hence the suggestion to identify demographic generations alongside the genetic ones. However, I accept, with Gorelick (2012), the implication that by identifying the origin of a new (genetic) generation with the occurrence of sex necessitates recognizing as a separate generation the life-cycle phase between meiosis and karyogamy.

This is not controversial in plants with alternation of gametophyte (starting with meiosis) and sporophyte (starting with karyogamy), while it may be odd in animals, where the generation between meiosis and karyogamy is limited to the gamete.

By accepting this view of generations, virtually all cycles become multigenerational, one of the included generations being commonly unicellular, as in the case of animal gametes. From the perspective of development, this provides a stronger reason to include unicells (unicellular stages, in this case) within the scope of developmental biology.

We may be hesitant to adopt multiple concepts of generation, but this strategy is not so different from the one often adopted with respect to the cognate concept of individual. Wilson (1999), for example, recognized many different and only partially overlapping notions of individual—as a historical entity, as a functional individual, as a genetic individual, as a developmental individual, and as a unit of evolution. To conflate too many notions within one term, be it individual or generation, does not help in addressing difficult questions; disentangling the semantic mess is one of the first steps towards a conceptual clarification of issues. This is relevant to our issue, because individuals and generations, although insufficient per se to define the scope of development, as discussed below, are nevertheless its main actors and its main units of periodization, respectively.

Multigenomic individuality in development

Another challenge to the naïve notion of individual as *the* unit of development is the multigenomic nature of most living organisms (Dupré, 2010). I do not refer here so much to the coexistence in eukaryotic cells of two or more genomes, as functional remnants of the long-stabilized events of endosymbiosis that gave rise to the mitochondria of (nearly) all eukaryotes and to the plastids of plants and many clades of protists. Let's only mention here that similar processes of complete integration of host and symbiont, at the cellular and, specifically, genomic level, still occur today, as shown for example by the recent integration of genes of the symbiotic/parasitic bacterium *Wolbachia* within the nuclear genome of the seed beetle *Callosobruchus chinensis* (Nikoh et al., 2008) and the mosquito *Aedes aegypti* (Klasson et al., 2009), as well as the similar transfer of genetic material from the bacterium *Buchnera* to the nuclear genome of host aphids (Nikoh & Nakabachi, 2009). Nuclear genes of the alga *Vaucheria litorea* were found in the nuclear genome of the sea slug *Elysia chlorotica*, which feeds on that alga (Rumpho et al., 2008).

I will also simply mention here two peculiar conditions of genome multiplicity, whose impact on development may deserve attention but is very likely minor. One of these conditions is the dikaryotic hypha, containing two independent and different nuclei deriving from the two parent hyphae, which represents a short (Ascomycetes) or lasting (Basidiomycetes) condition of the mycelia of two major groups of fungi. The other condition is heteroplasmy (e.g. Lane, 2012; Passamonti & Scali, 2001), i.e. the presence in the same cells of two populations of mitochondria of different origin: this happens when the sperm cell contributes more than just its nucleus to the zygote, as in several bivalves. Worth mention here is also the multigenomic and multigenerational nature of the seed (of seed plants generally), with tissues deriving from the maternal diploid sporophyte, plus remnants of the haploid megagametophyte, accompanying the tissues of the growing embryo (the new diploid sporophyte); in addition, there is also the embryo's triploid 'brother', i.e. the endosperm, fated to be consumed by the embryo.

Much more relevant here are systems such as lichens and insect-induced galls, as well as the developmental units represented by an animal plus its microbiome (see Nyholm and McFall-Ngai, this volume). Very sensible in this respect is Gilbert et al.'s (2012) notion of the individual animal as that which proceeds from ovum to ovum but only as a consortium of animal cells and microbes (cf. Fraune & Bosch, 2010; Gilbert & Epel, 2009; McFall-Ngai, 2002; Pradeu, 2011).

From the perspective of a theory of development, a most interesting feature is that these multigenomic systems are created de novo at each 'generation'. In lichens, for example, fungal hyphae and algal cells or filaments are initially independent but eventually join together in development to produce the lichen thallus (Sanders, 2006).

Galls produced by the interactions of cynipid wasps with oak (*Quercus*) and rose (*Rosa*) species have been described as novel plant organs (Harper et al., 2004). However, there are structural and developmental features that would justify treating them as multigenomic organisms rather than as peculiar plant organs. The fact that a gall grows attached to the plant would not be a reason for rejecting this suggestion. Indeed, there are other examples of morphologically and genetically distinct individuals, one of which is attached to the other and nutritionally dependent on it; think of mother and offspring in viviparous animals, or the diploid sporophytes of mosses growing attached to the haploid gametophytes. Moreover, similar to lichens, whose species-specific morphology is mainly attributable to the fungal symbiont, galls also exhibit species-specific morphology, mainly dependent on the insect, although the gall tissues are definitely vegetal, while the insect essentially intervenes as a genetically specific 'manipulator'. Manipulation, however, is not simply metabolic, but extends to the organization of the plant's genome. In the inner part of the gall, cell nuclei are often polyploid, sometimes to a very high degree. Hesse (1968) recorded degrees of polyploidy up to 1024*n* in galls induced by *Andricus quadrilineatus* (= *A. marginalis*); this polyploidization by endoreduplication is regarded as induced by the wasp larva growing in the middle of the gall, although the mechanisms of this induction are apparently unknown (Nagl, 1978). Gall induction results from secretions derived from the egg and larva and not from any maternal secretion (Stone et al., 2002).

Even if we consider only the nuclear genome of the 'dominant' multicellular organism, genetic homogeneity within an individual exists only in theory. Point mutations accumulated over the years cause our cells to be a mosaic of slightly different genetic identities. More dramatic, and more or less extensively systemic, are the instances of mosaicism in which a multicellular organism contains two or more populations originating from different zygotes. For example, fusion of separate larvae has been observed in the freshwater sponges *Ephydatia* and *Spongilla* (Brien, 1973). In various forms, mosaicism is widespread among colonial animals (sponges, cnidarians, bryozoans, tunicates), but it has been occasionally documented even in humans and is, amazingly, the rule in a group of small South American primates (marmosets and tamarins). In these monkeys, reciprocal chimerism between twins is the rule, especially in the haematopoietic tissue but also in other somatic tissues and even in the germ line (Haig, 1999; Ross et al., 2007).

Development of unicells

To some extent, most unicellular organisms, be they bacteria (e.g. Vlamakis et al., 2008), fungi (e.g. *Candida*; Shapiro et al., 2009, 2012), protists, or algae (e.g. *Chlamydomonas*; Pan & Snell, 2000), exhibit alternative states of differentiation, reversible or irreversible, under different conditions. It seems sensible to describe the transitions from one state to another as developmental processes (see also Griesemer, this book, for an example).

In the recent past, up to 24 different phenotypes have been described as representing as many different forms of a unicellular 'alga', the dinoflagellate *Pfiesteria piscicida* (e.g. Burkholder et al., 2001), but more recent studies have shown that most of these were indeed culture contaminants that have nothing to do with *Pfiesteria*. The latter's cycle does apparently include a couple of forms only, as in the majority of dinoflagellates (Litaker et al., 2002; Peglar et al., 2004). Complex cycles with several distinct phenotypes do anyway exist, e.g. in the malaria parasite *Plasmodium* and many other apicomplexans, but also in free-living protists, including several ciliates (e.g. *Tetrahymena*; Ryals et al., 2002; *Bromeliothrix*; Foissner, 2010).

A good catalogue of developmental events in unicells is not available to date. The relevant literature is not simply scattered but is often difficult to interpret. For example, papers describing multiple phenotypes in a clonal culture of bacteria or protists are very often silent about whether the transition from one phenotype to the other occurs with cell division or not. Certainly, mitosis does not accompany the transition from trypomastigotes to amastigotes in *Trypanosoma cruzi* (Contreras et al., 2002; Sooksri et al., 1991). An intrinsic ability to change from one cell type to another is possibly the default condition of cells generally, either isolated or within a multicellular organism. If so, a cell maintains a stable phenotype only by continuously receiving and integrating information, from both internal and external sources—its way 'to remember how to behave within an organ', to use Bissell's (2003: 103) colourful expression.

Adultocentrism

Sterelny (2000) remarked that evolutionary patterns and evolutionary trends are generally discussed based on the form of adult organisms, although embryology, per se, has its generalizations, e.g. von Baer's (1828) law and the notions of the zootype (Slack et al., 1993) and the phylotypic stage (Sander, 1983) with the associated egg-timer metaphor (Duboule, 1994). Sterelny's remark extends to well-known analyses of morphological complexity, disparity, and stasis (e.g. Conway Morris, 1998; Gould, 1989, 1996; McShea, 1996). Ironically, especially with reference to Stephen J. Gould (think of the spandrel paper by Gould & Lewontin, 1979), adultocentrism is fostered by adaptationism, as you cannot measure Darwinian fitness of non-reproductive stages.

As mentioned before, most biologists are probably reluctant to use the notion of development to describe morphological changes in unicells. To a large extent, this is another consequence of an adultocentric view of living organisms, according to which the adult is the 'real' organism, to which the faculty of reproduction (or, at least, of sexual reproduction) is nearly universally limited. As a consequence, all pre-reproductive stages are seen as merely preparatory, and development is thus literally limited to the sequence of events through which the adult is finally produced. However, this is a narrow view (Minelli, 2003, 2009a, 2011b).

First, sexual reproduction is not necessarily limited to the conventional, unique, adult phase. For example, some ctenophores have two widely separated reproductive periods, between which their body organization changes substantially (dissogony). Second, the egg, far from being 'the primordial animal cell', a direct descendant of the unicellular ancestors of animals, is one of the most 'modern', elaborate, and still rapidly evolving cell types (Boyden & Shelswell, 1959), thus hardly representing an evolutionary starting point upon which development would have been grafted with the advent of multicellularity. Newman (2011), based on the idea that different body plans may have remotely originated by self-organizing physical processes operating within clusters of cells, suggests that eggs may represent instead a set of independent evolutionary innovations grafted at some point onto the developmental trajectories of such aggregates.

A consequence of the identification of development with the series of changes occurring along an organism's life cycle is the often unjustified identification of the achievement of the adult somatic phenotype with the achievement of sexual maturity (Minelli & Fusco, 2013). The decoupling of these two aspects of development is amply demonstrated by the occurrence of neoteny and other forms of heterochrony (e.g. de Beer, 1958; Gould, 1977; McKinney & McNamara, 1991; McNamara, 1986, 1995). A particularly intriguing feature is periodomorphosis, i.e. the occurrence, during an individual's life, of two or more reproductive periods between which a non-reproductive stage is intercalated. Distinct morphological changes separating the latter from both the preceding and the subsequent reproductive stages are not restricted to the organs of reproduction, but additionally involve different somatic features. This phenomenon has been described in several julid millipedes (Sahli, 1985, 1989, 1990; Verhoeff, 1923, 1933, 1939) and in some springtails (Cassagnau, 1985). In female isopods (such as woodlice), there is a regular alternation between mating-ready stages, with fully developed egg-bearing appendages, and intervening stages, in which those appendages are reduced.

Non-adaptive aspects of development

The adultocentric view of development is arguably the most obvious consequence of a conception of development as restricted to positively selected, adaptive changes exhibited by an individual organism before sexual maturity or at most until the end of the individual's reproductive age. This restriction, however, is difficult to defend. First of all, it is unnecessarily finalistic: who can predict whether an organism's long-term response to an external disturbance will eventually increase its fitness? Second, its strict application would cause the establishment of unwarranted divides.

Regeneration events, although distinct from the 'normal' sequence of events in a species' life cycle, are often categorized as developmental (a choice not shared, however, by Vervoort, 2011) because of their expected adaptive value. However, we may sometimes question the adaptive significance of regeneration. At least, it imposes trade-offs that may be heavy to sustain. In sponges and corals, regeneration has negative effects, often strongly so, on 'normal' processes such as somatic growth, sexual reproduction, and the ability of these animals to defend themselves against predators, to face competition, and to recognize conspecifics (Henry & Hart, 2005).

If regeneration is not necessarily adaptive, what about ageing? If senescence is simply due to organic failures accumulating in a part of individual life which is not subject to natural selection, because its target is the post-reproductive segment of life, this amounts to saying that this phenomenon is intrinsically non-adaptive. If non-adaptive organic changes should not be regarded as developmental, then the study of senescence would be outside the scope of developmental biology. But this decision would make little sense, for example, in a social species like *Homo sapiens*, where there can be an adaptive social or cultural role for individuals in the post-reproductive age. On the other hand, would we say that an individual has not undergone development if, when adult, it is eventually sterile (cf. Griesemer, this volume)?

Thus, if by life cycle we mean all the events between two subsequent occurrences of a given stage, e.g. the egg (if present in the cycle) or the adult stage (to the extent that this is uniquely identifiable in the species considered), then development is not restricted to the 'normal' events occurring along the life cycle itself, but extends (i) to the senescent offshoots spinning off from the cycle and to (ii) a diversity of other-than-normal events, only some of which (regeneration) are quite reasonably adaptive; others (senescence) may be selectively irrelevant, although eventually rescued by exaptation under peculiar circumstances such as sociality in animals with life expectancy long extended beyond the reproductive age; and still others (such as carcinogenesis) are definitely pathological.

I agree with Oyama (2000), Bateson and Gluckman (2011), and a few others in regarding development as a process continuing until death (see also Minelli, 2011b). Moreover, in terms of the mechanisms involved it is also difficult, if even possible, to fix a strict divide between adaptive developmental changes and events of pathological morphogenesis such as carcinogenesis. I am inclined to regard these too as legitimately pertaining to developmental biology. In a similar vein, Ramos (2012) has recently proposed to treat even the inflammatory events typically described in immunological, pathological, and pharmacological contexts as mechanisms of animal development.

Ageing

According to some authors (e.g. Martin, 2011), there may be general common principles behind the different phenomena associated with ageing; but a broad, comparative survey of relevant evidence invites caution (Fahy, 2010).

Fahy (2010) divides organisms into four classes, according to the occurrence of ageing: (i) occurring only after the achievement of sexual maturity (zero ageing prior to sexual maturity); (ii) occurring throughout life; (iii) occurring before, but not after achieving sexual maturity; and (iv) not occurring at all.

The last two groups, and group (iii) especially, do not fit easily within the popular interpretation of ageing as being due to the accumulation of deleterious mutations whose phenotypic expression is limited to the post-reproductive age. Evidence for zero ageing limited to juvenile stages is indeed scanty and uncertain: Fahy's (2010) only example is the decline of the mortality rate of the medfly (*Ceratitis capitata*) with the adult's advancing age, as reported by Carey et al. (1992). However, organisms without evidence of 'actuarial ageing' (Fahy's group (iv)) are more numerous and include sponges; sea anemones and hydras; the queens of ants, bees, and probably some termites and wasps; some sea urchins; perhaps, tubeworms (*Lamellibrachia* sp.); and certain clams (*Arctica*) (Fahy, 2010; Jones, 1983; Martinez, 1997; Sebens, 1987).

Relevant to an effort to define the nature of development and, in particular, the limits of individual development are the correlations between lifespan, growth, ageing, and death. Once more, browsing through the data available for representatives of the different kinds of organism shows how diverse are these correlations, even between closely related taxa. A few examples follow.

(i) There are animals (and plants) with indeterminate growth, but these are not limited to animals (and plants) with simple organization; neither is this condition generally shared by all the members of a larger clade. Examples of animals with indeterminate growth are found among the annelids (e.g. the small freshwater oligochaete *Pristina*), the molluscs (e.g. the giant clam *Tridacna*), the echinoderms (e.g. the sea urchin *Strongylocentrotus*), the decapod crustaceans, the bony fishes, the reptiles, and also mammals, e.g. the bison, the giraffe, and the elephants (Karkach, 2006).

(ii) Total lifespan can be prolonged by regeneration, sometimes to a very high extent, as in the small flatworm *Macrostomum lignano* (Egger et al., 2006). This suggests that we should reverse the popular, finalistic view, that many animals regenerate in order to live longer, to say instead that some animals live longer because they are able to regenerate (Minelli, 2009a).

(iii) Total lifespan can be also prolonged by the prolonged absence of a specific environmental cue required by the organism to switch to a more advanced developmental stage. Temporary lack of the specific signals required to metamorphose will eventually increase total lifespan in the nudibranch gastropod *Phestilla sibogae* by as many days as its metamorphosis has been delayed (Miller & Hadfield, 1990). The marine snail *Fusitriton oregonensis* can spend up to four and a half years in the active larval state with no sign of senescence, and the post-larval segment of its life history exhibits normal growth and requires the usual time to achieve reproductive maturity (Strathmann & Strathmann, 2007).

(iv) Similarly, starvation or excessive crowding causes a second instar larva of *Caenorhabditis elegans* to enter the non-feeding dauer stage, in which it may remain up to six times the total length of the normal adult lifespan of the worm; but the time spent as a dauer larva will be simply added to the individual's lifespan, as its adult stage will have the same length as in an individual whose development was not arrested as a dauer larva (Klass & Hirsh, 1976).

(v) In Beck and Bharadwaj's (1972) experiments, under conditions of strict starvation, nearly mature (sixth instar) larvae of the dermestid beetle *Trogoderma glabrum* underwent retrogressive moults that led to progressively smaller larvae. Following 20–36 weeks of starvation, these larvae were fed again and regained their previous size and weight through a series of moults. This cycle of retrogression and regrowth could then be repeated. The normal lifespan of *T. glabrum* is 8 weeks; but in this experiment, the life of some larvae was prolonged up to more than two years.

(vi) Many perennial plants have indeterminate growth; nevertheless, some of them are subject to mechanisms determining their sudden death immediately after their only flowering season. This is best known for bamboos, with a species-specific lifespan often fixed at 30, 60, or 120 years. Another example is the legume *Tachigali versicolor*, popularly known as the 'suicidal tree', which can grow to a height of up to 40 meters and then flower, produce fruit, and die within a year (Foster, 1977).

(vii) Senescence is evident in many unicells, for example in hypotrichous ciliates. Following genome 'rejuvenation' through conjugation, the progeny obtained by continuing divisions progresses through (i) a juvenile phase, in which it does not undergo conjugation, (ii) a 'mature', mating-prone phase, and (iii) a senescent phase, eventually ending with the clone's extinction after a more or less predictable, lineage-specific number of mitoses (Duerr et al., 2004). Senescence, however, does not seem to be universal, even among the ciliates. *Tetrahymena pyriformis* has been cultured for more than 50 years without performing conjugation and without showing signs of senescence (Nanney, 1986).

Ageing may well be maladaptive, in general, but it encompasses an amazing diversity of predictable changes that cannot escape attention from developmental biologists; and, thus, it can be part of a satisfactory theory of development.

Canalization and reversibility of development

In introducing development in unicells, I used the term polyphenism to refer to the alternative phenotypes produced by an organism, at different times or in different conditions, without any related difference in the genotype. More precisely, the trophozoite and merozoite stages of *Plasmodium*, or the amastigote and trypomastigote stages of *Trypanosoma*, are instances of sequential, or individual polyphenism, whereas, e.g. queen and worker bee are an example of parallel, or population-level polyphenism.

Polyphenism must not be confused with polymorphism, where different phenotypes are produced, even under the same environmental conditions, due to genetic differences, as between ladybirds with different patterns of spots on the elytra within a local population (Fusco and Minelli 2010).

Both polyphenism and polymorphism occur sometimes in the same species, or in closely related species, thus offering a precious opportunity to look at possible ways in which an environmentally controlled polyphenism can eventually transform into a genetically controlled polymorphism. In the pea aphid *Acyrthosiphon pisum*, the two phenomena coexist within the species (e.g. Brisson, 2010). In females, presence or absence of wing is an environmentally controlled polyphenism, while in males the development of wings is genetically controlled and determined by polymorphism at a locus (*aphicarus*) on the X chromosome. Interestingly, irrespective of this fundamental difference in the mechanism of release of the winged vs wingless phenotype, the genes actually involved in accomplishing the developmental switch between the two alternative forms appear to be the same.

The increased canalization of development depending on the increased degree of genetic control causes the same phenotype to be more predictably expressed under a precise scenario of external conditions. As a consequence, the expression of that phenotype will more likely recur, in developmental time, subsequent to another, precisely defined developmental step. In its turn, the same robustly controlled phenotype is liable to offer the specific conditions for the predictable expression of another phenotype, or developmental step, to likely recur. This way, different genetically controlled alternative phenotypes compatible with a given genome, but separately expressed in the presence of different transcriptome patterns, will increasingly evolve towards a strictly controlled temporal sequence of developmental stages. As a further consequence, the probability that this developmental sequence will be occasionally reversed will likely vanish soon, as is in fact demonstrated by the life cycles of the vast majority of animal species.

Exceptions to this rule are indeed extremely rare. Best known is arguably the case of the hydrozoan *Turritopsis nutricula*: under certain conditions, a medusa is able to revert to polyp (Bavestrello et al., 1992; Piraino et al., 1996). A less adequately investigated case is provided by the cercariae and metacercariae of some gymnophalline digeneans, which do not transform into adult maritae, as expected, but revert to germinal sacs similar to those from which they have been produced (Galaktionov & Dobrovolskij, 2003).

At the cell level, to reverse a developmental sequence will generally mean to go back from a differentiated to an undifferentiated (or less differentiated) state. This may occur following injury, as an early phase of the regeneration process. Dedifferentiation may even renegotiate the divide between germinal and somatic cells. Following cuts through the region of gonads, specimens of the free-living flatworm *Dugesia lugubris* produced blastemas primarily derived from somatic cells but also from germ cells. Eventually, these blastema cells produced somatic tissue, thus demonstrating that cells committed to germ-line fate can dedifferentiate and take a new, somatic differentiation (Gremigni et al., 1980, 1982; Gremigni & Miceli, 1980).

The discovery of the potential reversibility of cell specification in animals has opened the way to the rapidly expanding field of somatic cloning, up to the production of new complete animals, even in vertebrates (Gurdon et al., 1958; Wilmut et al. 1997; and a huge subsequent literature), with the associated bioethical debate.

In plants, the reversibility of cell differentiation is unproblematic and has been long known. It is currently exploited commercially, for example in the practice of micropropagation, with strictly uniform clones being produced starting with callus cells, i.e. from cells produced by a growing culture of dedifferentiated cells.

Development vs metabolism

In general terms, development and metabolism are different kinds of processes. For example, there is a difference between, on the one hand, the growth and differentiation of that part of the digestive tract that eventually becomes our stomach, and, on the other, the functional changes our stomach wall's cells undergo every time we require them to digest a meal. However, the divide between metabolism and development is not always clear-cut. The involvement in organogenesis of what would be otherwise regarded as 'housekeeping genes' has been demonstrated by the specific defects correlated to a number of mutations in those genes encoding proteins that work in a diversity of basic metabolic and cellular functions (Tsukaya et al., 2013).

Many curious facts are clustered on either side of the uncertain divide between metabolism and development and contribute to demonstrating its often arbitrary nature. A good example is offered by the Burmese python (*Python molurus*), whose bulky prey items require a lot of physiological and even morphological changes to be efficiently digested. Within two days after feeding, the muscular mass of the snake's heart ventricle increases by 40% (Andersen et al., 2005), and the length of its intestinal villi increases fivefold (Secor, 2008). Why not regard these changes as developmental phenomena, despite the fact that the heart mass increase is not due to cell division, but to the enhanced expression of genes coding for contractile proteins? The reversibility of the process should also not be regarded as an argument to deny the developmental nature of these dramatic changes in the python's heart mass; a similar argument would delete from developmental biology whole chapters such as growth and even cause serious trouble in relation to instances of dedifferentiation, as mentioned in 'Canalization and reversibility of development' .

A further developmental feature of the python's heart story is that it involves the whole multigenomic community of the snake's gut microbiome. Following the ingestion of the prey, bacterial species diversity increases significantly, and the abundance of individual taxa also change, with *Clostridium* and *Lactobacillus* species, among others, progressively obtaining dominance over different bacterial taxa previously dominating in the gut of the fasting snake (Costello et al., 2010).

Also related to feeding is the merging of developmental with metabolic events observed during the pupal stage in the ant lion and its relatives. In the larva of these insects, the midgut is a blind-end sac, not communicating with the hindgut. Therefore, throughout its larval life, the ant lion accumulates in the midgut a large amount of waste materials, of which it can finally dispose only after completing metamorphosis, when the midgut eventually opens into the hindgut (Jepp, 1984; McDunnough, 1909).

Temporal phenotypes—development beyond morphological change

The phenomena we generally take into account as within the scope of development are changes in morphology, of the whole organism or of one of its parts. However, development includes also other, non-morphological aspects. Systemic and long-lasting (or definitive) functional changes such as the achievement of sexual maturity, or the regular massive falling of leaves from broad-leaved trees in autumn, are developmental events that are more pervasive than the associated morphological changes the organism undergoes. But these changes may be also considered from a perspective still more distant from morphology, i.e. the temporal one.

In a different context, it has been recently remarked that temporal phenotypes deserve to be studied by evolutionary developmental biology, as a legitimate companion to morphological phenotypes (Minelli & Fusco, 2012). A similar message is flagged here in the context of developmental biology, with the hope that it will be fleshed out before long.

One of these temporal phenotypes is the total life span, whose length may be controlled independently from the processes causing ageing (e.g. Gladyshev 2012). More generally, beyond its basic, mechanistic aspects at the level of metabolism (up to the cycling of the pulsating vacuole of a *Paramecium*, or our circadian wake-and-sleep periodicity), the temporal organization of life includes many aspects pertaining to development. Obvious examples are (i) arthropod moults, in terms of their succession, the temporal spacing between two subsequent moults, and the control—if any—of their number throughout an individual's life; and (ii) a plant's transition from growing/vegetative phase to the flowering phase.

Size and cellularity of developmental systems

I have recently suggested (Minelli, 2011a) that a useful starting point towards a theory of development could be a 'null model' defining a kind of inertial behaviour of cells, such that any systematic deviation from that behaviour would define a class of developmental phenomena. According to Griesemer (this volume), a theory of development should not be restricted to systems made of cells. In the light of the evidence summarized in this chapter, I must admit that his claim is justified by empirical reasons, besides the epistemic advantages outlined in his chapter.

Development below the level of the cell

As we have seen before, developmental phenomena are not restricted to multicells, and their scope at the one-cell level goes far beyond the vicissitudes of gametes and zygotes. Furthermore, there are developmental phenomena even at the subcellular level, for example the predictable (shall we say programmed?) changes in genome organization in ciliates (Bracht et al., 2013) and the less dramatic but somehow comparable genome rearrangements found in the somatic cells of lampreys (Smith et al., 2009, 2012) and *Ascaris* (Wang et al., 2012). The uniquely extensive and complex changes in genome architecture observed in ciliates are so exceptional, and so little known outside the world of protistologists, that a short summary should be given here, following Bracht et al. (2013).

According to textbook descriptions, a ciliate's cell includes a macronucleus and one or more micronuclei; in fact, a single cell can contain up to hundreds of nuclei, in the different phases of genome fragmentation, degradation, synthesis, and amplification (Prescott, 1994). Genome copies eventually forming the macronucleus are cut into pieces, scrambled and united again into functional coding sequences. In some species, just 2% of the genome is saved into these functional genes, and the tiny macronuclear 'chromosomes' are often reduced to one gene each and lack centromeres (Prescott, 1994; Swart et al., 2013).

Steps in developmental evolution: unicellular to multicellular systems

To bridge the gap between unicells and multicells from the perspective of developmental biology, the most obvious problem to be addressed is how the transition to multicellularity is actually accomplished. During the last decade, much progress has been made in understanding the emergence of key mechanisms of developmental pattern formation from mechanisms (e.g. Knoll, 2011; Newman, 2010; Newman & Bhat, 2008; Newman et al., 2006) and molecules (e.g. Abedin & King, 2008, 2010; King, 2004; King et al., 2008; Rokas et al., 2008; Sebé-Pedrós et al., 2010) already present in the unicellular world.

However, from the perspective of a theory of development, much more interesting is the question of what the developmental processes occurring in unicells (rather than the proteins they synthesize) have in common with those occurring in multicells. Of course, we should not expect to find among the unicells those aspects of development that are strictly related to multicellular organization but (at least) those that involve the production of alternative, stable cellular phenotypes.

Indeed, we may regard cell differentiation as a predictable form of polyphenism. All the diverse kinds of cells in an organism share (with a few marginal exceptions) an identical genome, and their different phenotypes are largely the result of the history of their interactions with their cellular environment.

This circumstance may suggest (Mikhailov et al. 2009; Minelli, 2009b; Minelli and Fusco, 2010; Sachwatkin, 1956 (originally published as Zakhvatkin, 1949); Schlichting, 2003; Valentine, 2004) that a first degree of cell differentiation within a multicellular organism may have originated by the stabilization, and increased predictability, of a pattern of coexistence of alternative cellular phenotypes. Thus, what at an earlier stage of evolution might have been different forms within the polyphenism of a unicell may have turned into the basis for cell differentiation within a multicell. This model may likely explain the origin of multicells with a very small number of coexisting cell types, but this in turn may have also provided the condition from which, and onto which, higher degrees of cell differentiation may have evolved.

Potentially different routes to differentiation could be expected among those organisms where the multicellular condition is obtained by aggregation of previously autonomous unicells rather than by the remaining together of the mitotic products of a founder cell, usually a spore or a zygote. However, the adventures of clades with aggregative multicellularity into the intricacies of cell differentiation are very limited, despite the fact that this way to becoming multicellular has evolved multiple times. Several examples are known in the domain Bacteria, most typically in myxobacteria such as *Myxococcus xanthus* (e.g. Hartzell & Youderian, 1995), but also in more 'normal' bacteria such as *Pseudomonas aeruginosa* (Klausen et al., 2003; for a review, see Webb et al., 2003). Among the protists, aggregative multicellularity has evolved at least six times. The case of the 'fruiting body' (sorocarp) of the cellular slime molds (Dictyostelida) is well known, to the extent that one representative of this little clade (*Dictyostelium discoideum*) long ago emerged as a model organism. Lesser known, independently evolved examples of aggregative multicellularity (cf. Wegener Parfrey & Lahr, 2013) have occurred among the rhizarians (the cercozoa *Guttulinopsis vulgaris*; Brown et al., 2012a), the labyrinthulids (*Sorodiplophrys*; Dykstra & Olive, 1975), amoebozoans (the tubulinean *Copromyxa*; Brown et al., 2011), the opisthokonts (the nucletmycean *Fonticula alba*; Brown et al., 2009), ciliates (*Sorogena stoianovitchae*; Lasek-Nesselquist et al., 2001), and heterolobosans (the acrasiids; Brown et al., 2012b).

On a different level, but still involving the sum of previously independent units eventually merging into a larger multicellular system, are phenomena like the 'blastomere anarchy' described in free-living flatworms of the genus *Dendrocoelum* (Hallez, 1887) and the split-embryo development of the freshwater fish *Cynolebias*, where one zygote often gives rise to two separate 'twin embryos' that subsequently merge to reconstitute a single embryo at gastrulation (Carter & Wourms, 1993).

Modularity and the effects of size

Within an organism comprising more than a handful of cells, development does not proceed uniformly and at the same pace throughout the whole system. Sooner or later, a number of more or less independent, more or less strictly circumscribed, and more or less durable units emerge, within which the developmental dynamics are different from what happens in the units nearby. These units have been called 'modules' (e.g. Callebaut & Rasskin-Gutman, 2005; Schlosser, 2002; Schlosser & Wagner, 2004).

This is not the place to discuss the merits of a description of development in terms of modules. Let's however remark that development is not simply a set of processes from the perspective of which living systems generally appear to articulate into a plurality of modules; it is also a set of processes through which a system's modularity is subject to change. Growing size is the most obvious reason for developmental change in modularity. An embryo's small size allows for effective cell-to-cell signalling followed by rapid changes in gene expression involving the whole organism or a large part of it; however, this becomes less and less feasible as the animal grows. This explains why embryonic development can proceed in the absence of a centralized control mechanism and is mainly or entirely self-organizing, whereas later, especially post-embryonic development becomes a largely local affair (Minelli, 2003; Nijhout, 1999). Regulation is now possible only within smaller regions, or developmental fields, which acquire their developmental autonomy; in other words, more and more modules emerge. Of course, this does not rule out the possibility of long-distance and even global coordination, mainly through hormones.

Conclusion

'Thus all morphological development is contained in the previous state. This work is pure repetition. [. . .] there is no morphology without predecessors. In reality we do not witness the birth of a new being:

we see only a periodic continuation [. . .] If things happen in this way, it is because the being is in some way imprisoned by a series of conditions which it cannot escape, since they are always repeated in the same way outside and inside it' (Bernard, 1878: 240–242).

This is still an excellent description of the individual life cycle. Moreover, we do not need to force it too much beyond its literal wording and meaning to accommodate for two different extensions, both of which are important in the perspective of a theory of development.

On the one hand, even if Claude Bernard clearly intended to apply his global view of morphological change to ontogeny, the same sentences would also essentially apply to phylogeny, thus opening the question of a possible holistic view of all these processes as aspects of one and the same chain of events (ontophylogenesis), as advocated by Kupiec (2009; and his chapter in this book), who recently called attention to the passage cited above from Bernard's *Leçons sur les phénomènes de la vie communs aux animaux et aux végétaux* (*Lectures on the Phenomena of Life Common to Animals and Plants*).

On the other hand, to say that 'there is no morphology without predecessors' and 'the being is in some way imprisoned by a series of conditions which it cannot escape' is just a way to point to the predictability of developmental change (without any obligation to see it as programmed); that is, to seeing in all and any stage the material memory of previous events (Minelli, 1971) that will bias, and even strictly canalize, the next possible developmental event. This will not necessarily be an adaptive event or one within the strict limits of an ever-repeating cycle.

The concept of development suggested in these pages is more extensive than the current notion of individual development or ontogeny and is definitely in contrast with its usual adultocentrism.

Box 15.1 summarizes the main differences between the 'disparity view' of development presented in this chapter and the 'common view'. which generally includes within the scope of developmental biology either only the individual development (egg to adult) or, a bit more liberally, the individual life cycle (egg to egg, or adult to adult).

To go back to the quotation with which I opened this chapter: 'Adult phenotypes do not evolve, life

Box 15.1 Properties and scope of development, according to the 'common view' and the 'disparity view' presented in this chapter.

	'Common view' of development		'Disparity view' of development
	Adultocentric perspective	**Other components**	
Development			
Runs from egg to adult	yes		not necessarily
Extends to post-reproductive events, especially ageing	no		yes
Is limited to adaptive traits	mostly, yes		no
Is strictly irreversible	yes		no
Includes pathological changes, especially carcinogenesis	no		yes
Is strictly distinct from metabolism		yes	sometimes very difficult to distinguish
Exists in unicells		no	yes
Concerns only morphological aspects		yes	no—add e.g. temporal phenotypes

cycles do; life *is* development' (Konner, 2010: 741). The study of individual development will arguably remain a privileged research target of developmental biology, but—from a broader and theoretically sensible perspective—the notion of individual development is only the intersection between the difficult notion of biological individuality (e.g. Bouchard & Huneman, 2013; Clarke, 2010, 2012; Folse & Roughgarden, 2010; Pradeu, 2010; Santelices, 1999; Wilson, 1999) and the not less difficult notion of development. I hope that considering development from the wider perspective suggested here may help to avoid formulating theories of development destined too soon to show their low fitness when confronted with the variegated expressions of the real world.

Acknowledgments

This chapter has strongly benefited from the critical comments of Wallace Arthur, Elena Casetta, Giuseppe Fusco, and Thomas Pradeu, even if many of the ideas presented here diverge anyway from their suggestions.

References

Abedin, M., and King, N. (2008). The premetazoan ancestry of cadherins. *Science*, **319**, 946–8.

Abedin, M., and King, N. (2010). Diverse evolutionary paths to cell adhesion. *Trends in Cell Biology*, **20**, 734–42.

Andersen, J.B., Rourke, B.C., Caiozzo, et al. (2005). Postprandial cardiac hypertrophy in pythons. *Nature*, **434**, 37–8.

Bateson, P., and Gluckman, P. (2011). *Plasticity, Robustness, Development and Evolution*. Cambridge University Press, Cambridge.

Bavestrello, G., Sommer, C., and Sarà, M. (1992). Bi-directional conversion in *Turritopsis nutricula*. *Scientia Marina*, **56**, 137–40.

Beck, S.D. and Bharadwaj, R.K. (1972). Reversed development and cellular aging in an insect. *Science*, **178**, 1210–11.

Bernard, C. (1878). *Leçons sur les phénomènes de la vie communs aux animaux et aux végétaux*. Baillière, Paris; trans. Hoff, E.H., Guillemin, R., and Guillemin, L. (1974) as *Lectures on the Phenomena of Life Common to Animals and Plants*. Thomas, Springfield.

Biella, S., Smith, M.L., Aist, J.R., et al. (2002). Programmed cell death correlates with virus transmission in a filamentous fungus. *Proceedings of the Royal Society of London B*, **269**, 2269–76.

Bissell, M.J., Mian, I.S., Radisky, D., et al. (2003). Tissue specificity: structural cues allow diverse phenotypes from a constant genotype. In G.B. Müller and S.A. Newman, eds., *Origination of Organismal Form. Beyond the Gene in Developmental and Evolutionary Biology*. MIT Press, Cambridge, pp. 103–117.

Bouchard, F., and Huneman, P., eds. (2013). *From Groups to Individuals*. MIT Press, Cambridge.

Boyden, A., and Shelswell, E.M. (1959). Prophylogeny: some considerations regarding primitive evolution in lower Metazoa. *Acta Biotheoretica*, **13**, 115–30.

Bracht, J.R., Fang, W., Goldman, A.D., et al. (2013). Genomes on the edge: programmed genome instability in ciliates. *Cell*, **152**, 406–16.

Brien, P (1973). Les démosponges. Morphologie et reproduction. In P.P. Grassé, ed., *Traité de Zoologie*, **Vol. 3**. Masson, Paris, pp. 133–461.

Brisson, J. (2010). Aphid wing dimorphism: linking environmental and genetic control of trait variation. *Philosophical Transactions of the Royal Society of London B*, **365**, 605–16.

Brown, M.W., Kolisko, M., Silberman, J.D., et al. (2012a). Aggregative multicellularity evolved independently in the eukaryotic Supergroup Rhizaria. *Current Biology*, **22**, 1123–7.

Brown, M.W., Silberman, J.D., and Spiegel, F.W. (2012b). A contemporary evaluation of the acrasids (Acrasidae, Heterolobosea, Excavata). *European Journal of Protistology*, **48**, 103–23.

Brown, M.W., Spiegel, F.W., and Silberman, J.D. (2009). Phylogeny of the 'forgotten' cellular slime mold, *Fonticula alba*, reveals a key evolutionary branch within Opisthokonta. *Molecular Biology and Evolution*, **26**, 2699–709.

Burkholder, J.M., Glasgow, H.B., and Deamer-Melia, N.J. (2001). Overview and present status of the toxic *Pfiesteria* complex. *Phycologia*, **40**, 186–214.

Callebaut, W., and Rasskin-Gutman, D., eds. (2005). *Modularity: Understanding the Development and Evolution of Natural Complex Systems*. MIT Press, Cambridge.

Carey, J.R., Liedo, P., Orozco, D., et al. (1992). Slowing of mortality rates at older ages in large medfly cohorts. *Science*, **258**, 457–61.

Carter, C.A., and Wourms, J.P. (1993). Naturally occurring diblastodermic eggs in the annual fish *Cynolebias*: implications for developmental regulation and determination. *Journal of Morphology*, **215**, 301–12.

Cassagnau, P. (1985). Le polymorphisme des femelles d'*Hydroisotoma schaefferi* (Krausbauer): un nouveau cas d'épitoquie chez les collemboles. *Annales de la Société entomologique de France, Nouvelle Série*, **21**, 287–96.

Clarke, E, (2010). The problem of biological individuality. *Biological Theory*, **5**, 312–25.

Clarke, E. (2012). Plant individuality: a solution to the demographer's dilemma. *Biology and Philosophy*, **27**, 321–61.

Contreras, V.T., Navarro, M.C., De Lima, A.R., et al. (2002). Production of amastigotes from metacyclic trypomastigotes of *Trypanosoma cruzi*. *Memórias do Instituto Oswaldo Cruz*, **97**, 1213–20.

Conway Morris, S. (1998). *The Crucible of Creation: The Burgess Shale and the Rise of Animals*. Oxford University Press, Oxford.

Costello, E.K., Gordon, J.I., Secor, S.M., et al. (2010). Postprandial remodeling of the gut microbiota in Burmese pythons. *ISME Journal*, **4**, 1375–85.

de Beer, G.R. (1958). *Embryos and Ancestors*, 3rd ed. Clarendon Press, Oxford.

Duboule, D. (1994). Temporal colinearity and the phylotypic progression: a basis for the stability of a vertebrate Bauplan and the evolution of morphologies through heterochrony. *Development*, **1994** *Supplement*, 135–42.

Duerr, H.P., Eichnera, M., and Ammermann, D. (2004). Modeling senescence in hypotrichous ciliates. *Protist*, **155**, 45–52.

Dupré, J. (2010). The polygenomic organism. *The Sociological Review*, **58**, 19–31.

Dykstra, M.J., and Olive, L.S. (1975). *Sorodiplophrys*—unusual sorocarp producing protist. *Mycologia*, **67**, 873–9.

Egger, B., Ladurner, P., Nimeth, K., et al. (2006). The regeneration capacity of the flatworm *Macrostomum lignano*—on repeated regeneration, rejuvenation, and the minimal size needed for regeneration. *Development Genes and Evolution*, **216**, 565–77.

Fahy, G.M. (2010). Precedents for the biological control of aging: experimental postponement, prevention, and reversal of aging processes. In G.M. Fahy, ed., *The Future of Aging. Pathways to Human Life Extension*. Springer, Dordrecht, pp. 127–225.

Foissner, W. (2010). Life cycle, morphology, ontogenesis, and phylogeny of *Bromeliothrix metopoides* nov. gen., nov. spec., a peculiar ciliate (Protista, Colpodea) from tank bromeliads (Bromeliaceae). *Acta Protozoologica*, **49**, 159–93.

Folse, H.J., and Roughgarden, J. (2010). What is an individual organism? A multilevel selection perspective. *Quarterly Review of Biology*, **85**, 447–72.

Foster, R.B. (1977). *Tachigalia versicolor* is a suicidal neotropical tree. *Nature*, **268**, 624–6.

Fraune, S. and Bosch, T.C.G. (2010). Why bacteria matter in animal development and evolution. *Bioessays*, **32**, 571–80.

Fryer, G. (1961). The developmental history of *Mutela bourguignati* (Ancey) Bourguignat (Mollusca: Bivalvia). *Philosophical Transactions of the Royal Society of LondonB*, **244**, 259–98.

Fusco, G., and Minelli, A. (2010). Phenotypic plasticity in development and evolution. *Philosophical Transactions of the Royal Society of London B*, **365**, 547–56.

Galaktionov, K.V., and Dobrovolskij, A.A. (2003). *The Biology and Evolution of Trematodes. An Essay on the Biology, Morphology, Life Cycles, and Evolution of Digenetic Trematodes*. Kluwer, Dordrecht.

Gilbert, S.F., and Epel, D. (2009). *Ecological Developmental Biology: Integrating Epigenetics, Medicine, and Evolution*. Sinauer Associates, Sunderland.

Gilbert, S.F., Sapp, J., and Tauber, A.I. (2012). A symbiotic view of life: we have never been individuals. *The Quarterly Review of Biology*, **87**, 325–41.

Gladyshev, V.N. (2012). On the cause of aging and control of lifespan. *BioEssays*, **34**, 925–29.

Gorelik, R. (2012). Mitosis circumscribes individuals; sex creates new individuals. *Biology and Philosophy*, **27**, 871–90.

Gould, S.J. (1977). *Ontogeny and Phylogeny*. The Belknap Press of Harvard University Press, Cambridge.

Gould, S.J. (1989). *Wonderful Life: The Burgess Shale and the Nature of History*. W.W. Norton, New York.

Gould, S.J. (1996). *Full House: The Spread of Excellence from Plato to Darwin*. Harmony Press, New York.

Gould, S.J., and Lewontin, R.C. (1979). The spandrels of San Marco and the Panglossian paradigm: A critique of the adaptationist programme. *Proceedings of the Royal Society of London B*, **205**, 581–98.

Gremigni, V., Miceli, C., and Picano, E. (1980). On the role of germ cells in planarian regeneration. II. Cytophotometric analysis of the nuclear Feulgen-DNA content in cells of regenerated somatic tissues. *Journal of Embryology and Experimental Morphology*, **55**, 65–76.

Gremigni, V., Nigro, M., and Puccinelli, I. (1982). Evidence of male germ cell redifferentiation into female germ cells in planarian regeneration. *Journal of Embryology and Experimental Morphology*, **70**, 29–36.

Gremigni,V., and Miceli, C. (1980). Cytophotometric evidence for cell 'transdifferentiation' in planarian regeneration. *Wilhelm Roux's Archives of Developmental Biology*, **188**, 107–13.

Gurdon, J.B., Elsdale, T.R., and Fischberg, M. (1958). Sexually mature individuals of *Xenopus laevis* from the transplantation of single somatic nuclei. *Nature*, **182**, 64–5.

Haig, D. (1999). What is a marmoset? *American Journal of Primatology*, **49**, 285–96.

Hallez, P. (1887). *Embryogénie des dendrocoeles d'eau douce*. Baillière, Paris.

Harper, L.I., Schönrogge, K., Lim, K.Y., et al. (2004). Cynipid galls: insect-induced modifications of plant development create novel plant organs. *Plant, Cell and Environment*, **27**, 327–35.

Hartzell, P.L., and Youderian, P. (1995). Genetics of gliding motility and development in *Myxococcus xanthus*. *Archives of Microbiology*, **164**, 309–23.

Henry, L.-A., and Hart, M. (2005). Regeneration from injury and resource allocation in sponges and corals—a review. *International Review of Hydrobiology*, **90**, 125–58.

Hesse, M. (1968). Karyologische Anatomie von Zoocecidien und ihre Kernstrukturen. *Österreichische Botanische Zeitschrift*, **115**, 34–83.

Janzen, D.H. (1977). What are dandelions and aphids? *American Naturalist*, **111**, 586–9.

Jepp, J. (1984). Morphology and anatomy of the preimaginal stages of Chrysopidae: a short survey. In M. Canard, Y. Semeria and T.R. New, eds., *Biology of Chrysopidae*. Junk, The Hague, pp. 9–18.

Jones, D.S. (1983). Sclerochronology: reading the record of the molluscan shell. *American Scientist*, **71**, 384–91.

Karkach, A.S. (2006). Trajectories and models of individual growth. *Demographic Research*, **15**, 347–400.

King, N. (2004). The unicellular ancestry of animal development. *Developmental Cell*, **7**, 313–25.

King, N., Westbrook, M.J., Young, S.L., et al. (2008). The genome of the choanoflagellate *Monosiga brevicollis* and the origin of metazoans. *Nature*, **451**, 783–8.

Klass, M., and Hirsh, D. (1976). Non-ageing developmental variant of *Caenorhabditis elegans*. *Nature*, **260**, 523–5.

Klasson, L., Kambris, Z., Cook, P.E., et al. (2009). Horizontal gene transfer between *Wolbachia* and the mosquito *Aedes aegypti*. *BMC Genomics*, **10**, 33.

Klausen, M., Aaes-Jorgensen, A., Molin, S., et al. (2003). Involvement of bacterial migration in the development of complex multicellular structures in *Pseudomonas aeruginosa* biofilms. *Molecular Microbiology*, **50**, 61–8.

Knoll, A.H. (2011). The multiple origins of complex multicellularity. *Annual Review of Earth and Planetary Sciences*, **39**, 217–39.

Konner, M. (2010). *The Evolution of Childhood: Relationships, Emotion, Mind*. Harvard University Press, Cambridge.

Kupiec, J.-J. (2009). *The Origins of Individuals*. World Scientific, Singapore.

Lane, N. (2012). The problem with mixing mitochondria. *Cell*, **151**, 246–8.

Lasek-Nesselquist, E., and Katz, L.A. (2001). Phylogenetic position of *Sorogena stoianovitchae* and relationships within the class Colpodea (Ciliophora) based on SSU rDNA sequences. *Journal of Eukaryotic Microbiology*, **48**, 604–7.

Litaker, R.W., Vandersea, M.W., Kibler, S.R., et al. (2002). Life cycle of the heterotrophic dinoflagellate *Pfiesteria piscicida* (Dinophyceae). *Journal of Phycology*, **38**, 442–63.

Manni, L., and Burighel, P. (2006). Common and divergent pathways in alternative developmental processes of ascidians. *BioEssays*, **28**, 902–12.

Martin, A.C.R. (2011). Change and aging senescence as an adaptation. *PLoS ONE*, **6**, e24328.

Martinez, D.E. (1997). Mortality patterns suggest lack of senescence in hydra. *Experimental Gerontology*, **33**, 217–25.

McDunnough, J. (1909). Uber den Bau des Darmes und seiner Anhänge von *Chrysopa perla* L. *Archiv für Naturgeschichte*, **75**, 313–60.

McFall-Ngai, M.J. (2002). Unseen forces: the influences of bacteria on animal development. *Developmental Biology*, **242**, 1–14.

McKinney, M.L., and McNamara, K.J. (1991). *Heterochrony. The Evolution of Ontogeny*. Plenum Press, New York.

McNamara, K.J. (1986). A guide to the nomenclature of heterochrony. *Journal of Paleontology*, **60**, 4–13.

McNamara, K.J., ed. (1995). *Evolutionary Change and Heterochrony*. Wiley, Chichester.

McShea, D.W. (1996). Metazoan complexity and evolution: is there a trend? *Evolution*, **50**, 477–92.

Mikhailov, K.V., Konstantinova, A.V., Nikitin, M.A., et al. (2009). The origin of Metazoa: a transition from temporal to spatial cell differentiation. *BioEssays*, **31**, 758–68.

Miller, S.E. and Hadfield, M.G. (1990). Developmental arrest during larval life and life-span extension in a marine mollusc. *Science*, **248**, 356–8.

Minelli, A. (1971). Memory, morphogenesis and behaviour. *Scientia*, **106**, 798–806.

Minelli, A. (2003). *The Development of Animal Form: Ontogeny, Morphology, and Evolution*. Cambridge University Press, Cambridge.

Minelli, A. (2009a). *Perspectives in Animal Phylogeny and Evolution*. Oxford University Press, Oxford.

Minelli, A. (2009b). When evolution invented development. In S. Casellato, P. Burighel, and A. Minelli, eds., *Life and Time: Selected Contributions on the Evolution of Life and its History*. Cleup, Padova, pp. 141–150.

Minelli, A. (2011a). A principle of developmental inertia. In B. Hallgrímsson and B.K. Hall, eds., *Epigenetics: Linking Genotype and Phenotype in Development and Evolution*. University of California Press, San Francisco, pp. 116–133.

Minelli, A. (2011b). Development, an open-ended segment of life. *Biological Theory*, **6**, 4–15.

Minelli, A., and Fusco, G. (2010). Developmental plasticity and the evolution of animal complex life cycles. *Philosophical Transactions of the Royal Society of London B*, **365**, 631–40.

Minelli, A., and Fusco, G. (2012). On the evolutionary developmental biology of speciation. *Evolutionary Biology*, **39**, 242–54.

Minelli, A., and Fusco, G. (2013). Arthropod post-embryonic development. In A. Minelli, G. Boxshall, and G. Fusco, eds., *Arthropod Biology and Evolution. Molecules, Development, Morphology*. Springer, Heidelberg, pp. 91–122.

Nagl, W. (1978). *Endopolyploidy and Polyteny in Differentiation and Evolution*. North-Holland, Amsterdam.

Nanney, D.L. (1986). Introduction. In J.G. Gall, ed., *The Molecular Biology of Ciliated Protozoa*. Academic Press, Orlando, pp. 1–7.

Newman, S.A. (2010). Dynamical patterning modules. In M. Pigliucci and G.B. Müller, eds., *Evolution: The Extended Synthesis*. MIT Press, Cambridge, pp. 281–306.

Newman, S.A. (2011). Animal egg as evolutionary innovation: a solution to the 'embryonic hourglass' puzzle. *Journal of Experimental Biology (Molecular and Developmental Evolution*, **316**, 467–83.

Newman, S.A., and Bhat, R (2008). Dynamical patterning modules: physico-genetic determinants of morphological development and evolution. *Physical Biology*, **5**: 015008.

Newman, S.A., Forgacs, G., and Müller, G.B. (2006). Before programs: the physical origination of multicellular forms. *International Journal of Developmental Biology*, **50**, 289–99.

Nijhout, H.F. (1999). When developmental pathways diverge. *Proceedings of the National Academy of Sciences USA*, **96**, 5348–50.

Nikoh, N., and Nakabachi, A. (2009). Aphids acquired symbiotic genes via lateral gene transfer. *BMC Biology*, **7**, 12.

Nikoh, N., Tanaka, K., Shibata, F., et al. (2008). *Wolbachia* genome integrated in an insect chromosome: evolution and fate of laterally transferred endosymbiont genes. *Genome Research*, **18**, 272–80.

Oyama, S. (2000). *The Ontogeny of Information: Developmental Systems and Evolution*, 2nd ed. Duke University Press, Durham.

Pan, J., and Snell, W.J. (2000). Signal transduction during fertilization in the unicellular green alga, *Chlamydomonas*. *Current Opinion in Microbiology*, **3**, 596–602.

Passamonti, M., and Scali, V. (2001). Gender-associated mitochondrial DNA heteroplasmy in the venerid clam *Tapes philippinarum* (Mollusca: Bivalvia). *Current Genetics*, **39**, 117–24.

Peglar, M.T., Nerad, T.A., Anderson, O.R., et al. (2004). Identification of amoebae implicated in the life cycle of *Pfiesteria* and *Pfiesteria*-like dinoflagellates. *Journal of Eukaryotic Microbiology*, **51**, 542–52.

Piraino, S., Boero, F., Aeschbach, B., et al. (1996). Reversing the life cycle: medusae transforming into polyps and cell transdifferentiation in *Turritopsis nutricula* (Cnidaria, Hydrozoa). *Biological Bulletin*, **190**, 302–12.

Pradeu, T. (2010). What is an organism? An immunological answer. *History and Philosophy of the Life Sciences*, **32**, 247–68.

Pradeu, T. (2011). A mixed self: the role of symbiosis in development. *Biological Theory*, **6**, 80–8.

Prescott, D.M. (1994). The DNA of ciliated protozoa. *Microbiological Reviews*, **58**, 233–67.

Ramos, G.C. (2012). Inflammation as an animal development phenomenon. *Clinical and Developmental Immunology*, **2012**, Article ID 983203.

Rokas, A. (2008). The origins of multicellularity and the early history of the genetic toolkit for animal development. *Annual Review of Genetics*, **42**, 235–51.

Ross, C.N., French, J.A., and Orti, G (2007). Germ-line chimerism and paternal care in marmosets (*Callithrix kuhlii*). *Proceedings of the National Academy of Sciences USA*, **104**, 6278–82.

Rumpho, M.E., Worful, J.M., Lee, J., et al. (2008). Horizontal gene transfer of the algal nuclear gene *psbO* to the photosynthetic sea slug *Elysia chlorotica*. *Proceedings of the National Academy of Sciences USA*, **105**, 17867–71.

Ryals, P.E., Smith-Somerville, H.E., and Buhse, H.E. Jr. (2002). Phenotype switching in polymorphic *Tetrahymena*: a single-cell Jekyll and Hyde. *International Review of Cytology*, **212**, 209–38.

Sachwatkin, A.A. (1956). *Vergleichende Embryologie der niederen Wirbellosen (Ursprung und Gestaltungswege der individuellen Entwicklung der Vielzeller)*. Deutscher Verlag der Wissenschaften, Berlin.

Sahli, F. (1985). Periodomorphose et mâles intercalaires des Diplopodes Julida: une nouvelle terminologie. *Bulletin scientifique de Bourgogne*, **38**, 23–31.

Sahli, F. (1989). The structure of two populations of *Tachypodoiulus niger* (Leach) in Burgundy and some remarks on periodomorphosis. *Revue d'Écologie et de Biologie du Sol*, **26**, 355–61.

Sahli, F. (1990). On post-adult moults in Julida (Myriapoda: Diplopoda). Why do periodomorphosis and intercalaries occur in males? In A. Minelli, ed., *Proceedings of the 7th International Congress of Myriapodology*. Brill, Leiden, pp. 135–156.

Sander, K. (1983). The evolution of patterning mechanisms: gleanings from insect embryogenesis and spermatogenesis. In B.C. Goodwin, N. Holder, and C.C. Wylie, eds., *Development and Evolution*. Cambridge University Press, Cambridge, pp. 124–137.

Sanders, W.B. (2006). A feeling for the superorganism: expression of plant form in the lichen thallus. *Botanical Journal of the Linnean Society*, **150**, 89–99.

Santelices, B. (1999). How many kinds of individual are there? *Trends in Ecology and Evolution*, **14**, 152–5.

Schlichting, C. (2003). Origins of differentiation via phenotypic plasticity. *Evolution & Development*, **5**, 98–105.

Schlosser, G. (2002). Modularity and the units of evolution. *Theory in Biosciences*, **121**, 1–80.

Schlosser, G., and Wagner, G.P., eds. (2004). *Modularity in Development and Evolution*. University of Chicago Press, Chicago.

Sebens, K.P. (1987). The ecology of indeterminate growth in animals. *Annual Review of Ecology and Systematics*, **18**, 371–407.

Sebé-Pedrós, A., Roger, A.J., Lang, F.B., et al. (2010). Ancient origin of the integrin-mediated adhesion and signaling machinery. *Proceedings of the National Academy of Sciences USA*, **107**, 10142–7.

Secor, S.M. (2008). Digestive physiology of the Burmese python: broad regulation of integrated performance. *Journal of Experimental Biology*, **211**, 3767–74.

Shapiro, R.S., Sellam, A., Tebbji, F., et al. (2012). Pho85, Pcl1 and Hms1 signaling governs *Candida albicans* morphogenesis induced by high temperature or Hsp90 compromise. *Current Biology*, **22**, 461–70.

Shapiro, R.S., Uppuluri, P., Zaas, A.K., et al. (2009). Hsp90 orchestrates temperature-dependent *Candida albicans* morphogenesis via Ras1-PKA signaling. *Current Biology*, **19**, 621–9.

Slack, J.M.W., Holland, P.W.H., and Graham, C.F. (1993). The zootype and the phylotypic stage. *Nature*, **361**, 490–2.

Smith, J.J., Antonacci, F., Eichler, E.E., et al. (2009). Programmed loss of millions of base pairs from a vertebrate genome. *Proceedings of the National Academy of Sciences USA*, **106**, 11212–17.

Smith, J.J., Baker, C., Eichler, E.E., et al. (2012). Genetic consequences of programmed genome rearrangement. *Current Biology*, **22**, 1524–9.

Song, J.L., Wong, J.L., and Wessel, G.M. (2006). Oogenesis: single cell development and differentiation. *Developmental Biology*, **300**, 385–405.

Sooksri, V., Nakazawa, S., Fukuma, T., et al. (1991). An observation of the transitional forms between trypomastigote and amastigote of *Trypanosoma cruzi* by scanning electronmicroscopy. *Tropical Medicine*, **33**, 35–40.

Steenstrup, J.J.S. (1845). *On the Alternation of Generation or the Propagation and Development of Animals through Alternate Generations*. Ray Society, London.

Sterelny, K. (2000). Development, evolution, and adaptation. *Philosophy of Science*, **67**, S369–S387.

Stone, G.N., Schönrogge, K., Atkinson, R.J., et al. (2002). The population biology of oak gallwasps (Hymenoptera: Cynipidae). *Annual Review of Entomology*, **47**, 633–68.

Strathmann, M.F., and Strathmann, R.R. (2007). An extraordinarily long larval duration of 4.5 years from hatching to metamorphosis for teleplanic veligers of *Fusitriton oregonensis*. *Biological Bulletin*, **213**, 152–9.

Swart, E., Bracht, J.R., Magrini, V., et al. (2013). The *Oxytricha trifallax* macronuclear genome: a complex eukaryotic genome with 16,000 tiny chromosomes. *PLoS Biology*, **11**, e1001473.

Tsukaya, H., Byrne, M.E., Horiguchi, G., et al. (2013). How do 'housekeeping' genes control organogenesis?–unexpected new findings on the role of housekeeping genes in cell and organ differentiation. *Journal of Plant Research*, **126**, 3–15.

Valentine, J.W. (2004). *On the Origin of Phyla*. University of Chicago Press, Chicago.

Verhoeff, K.W. (1923). Periodomorphose. *Zoologischer Anzeiger*, **56**, 241–54.

Verhoeff, K.W. (1933). Wachstum und Lebensverlängerung bei Blaniuliden und über die Periodomorphose. *Zeitschrift für Morphologie und Ökologie der Tiere*, **27**, 732–48.

Verhoeff, K.W. (1939). Wachstum und Lebensverlängerung bei Blaniuliden und über die Periodomorphose, II Teil. *Zeitschrift für Morphologie und Ökologie der Tiere*, **36**, 21–40.

Vervoort, M. (2011). Regeneration and development in animals. *Biological Theory*, **6**, 25–35.

von Baer, K.E. (1828). *Ueber Entwickelungsgeschichte der Thiere, Beobachtung and Reflexion*. Bornträger, Königsberg.

Vlamakis, H., Aguilar, C., Losick, R., et al. (2008). Control of cell fate by the formation of an architecturally complex bacterial community. *Genes & Development*, **22**, 945–53.

Wang, J., Mitreva, M., Berriman, M., et al. (2012). Silencing of germline-expressed genes by DNA elimination in somatic cells. *Developmental Cell*, **23**, 1072–80.

Webb, J.S., Givskovy, M., and Kjelleberg, S. (2003). Bacterial biofilms: prokaryotic adventures in multicellularity. *Current Opinion in Microbiology*, **6**, 578–85.

Wegener Parfrey, L., and Lahr, D.J.G. (2013). Multicellularity arose several times in the evolution of eukaryotes. *BioEssays*, 35, 339–47.

Williamson, D.J. (2006). Hybridization in the evolution of animal form and life-cycle. *Biological Journal of the Linnean Society*, **148**, 585–602.

Wilmut, I., Schnieke, A.E., McWhir, J., et al. (1977). Viable offspring derived from fetal and adult mammalian cells. *Nature*, **385**, 810–13.

Wilson, J. (1999). *Biological Individuality: The Identity and Persistence of Living Entities*. Cambridge University Press, Cambridge.

Zakhvatkin, A.A. (1949). *The Comparative Embryology of the Low Invertebrates. Sources and Method of the Origin of Metazoan Development* [in Russian]. Soviet Science, Moscow.

CHAPTER 16

Identifying some theories in developmental biology: the case of the cancer stem cell theory

Lucie Laplane

To know whether there are theories of development in developmental biology, one will need to know how to recognize them. This is, in fact, far from a trivial task. Biologists use the concept of theory loosely. This makes any intuitive distinction between 'theories', 'hypotheses', or 'models' very difficult (see Pradeu, 2009: 206). Furthermore, as we will see, syntactic and semantic conceptions of scientific theories developed by philosophers of science provide no tools for an attempt to identify theories in any particular field of science. Thus the question 'are there theories of development in developmental biology?' can be subdivided into three independent questions:

(i) What is a scientific theory?
(ii) How can existing scientific theories be identified?
(iii) Is there, in fact, such a thing as a scientific theory in developmental biology?

Thomas Pradeu's chapter in the present book deals mainly with the first and third questions, while this chapter will focus on the second and third questions. My aim is primarily practical rather than conceptual. After showing that we lack means to distinguish theories from non-theories, I will attempt to provide some tools that will allow us to make the distinction. I will argue that a particular interpretation of the old vera causa principle can provide such a tool. Through its use, I will then answer positively to the third question and set out what I consider a good example of a theory in development biology: the cancer stem cell (CSC) theory.

Classical conceptions of scientific theory provide no tools for the identification of theories

Scientific theories are commonly described as sets of explanations and predictions. This cannot be taken as a satisfying definition of a scientific theory given that a scientific hypothesis, defined as 'plausible propositions on how a phenomenon occurs' (Pradeu, 2009: 203), can also provide a set of explanations and predictions. During the 20th century, philosophers of science of the analytic tradition proposed two major conceptions of scientific theories: the syntactic and the semantic. These conceptions give precise formal characterizations of scientific theories. Our aim does not require us to examine these conceptions in detail but just to recall how they describe theories. The syntactic conception, also called the 'received view', regroups several more or less heterogeneous views that describe theories as 'networks of principles and laws', to use Vorms' (2011: 122) metaphor (for insight on this conception, see Hempel, 1965). The semantic conception, which was elaborated in response to the syntactic views, conceives scientific theories as 'classes (or families) of models' (Vorms, 2011: 128; for insight on this conception see Suppe, 1977 and van Fraassen, 1987). Of critical importance for us is that they both aim to provide formal reconstructions of theories rather than any precise definition of scientific theories that would enable us to identify them in a particular

Towards a Theory of Development. Edited by Alessandro Minelli and Thomas Pradeu

field of science. Suppes illustrates this point very well when writing:

> If someone asks, "what is a scientific theory?" it seems to me there is no simple response to be given. [. . .]. It does not seem to me important to give precise definitions of the form: X is a scientific theory if and only if, so-and-so. (Suppes, 1967: 63–64).

When trying to know if there are theories of development, it is precisely 'so-and-so' criteria that are needed to determine if 'X' is a theory. It is important to acknowledge that our goal is fundamentally different from the aims of syntactic and semantic conceptions of theories. But even if the goals differ, one can ask whether their descriptions of theories provide tools that could be used for the identification of theories. We will now address this question.

The syntactic conception is unsuited to biology

From the syntactic perspective, the major characteristic of theories is the enunciation of laws. Does that provide us with a criterion to identify theories? Logical empiricists define laws as 'true generalizations that are "purely qualitative," meaning that they do not refer to any place, time, or individual' (Sober, 2006: 249). Such laws cannot exist in biology (see Gayon, 1993; Hull, 1974; Smart, 1963). Every biological entity is the outcome of an evolutionary history. First of all, evolutionary history could have been different (Gould, 1990). Second, generalizations that are true on earth can be false elsewhere in the universe (see Morange, 2003 or 2008 on exobiology). Third, biology is characterized by its countless exceptions (for examples see Gayon, 1993). True generalizations in biology refer to history, locations, and/or individuals. This general argument against the syntactic conception of scientific theories became particularly popular with Beatty's 'evolutionary contingency thesis' (Beatty, 1995).

Moreover, the adequacy of the syntactic conception of scientific theories based on such laws has also been questioned in physics. On the one hand, some fields of research in physics are as historical as biology can be, such as astrophysics. On the other hand, and even more problematic, classical physics itself could rest on contingently true generalizations, an outcome of the Big Bang (Rosenberg & McShea, 2008; Ruse, 1973; more references in Pradeu, 2009). The syntactic conception not only does not apply to all sciences, which is problematic for a general conception of 'scientific' theories, but worse, it could apply to no science at all.

Insofar as the characterization of scientific theories as sets of laws is not suitable for all sciences, it does not provide relevant criteria for the identification of theories, in particular in biology. Some philosophers have argued that this problem could be overcome by revising the notion of law (see Gayon, 1993; Hull, 1977; Pradeu, 2011; Sober, 2006). But, empirically, every attempt to axiomatize the most famous biological theory (evolution by natural selection theory) has fallen short (Lloyd, 1988; Williams, 1981; see also Duchesneau, 1997). This has led to a consensus that the syntactic conception of scientific theories is too exclusive. Frederick Suppe has famously made a similar criticism, indicating that 'not all scientific theories admit the canonical axiomatic formulation required by the Received View' (Suppe, 1977: 63). He concluded that 'the Received View is not plausible as an analysis of the structure of *all* scientific theories' (Suppe, 1977: 66).

A historical lesson about induction from one science to another

Debates over the received view of theories have provoked severe criticisms of the predominant status of physics in philosophy of science. For example, Beatty (1980: 398) argues:

> In comparing biology and physics in light of the received view of theories, the issue certainly has not been the adequacy of the received view. The issue has been the adequacy of biology.

Indeed, scientific theories have been (and still are to some extent) conceived with theories of classical physics as a model. But because theories in physics enunciate laws, do we have to conclude that every scientific theory should do so? How ironic it is that a conception of theories willing to ascertain the logical basis of particular theories is itself based on a very loose induction from physics to science in general!

The particular history of the emergence of the philosophy of biology as a response to the physics-driven philosophy of science taught us a major lesson: transfers of criteria from one science to another and inductions from one particular science to science in general should be done with caution. Any attempt to characterize scientific theories should pay attention to this lesson.

The semantic conception is not specific to theories

The semantic conception of scientific theories has been described as far more liberal than the syntactic conception, leaving open the possibility of theories in most sciences, including biology but also economics and social sciences (e.g. Pradeu, 2009). This raises the hope of finding in this conception a description of theories that would enable us to identify them in a particular field of science.

The semantic conception of theories describes them as families of models. The notion of model is highly polysemous and is subject to numerous debates. In the semantic conception, the concept of model has to be understood in the sense of formal logic: 'it is a structure that satisfies a theory' (Vorms, 2011: 128). It must not be confused with its meaning in biology, which is fuzzy and multiple but clearly different, even if Suppes and van Fraassen both argued for compatibility between those meanings (Suppes, 1960; van Fraassen, 1980). Nonetheless, proponents of the semantic conception have shown that it is possible to formalize biological theories, like the theory of evolution, with models (in the sense of logic).

The question then would be: is the possibility of such a formal reconstruction a proof of the presence of a scientific theory? According to Peter Godfrey-Smith, this is not the case: non-theory, 'perhaps any representations with indicative content at all', like 'the sports pages of the *New York Times*', can be formalized with mathematical models:

> What we have here is a general way of thinking about meaning; we think about the meaning of a representation by thinking about the formal characteristics of structures that the representation can be seen as true of (Godfrey-Smith, 2006: 727).

Formalization by mathematical model might be a good way to represent and analyse scientific theories, but it isn't a suitable way to distinguish theories from non-theories. Contrary to the syntactic conception (which is too restrictive), the semantic conception would be too inclusive as an approach to identify scientific theories.

Thus, a first conclusion may be drawn: neither the syntactic nor the semantic conception of scientific theories provide adequate tools for an attempt to distinguish between theories and non-theories in science (which is fine, considering that this is not their aim). Consequently, we have to find other criteria to identify scientific theories. We will argue that a particular interpretation of the vera causa principle can provide such criteria.

The vera causa principle can provide useful criteria for the identification of theories

Before the syntactic and semantic conceptions of scientific theories, from Newton to Darwin, there have been two centuries of discussion of the so-called 'vera causa principle' (for a historical overview, see Kavaloski, 1974). The vera causa principle is a statement of scientific methodology. It is aimed at discriminating true causes from fictitious ones in order to guarantee the scientificity of explanations.

According to the historical analysis of Vincent Kavaloski, the vera causa principle has neither been formulated nor has it been used to identify theories. Indeed, Kavaloski (1974) distinguishes five functions of the vera causa principle:

> F1. It functions as a rule in the formation of hypotheses: viz., only *verae causae* are to be employed as explanatory factors therein.
>
> F2. It acts as a criterion of evaluation for hypotheses already in existence.
>
> F3. It plays a part in the definition of 'scientific explanation'.
>
> F4. It constitutes a demarcation criterion between science and non-science.
>
> F5. It acts as a requirement for something being a *plausible* scientific explanation, i.e., one worthy of investigation and, more particularly, of testing. (Kavaloski, 1974: 50)

Two conclusions can be drawn from the description of these five functions. First, the vera causa principle does not seem specific to theories. In this citation Kavaloski refers to 'hypotheses', not 'theories', but more generally he uses 'theory' and 'hypothesis' interchangeably. Second, the vera causa principle appears rather heterogeneous. Furthermore, in addition to these five functions, Kavaloski highlighted the existence of six interpretations of the principle (Kavaloski, 1974). This, one might think, would preclude the use of this principle for the aim of identification of theories in science. Besides, the vera causa principle was abandoned by philosophers of science during the 20th century. However, we will show that one particular interpretation of the vera causa principle escapes these pitfalls and could be useful for the specific aim of distinguishing theories from non-theories. This is Darwin's interpretation, or more precisely, Gayon's interpretation of Darwin's interpretation (or even my interpretation of Gayon's interpretation of Darwin's interpretation of the vera causa principle).

As recalled by Kavaloski, in his *Origin of Species* Darwin 'explicitly appeals to the *vera causa* principle as a desideratum favoring his own theory vis-à-vis its chief rival, special creationism' (Kavaloski, 1974: 105). He 'never explicitly defines what he means by the term *vera causa*' (Kavaloski, 1974: 105), but in response to criticisms that have been addressed by philosophers of science (i.e. Herschel, Whewell, Mill) on the methodology of his *Origin of Species*, Darwin argued:

> In scientific investigations, it is permitted to invent any hypothesis, and if it explains various large and independent classes of facts it rises to the rank of a well-grounded theory (Darwin, 1875: 9).

This citation provides a synthetic sketch of Darwin's interpretation of the vera causa principle (Kavaloski, 1974; Gayon, 2009), revealing the criteria that we think useful for the identification of theories—namely the explanation of 'independent classes of facts'. The Darwinian interpretation of the vera causa principle has been extensively analysed (Gayon, 1997, 2009; Hodge, 1977; Hull, 2003; Kavaloski, 1974; Ruse, 1975, 1992). We focus on Gayon's analysis because its two-level interpretation (see Figure 16.1) provides great insight into the question of the difference between hypotheses and theories. Note that this question has never been part of the debates over Darwin and the vera causa principle, nor has it been the subject of a direct argumentation by Gayon (who could disagree in some respect with my interpretation).

Gayon describes Darwin's argumentation as a two-level method relying on three criteria (illustrated in Figure 16.1). The two levels are those of 'hypothesis' and 'theory'. At the first level is the hypothesis (upper portion of the diagram). The scientificity of the hypothesis relies on the following criteria: it must be derived from 'empirical premises', or more precisely, from 'well-established classes of facts' (Gayon, 2009: 330). At the second level is the theory (bottom of the diagram). A theory, according to Gayon's analysis of Darwin, relies on a scientific hypothesis of the first level meeting a second criterion, namely it explains 'independent classes of facts' (the eight boxes at the bottom). This second criterion allows one to distinguish a scientific theory from a scientific hypothesis. Thus it gives us a way to identify theories in science. By '*independent* classes of facts,' we understand 'facts that are caused by independent causes'. This leads us to a historical interpretation of scientific theories, since the independent facts are no longer independent once the theory has fitted them with a unifying explanation. By '*classes* of facts', we understand 'groups of facts sharing particular properties'. Facts are always singular. Thus the classes of facts, like 'extinction' or 'divergence', designate a particular kind of singular fact.

The third criterion is transversal. It is the assistance of an analogical model (top right of Figure 16.1). The principle of an analogical model (or isomorphic model) is to consider a specific case as a model for the theory: in a specific case, particular causes are known to have particular effects, and in the theory, analogous causes are considered to have analogous effects. Proof by analogy was of major importance for several proponents of the vera causa principle, such as Herschel and Whewell. Newton also made use of it by taking fronds as models of the movement of the moon around the earth. It has been shown to have been of major importance for Darwin (Gayon, 2009; Kavaloski, 1974; Lloyd, 1983; Recker, 1987; Ruse, 1975; Thagard, 1978; Waters, 1986). This criterion is transversal to the levels of hypothesis and theory because (i) it increases the likelihood and scientificity

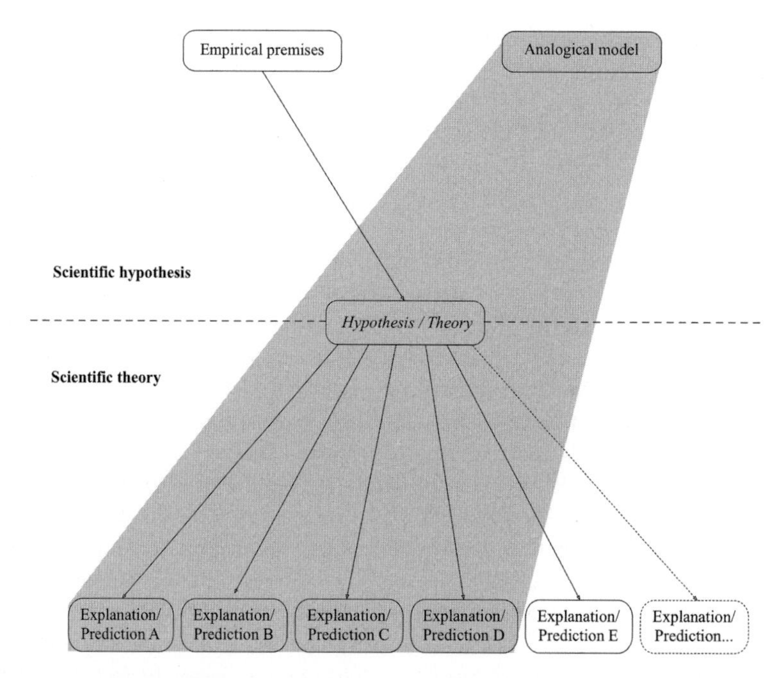

Figure 16.1 Schematic representation of the Darwin–Gayon view of scientific theories as interpreted in this chapter. This diagram represents an abstraction from, and analysis of, Gayon's (1997, 2009) two-level diagram. In the upper part of the diagram is the scientific hypothesis. A hypothesis is scientific if some empirical premises make it plausible, and even more valuable if any analogical model of it is available. At the bottom is the representation of the conditions for any scientific hypothesis to be considered as a scientific theory. There is only one requirement, which is the ability to explain (or predict) independent facts (represented by the boxes at the bottom). The availability of an analogical model reinforces the explanatory power of the theory and contributes to guarantee its scientificity; however, such a model is not required.

of the hypothesis, and (ii) it supports the theory by providing an empirical demonstration of the causal explanation (the grey area in Figure 16.1), e.g. artificial selection shows how selection can provide a causal explanation to the origin of a new variety of pigeon. One might ask if it is a requirement to be fulfilled by any theory. According to Gayon, only the first two criteria are required. The analogy clearly improves the scientificity of a theory but is not to be understood as a necessary criterion per se (Gayon, personal communication).

Domain of extension

As in the case of the syntactic and semantic conceptions of scientific theories, the vera causa principle was not designed to identify existing theories in particular fields of science. Unlike these conceptions, however, a particular interpretation of the vera causa principle can provide a useful criterion to achieve this aim. The study of the syntactic conception of scientific theories has revealed a methodological difficulty in the conceptualization of scientific theories: theories might have specific characteristics in specific fields of science (e.g. enunciation of laws). These specific characteristics must not be generalized to theories in science. Thus, we must raise the issue of the scope of application of our criterion. Two comments can be made on this subject. First, our main aim in the context of the present volume is to identify theories in (developmental) biology. Our criterion fulfills this specific aim (see also Pradeu, this volume). Second, other criteria might be better suited to other fields of science; however, there is evidence advocating the applicability of our criterion to other sciences as well. Notably, the vera causa principle and its interpretation by Darwin were first elaborated by Newton in the context of classical physics (Newton, 1960; also see the analysis in Kavaloski, 1974: Ch. 2). Furthermore, Lyell also applied it to geology (Lyell, 1830–1833; see the analysis of Kavaloski, 1974: Ch. 4). To be conclusive, a more general study should be carried out, but one might reasonably expect that this criterion can be generalized.

We can now draw a second conclusion—a theory can be identified through its ability to explain independent facts—and use it to address the very question 'are there theories of development in developmental biology today?'

The CSC theory

I will argue that there are theories of development in developmental biology. The CSC theory is a convincing proof of concept. Though it is almost never presented as such, I will show that:

(i) It fits the usual description of a scientific theory as a set of explanations and predictions.
(ii) It satisfies my identification of theories' criterion. That is to say, it explains independent classes of facts.

'CSC' is a concept that has become majorly influential in oncology in the course of the past two decades (for a precise historical account see Laplane, 2013). A CSC has been defined as a cancer cell that possesses stemness properties, that is to say, the capacity for long-term self-renewal and differentiation potency (ability to give rise to different kinds of cells). In other words, CSCs are cells that are both cancerous and stem-cell like.

> The consensus definition of a cancer stem cell that was arrived at in this Workshop is a cell within a tumor that possesses the capacity to self-renew and to cause the heterogeneous lineages of cancer cells that comprise the tumor (Clarke et al., 2006: 9340).

Concomitantly to the concept of CSC, biologists have formulated what I argue to be a theory of carcinogenesis and, more generally, of the developmental biology of cancers. However, researchers do not demand recognition of its status as a theory. Indeed it has been variously described as a theory (Dewi et al., 2011; Gil et al., 2008; Li et al., 2011; Tai et al., 2005; Yoo et al. 2008), a model (Dalerba et al., 2007; McClellan & Majeti, 2012; Vermeulen et al., 2012), and a hypothesis (Gespach, 2010; Hemmings, 2010; Morrison et al., 2011; Rahman et al., 2011; Shipitsin & Polyak, 2008; Tan et al., 2006). I will show that it is definitely a theory.

The CSC theory is composed of three models (representational models of the empiric sciences as defined by Giere, 1988; represented in Figure 16.2):

(i) a model of carcinogenesis;
(ii) a model of relapse; and
(iii) a model of therapy.

These models are based on a central two-fold premise: cancers comprise CSCs (cancer cells with stemness properties) but not every cancer cell is a CSC. In fact, CSCs are considered to make up a very small subpopulation of cancer cells. Evidence showing the coexistence of numerous non-CSCs and rare CSCs in a growing number of cancers is building up (for a review, see Nguyen et al., 2012). The three models are abstracted representations of the explanations and predictions of the CSC theory based on this two-fold premise.

The CSC theory differs from the classical view in that it ascribes stemness properties to a discrete population of cells (i.e. the CSCs), whereas the classical view considers that self-renewal and differentiation potency are stochastic properties that virtually any cancer cell can express (with a low but approximately equal probability).

The CSC theory is a set of explanations and predictions

According to this premise, among the cancer cells only the CSCs are capable of long-term self-renewal and of generating all the different types of cells that compose the malignant tumour. This provides explanations for:

(i) low clonogenicity of cancer cell populations;
(ii) heterogeneity of the cells composing the tumours;
(iii) presence of non-metastatic cancer cells at a distance from the primary tumour; and
(iv) relapses after apparently successful therapies;

as well as predictions for:

(v) long-term therapeutic failure of drugs that do not eliminate all the CSCs; and
(vi) long-term success of CSCs targeting therapies.

I will now detail these six aspects.

Low clonogenicity. Clonogenicity refers to the ability of a cell to divide and create an entire population,

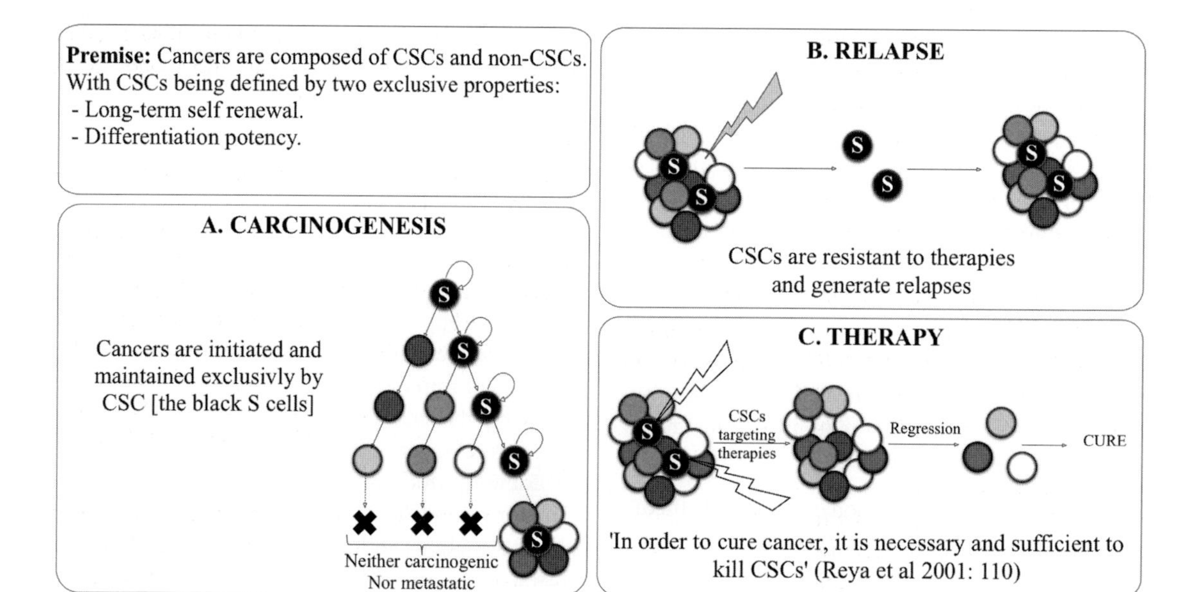

Figure 16.2 The CSC theory. The CSC theory comprises three models, which are abstracted representations of explanations and/or predictions: how carcinogenesis takes place, why relapse can occur after apparently successful therapy, and the only efficient therapy possible (according to the CSC theory). This theory relies on one premise (upper left box): the very existence of CSCs. Black circles containing an 'S' represent CSCs. The remaining circles (in dark grey, light grey, and white) are heterogeneous cancer non-stem cells. The grey thunderbolt represents classical radio- or chemotherapy. The white thunderbolts represent hypothetical CSC-targeting therapies. The drawings in the figure are inspired by the numerous ones published by the CSC theory proponents.

named a 'clone' or 'clonal population'. Cancer cells are known to be poorly clonogenic. When populations of cancer cells are cultured, only a few of them produce clones. A low stochastic ability to produce clones at the cell level classically explained this characteristic:

> Cancer cells of many different phenotypes have the potential to proliferate extensively, but any one cell would have a low probability of exhibiting this potential in an assay of clonogenicity or tumorigenicity (Reya et al., 2001: 109).

The CSC theory explains this low clonogenicity differently. First, only the CSCs are capable of long-term self-renewal and differentiation. Therefore, only these cells are really clonogenic. Second, CSCs are very few in number, which explains the low clonogenicity at the population level.

Heterogeneity. Although differentiation is often partial in cancers, cancer cell populations are heterogeneous. This heterogeneity is usually explained by mutations and selection, through the 'clonal evolution' model developed by Nowell (1976). In the CSC theory, the heterogeneity of the cancer cells is a direct consequence of the differentiation potency of the CSCs, as is the case in non-pathologic development. This idea results from two research pathways: studies on teratocarcinomas, which have shown that embryonic carcinoma cells (the CSCs of teratocarcinomas) are pluripotent and differentiate to give rise to the different cells that constitute the tumour (Fekete & Ferrigno, 1952; Kleinsmith & Pierce, 1964; Stevens, 1960); and research into blood cancers, which have shown that leukemic stem cells can give rise to all the leukemic blood cells and thus are multipotent (Bonnet & Dick, 1997; Fialkow, 1980).

Non-metastatic cells. Since the 1950s, researchers have shown that 'cancer cells exfoliate into body fluids, including blood and lymph, in a very high percentage of patients', but that 'not all disseminated cells develop into metastases and [. . .] in many patients none produce metastatic lesions' (Southam et al., 1961: 971). The frequent failure of disseminated cancer cells to develop metastases is usually explained by the postulated efficacy of the immune system against a limited number of cancer cells (see Reya et al., 2001; Southam et al., 1961). Under the CSC theory, this phenomenon falls under the same explanation as that for low clonogenicity: an exfoliated cancer cell can produce a metastasis if and only if it is a CSC. But many disseminated cells are not CSCs, as the latter are rare (Reya et al., 2001).

Relapses. After apparently successful therapies, patients are declared in 'remission'. Regrettably, relapses often occur. Even a very small number of primary cancer cells having escaped the therapies may cause these relapses. It is claimed that their escape is due to an evolutionary phenomenon of variation and selection (mutations being frequent in cancers). Drugs exert a selective pressure from which a few cells might stochastically escape through the accumulation of favorable mutations that confer upon them 'multiple drug resistance' (Persidis, 1999). Note that explanation (iv) is incompatible with (i) and (iii) in the classical view. According to (i), the probability that one or a few cells initiate a tumour is very low. Thus relapses should be very rare, which, unfortunately, is not the case. Therefore, (i) and (iv) are mutually exclusive. General data are lacking because cancer registries do not collect recurrences, but cancer survivors are considered 'at risk for recurrence'. For example, 40% of patients treated for colorectal cancers experience recurrence, and the rate of recurrence among bladder cancer patients ranges from 50% to 90% (see American Cancer Society, 2012). According to (iii), the immune system is very efficient against one or a few cancer cells. Thus, one should predict that if a few cells remain after an ostensible effective radio- or chemotherapy, the immune system should eliminate them. Relapses, then, should be very rare after effective therapies, which, again, is not the case. Therefore, (iii) and (iv) also contradict each other. Thus, there is an internal incoherence in the classical view between the explanation of relapses and the explanations of non-metastatic disseminated cancer cells and low clonogenicity of cancer cells.

Under CSC theory, the resistance of CSCs explains relapses. CSCs are (the only cells) able to initiate, maintain, and propagate cancers. Thus the resistance of one CSC is sufficient for a relapse to occur. The explanation for the higher resistance of CSC was first only hypothetical and relied on an analogy with normal stem cells.

> It seems that normal stem cells from various tissues tend to be more resistant to chemotherapeutics than mature cell types from the same tissues. The reasons for this are not clear, but may relate to high levels of expression of anti-apoptotic proteins or ABC transporters such as the multidrug resistance gene. If the same were true of cancer stem cells, then one would predict that these cells would be more resistant to chemotherapeutics than tumour cells with limited proliferative potential. Even therapies that cause complete regression of tumours might spare enough cancer stem cells to allow regrowth of the tumours (Reya et al., 2001: 110).

Since then, biologists have ascertained that CSCs are more resistant than non-CSCs to radio- and chemotherapies (Bao et al., 2006; Dylla et al., 2008; Li et al., 2008; Ma et al., 2008; Woodward et al., 2007; for reviews, see Alison et al., 2011; Dean et al., 2005). Nonetheless, the explanation of this phenomenon often remains unclear (e.g. Buczacki et al., 2011). However, if CSCs are resistant to therapies, then relapses become a very predictable outcome.

Let us now turn to predictions.

Failure. The CSC theory predicts that any therapy that fails to eradicate all CSCs runs a high risk of relapse in the long term. This prediction is directly linked to a reciprocal prediction on the efficiency of any therapy against cancers.

Note that the prediction of therapeutic failure by the CSC applies to 'relapses after apparently successful therapies'. Direct failures of therapies to cure cancers are very distinct from long-term failures after a period of remission. The apparent successfulness of therapies is critical to both (iv) and (v), since multiple other factors, independent (at least to a certain degree) from CSCs, can explain direct failures (e.g. multidrug resistance, gene amplification, inherent resistance of a patient, alteration in membrane transport, drug interaction, etc . . .).

Successful therapies. The proponents of the CSC theory argue that 'in order to cure cancer, it is necessary and sufficient to kill cancer stem cells' (Reya et al., 2001: 110). Indeed, CSCs are the only cells able to self-renew and differentiate and thus the only cells able to replenish the tumour. Since the 1950s, we know that there are turnovers and cell-death in cancers, as in non-pathological tissues (Astaldi & Mauri, 1953; Killmann et al., 1962). Hence, if all the CSCs of a given cancer are eliminated, then the cancer will not be able to replenish and will consequently disappear progressively through the death of the cancer non-stem cells. This, in turn, will guarantee a definitive healing (for a criticism of this part of the theory see Laplane, forthcoming).

The long-term aspect of predictions (v) and (vi) is of critical importance. First, healing is arbitrarily declared after a five-year period of remission, and statistics on survival are generally based on the same duration. Yet CSCs, in particular if they are quiescent, could lead to a relapse much later. Second, the efficiency of therapies against solid cancers is evaluated on the basis of the therapy's ability to *rapidly* shrink the tumour (Eisenhauer, 2009; Therasse et al., 2000). Yet, a drug that rapidly shrinks a tumour can fail to kill the rare CSCs that can later cause a relapse. Conversely, 'potentially potent agents are discarded when they fail to induce rapid tumour shrinkage even if they have successfully eliminated the CSC population' (Vermeulen et al., 2012: e87; see also Rivera et al., 2011).

I have shown that the CSC theory provides a set of explanations and predictions, but this is not sufficient to establish that we are in the presence of an actual theory rather than a mere scientific hypothesis. According to my argumentation, a scientific theory differs from a scientific hypothesis by its ability to explain 'independent classes of facts'. The explanations and predictions provided by the CSC theory have this ability.

The CSC theory explains independent facts

My description of the explanations and predictions of the CSC theory shows that they focus on independent classes of facts (as interpreted above). Indeed, before the elaboration of the CSC theory, distinct causes explained the low clonogenicity of cancer cell populations, the heterogeneity of the cells composing the tumours, the presence of non-metastatic cells at distance from the site of the primary tumour, and the relapses that occur after apparently successful therapies. Clonogenicity is stochastic. Heterogeneity is due to mutations and selection. Non-metastatic distant cancer cells are explained by the activity of the immune system, and relapses after apparently successful therapies are due to resistant sub-populations. This last explanation can be linked to the explanation of heterogeneity through the clonal evolution model developed by Nowell. Hence, classical oncology dealt with three to four apparently independent classes of facts. The CSC theory suggests a common causal explanation for all these classes of facts. A low level of clonogenicity and different patterns of division between cancer cells are due to stemness (i.e. the ability to self-renew and differentiate). Heterogeneity between cancer cells in a given cancer is due to stemness, and more precisely, to the ability of stem cells to differentiate. The presence of non-metastatic cancer cells away from the primary tumour is due to a lack of stemness. Finally, relapses are due to the inherent resistance of stem cells to therapies and their ability to self-renew and differentiate. Thus, the CSC theory fulfills my Darwin–Gayon main criterion—it explains independent classes of facts—ensuring us that it is a scientific theory (illustrated in Figure 16.3).

The CSC theory benefits from the powerful transversal criterion of the analogical model

The CSC theory is strongly linked, at all levels, with an analogical model, namely the non-pathologic stem cell. Stem cells, as CSCs, are capable of long-term self-renewal and differentiation. These features empower them with high clonogenicity, whereas non-stem cells are poorly clonal. Low clonogenicity of cells at the cell population level has been proven to be due to this deterministic (and not stochastic) distribution of the ability to produce clones. This provides a strong analogy for (i) and (iii), since the ability to produce a metastasis requires the ability to produce a clone. The differentiation potency of the stem cells is the source of the production of the different kinds of cells that make up tumour tissues (rather than mutations). This provides an analogical model of (ii). Stem cells also provide a powerful analogical model of (iv), (v), and (vi) through their ability to self-renew and differentiate and their capacity to escape radio- and chemotherapies. First, by their capacity to self-renew and differentiate, stem cells are able to reproduce their tissues. This

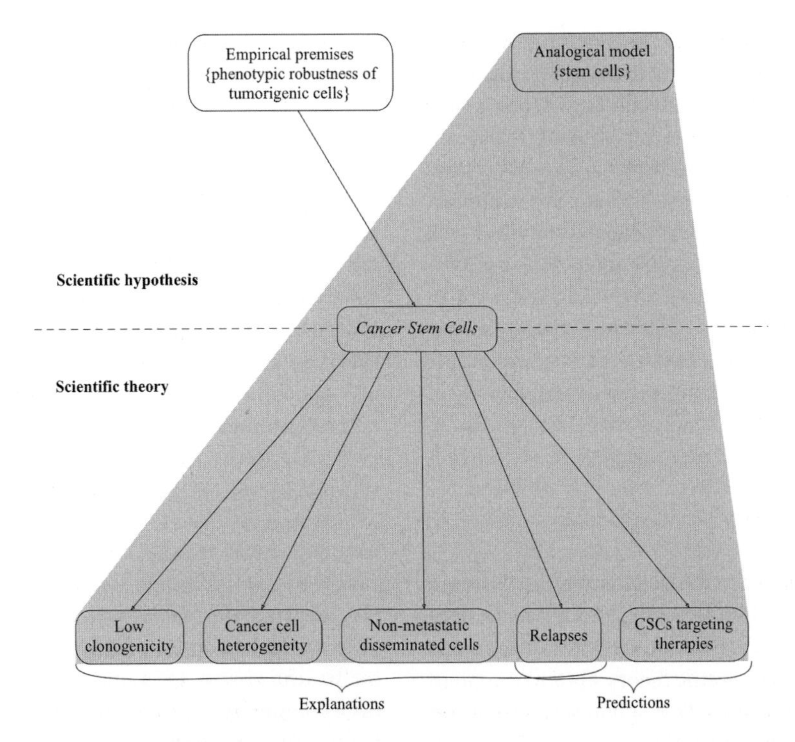

Figure 16.3 The CSC theory fits the interpretation of the Darwin–Gayon view of scientific theories presented in this chapter.

usually explains the homeostasis of tissues with high turnover, reparations of tissues after little to huge injuries (depending on the species and types of stem cells), and transplantations. This provides an analogy for the relapse phenomenon. Second, stem cells are resistant to radio- and chemotherapies, a mandatory condition, since the patient would die if therapies eliminated normal stem cells. Scientists have shown that stem cells have mechanisms allowing them to resist therapies such as the production of anti-apoptotic proteins, quiescence, increased capacity for DNA damage, high drug efflux capacity, and low levels of reactive oxygen species (Ahmed et al., 2011). This analogical model plays a major role in the understanding of relapses and long-term failures of classical therapies from the CSCs perspective.

The analogy between stem cells and CSCs can be considered stronger than the analogy between natural selection and artificial selection, or between the frond and the movement of the moon around the earth. It is more than an analogy. In fact, one could even argue that it is not an analogy at all, since CSCs are a subtype of stem cell. However, normal stem cells do play the role of an analogical model in the CSC theory. First, they represent specific cases that are taken as models for the theory. Second, these models are considered as no more than analogies, since CSCs are the pathological counterparts of normal stem cells and thus might present some differences.

According to this analysis, the CSC theory is a genuine scientific theory, and it benefits from a particularly powerful analogical model (Figure 16.3). The question that remains to be answered is whether this theory can be considered a theory of developmental biology.

The CSC theory is a theory of developmental biology

That the CSC theory belongs to developmental biology seems quite straightforward to me. First, the CSC theory is a theory on the *development* of cancers. It is a theory on how cancers start to develop, grow, differentiate, and maintain themselves. Second, cancers are neoformations; they are pathological developments. Besides, they are mentioned in all the major textbooks of developmental biology. Thus cancer development is development, and the CSC theory is a theory that pertains to developmental biology. One could argue that it is far from the best example. Better examples are surely available to highlight the existence of theories in developmental biology. Nevertheless, this one provides a powerful illustration of the conception of theories presented in this chapter. It should therefore be considered both as an example for the Darwin–Gayon conception of scientific theories and as a proof of concept that theories can be identified in developmental biology.

Conclusion

Discussions over the actual and potential existence of theories of development (and more generally of theories in biology) are biased by (i) the vagueness of the definition of scientific theory, and (ii) the latent idea that the syntactic and/or semantic conceptions of scientific theories provide us with good definitions of what a scientific theory is. However, there is no rigorous definition of what a scientific theory is. I have shown that the syntactic and semantic conceptions of theories do not aim to provide any specific definitions of theories. This chapter has also shown that they cannot provide adequate tools to distinguish theories from non-theories in sciences. My aim was not to produce an adequate definition of scientific theories (for this aim, see Pradeu's contribution in this volume), but to find empirically useful criteria to identify existing theories in a particular field of biology. I claim that a particular interpretation of the vera causa principle offers such a criterion. According to Gayon's analysis of Darwin's use and interpretation of the vera causa principle, a true theory can be distinguished from a mere hypothesis by its ability to explain independent facts. This leads me to a minimal conception of scientific theories as a scientific hypothesis derived from empirical premises (or more precisely, from well-established classes of facts) that has been proven able to explain independent classes of facts. A more precise definition and characterization of theories might be valuable. However, this one has the advantage of being simple and operational.

Through the use of this theory identification criterion, I have shown that there is already at least one theory of development, which is the CSC theory of development of cancers. This theory basically matches the fuzzy conception of theories as sets of explanations and predictions. It also answers positively to the specific criterion suggested in this chapter. Indeed, it explains classes of facts as different as the low clonogenicity of cancer cells at the population level, the heterogeneity of the cells composing the tumours, the presence of non-metastatic cells at a distance from the sites of the primary tumours, and relapses after apparently successful therapies.

Acknowledgments

I would like to thank Thomas Pradeu, Michel Morange, Michel Vervoort, Francesca Merlin, Thierry Hoquet, and Karine Prévot, members of the group 'Theory and Definition of Development' (IHPST), for fruitful discussions on the question of theories of development. I also want to thank Jean Gayon for his useful comments during the workshop 'Theory and Definition of Development', Melinda Fagan, Francesca Merlin, Alessandro Minelli, Thomas Pradeu, Davide Vecchi, and Marion Vorms for their useful comments on the text, as well as Alexandre Guay for discussions on theories in physics. I am also thankful to Joyce Kettering for her help.

References

Ahmed, A.U., Thaci, B., Wainwright, D.A., et al. (2011). Therapeutic strategies targeting cancer stem cells. In S. Shostak, ed., *Cancer Stem Cells—The Cutting Edge*. In-Tech, Croatia, pp. 375–88.

Alison, M.R., Lim, S.M., and Nicholson, L.J. (2011). Cancer stem cells: problems for therapy? *Journal of Pathology*, **223**, 147–61.

American Cancer Society (2012). *Cancer Treatment and Survivorship Facts & Figures 2012–2013*. American Cancer Society, Atlanta.

Astaldi, G., and Mauri, C. (1953). Recherches sur l'activite proliferative de l'hemocytoblaste de la leucémie aigue. *Revue belge de pathologie et de médecine expérimentale*, **23**, 69–82.

Bao, S., Wu, Q., McLendon, R.E., et al. (2006). Glioma stem cells promote radioresistance by preferential activation of the DNA damage response. *Nature*, **444**, 756–60.

Beatty, J. (1980). What's wrong with the received view of evolutionary theory? *PSA Proceedings of the Biennial Meeting of the Philosophy of Science Association*, **2**, 397–426.

Beatty, J. (1995). The evolutionary contingency thesis. In G. Wolters and J. Lennox, eds., *Concepts, Theories, and Rationality in the Biological Sciences: The Second Pittsburgh-Konstanz Colloquium in the Philosophy of Science*. University of Pittsburgh Press, Pittsburgh, pp. 45–81.

Bonnet, D., and Dick, J.E. (1997). Human acute myeloid leukemia is organized as a hierarchy that originates from a primitive hematopoietic cell. *Nature Medicine*, **3**, 730–7.

Buczacki, S., Davies, R.J., and Winton, D.J. (2011). Stem cells, quiescence and rectal carcinoma: an unexplored relationship and potential therapeutic target. *British Journal of Cancer*, **105**, 1253–9.

Clarke, M.F., Dick, J., Dirks, P.B., et al. (2006). Cancer stem cells—perspectives on current status and future directions: AACR Workshop on cancer stem cells. *Cancer Research*, **66**, 9339–44.

Dalerba, P., Cho, R.W., and Clarke, M.F. (2007). Cancer stem cells: models and concepts. *Annual Review of Medicine*, **58**, 267–84.

Darwin, C. (1875). *The Variation of Animals and Plants Under Domestication*, 2nd ed. John Murray, London.

Dean, M., Fojo, T., and Bates, S. (2005). Tumour stem cells and drug resistance. *Nature Reviews Cancer*, **5**, 275–84.

Dewi, D.L., Ishii, H., Kano, Y., et al. (2011). Cancer stem cell theory in gastrointestinal malignancies: recent progress and upcoming challenges. *Journal of Gastroenterology*, **46**, 1145–57.

Duchesneau, F. (1997) *Philosophie de la biologie*. PUF, Paris.

Dylla, S.J., Beviglia, L., Park, I.K., et al. (2008). Colorectal cancer stem cells are enriched in xenogeneic tumors following chemotherapy. *PLoS One*, **3**, e2428.

Eisenhauer, E.A., Therasse, P., Bogaerts, J., et al. (2009). New response evaluation criteria in solid tumours: revised RECIST guideline (version 1.1). *European Journal of Cancer*, **45**, 228–47.

Fekete, E., and Ferrigno, M.A. (1952). Studies on a transplantable teratoma of the mouse. *Cancer Research*, **12**, 438–40.

Fialkow, P.J. (1980). Clonal and stem cell origin of blood cell neoplasms. In J. Lobue, A.S. Gordon, R. Silber and F.M. Muggia, eds., *Contemporary Hematology and Oncology*, Vol. 1. Plenum, New York, pp. 1–46.

Gayon, J. (1993). La biologie entre loi et histoire. *Philosophie*, **38**, 30–57.

Gayon, J. (1997). The paramount power of selection: from Darwin to Kauffman. In M.L. Dalla Chiara, K. Doets, D. Mundici et al., eds. *Structures and Norms in Science, Volume Two of the Tenth International Congress of Logic, Methodology and Philosophy of Science*. Kluwer, Dordrecht, pp. 265–82.

Gayon, J. (2009). Mort ou persistance du darwinisme? Regard d'un épistémologue. *Comptes Rendues Palevol*, **8**, 321–40.

Gespach, C. (2010). Stem cells and colon cancer: the questionable cancer stem cell hypothesis. *Gastroentérologie Clinique Biologique*, **34**, 653–61.

Giere, R.N. (1988). *Explaining Science: A Cognitive Approach*. Chicago University Press, Chicago.

Gil, J., Stembalska, A., Pesz, K.A., et al. (2008). Cancer stem cells: the theory and perspectives in cancer therapy. *Journal of Applied Genetics*, **49**, 193–9.

Godfrey-Smith, P. (2006). The strategy of model-based science. *Biology and Philosophy*, **21**, 725–40.

Gould, S.J. (1990). *Wonderful Life. The Burges Shale and the Nature of History*. Norton, New York-London.

Hemmings, C. (2010). The elaboration of a critical framework for understanding cancer: the cancer stem cell hypothesis. *Pathology*, **42**, 105–12.

Hempel, C.G. (1965). *Aspects of Scientific Explanation*. Free Press, New York.

Hodge, M.J.S. (1977). The structure and strategy of Darwin's 'long argument'. *British Journal for the History of Science*, **10**, 237–46.

Hodge, M.J.S. (1992). Darwin's argument in the Origin. *Philosophy of Science*, **59**, 461–4.

Hull, D. (1974). *Philosophy of Biological Science*. Prentice-Hall, Englewood Cliffs.

Hull, D. (1977). A logical empiricist looks at biology. *British Journal for the Philosophy of Science*, **28**, 181–9.

Hull, D. (2003). Darwin's science and Victorian philosophy of science. In J. Hodge and G. Radick, eds., *The Cambridge Companion to Darwin*. Cambridge University Press, Cambridge, pp. 168–191.

Kavaloski, V. (1974). *The Vera Causa Principle: A Historico-philosophical Study of a Metatheoretical Concept from Newton through Darwin*. PhD dissertation, University of Chicago.

Killmann, S.A., Cronkite, E.P., Fliedner, T.M., et al. (1962). Cell proliferation in multiple myeloma studied with tritiated thymidine in vivo. *Laboratory Investigation*, **11**, 845–53.

Kleinsmith, L.J., and Pierce, G.B. (1964). Multipotentiality of single embryonal carcinoma cells. *Cancer Research*, **24**, 1544–51.

Laplane, L. (2013). *Cancer stem cells: ontologies and therapies*. PhD dissertation, Paris Ouest University.

Laplane, L. (forthcoming). Cancer stem cells: Off with their heads. In B. Bensaude-Vincent, S. Loeve, A. Nordmann and A. Schwarz, eds., *Attractive Objects. The Furniture of the Technoscientific World*. Pickering & Chatto, London. [YEAR to be updated]

Li, X., Lewis, M.T., Huang, J., et al. (2008). Intrinsic resistance of tumorigenic breast cancer cells to chemotherapy. *Journal of the National Cancer Institute*, **100**, 672–9.

Li, Y., Wicha, M.S., Schwartz, S.J., et al. (2011). Implications of cancer stem cell theory for cancer chemoprevention by natural dietary compounds. *The Journal of Nutritional Biochemistry*, **22**, 799–806.

Lloyd, E. (1983). The nature of Darwin's support for the theory of natural selection. *Philosophy of Science*, **50**, 112–29.

Lloyd, E. (1988). *The Structure and Confirmation of Evolutionary Theory*. Princeton Universtity Press, Princeton.

Lyell, C. (1830–1833) *Principles of Geology, Being an Attempt to Explain the Former Changes of the Earth's Surface by Reference to Causes Now in Operation*. John Murray, London.

Ma, S., Lee, T.K., Zheng, B.J., et all. (2008). CD133 + HCC cancer stem cells confer chemoresistance by preferential expression of the Akt/PKB survival pathway. *Oncogene*, **27**, 1749–58.

McClellan, J.S., and Majeti, R. (2012). The cancer stem cell model: B cell acute lymphoblastic leukaemia breaks the mould. *EMBO Molecular Medicine*, **5**, 7–9.

Morange, M. (2003). *La vie expliquée? 50 ans après la double hélice*. Odile Jacob, Paris.

Morange, M. (2008)). *Life explained*. (trans. Cobb, M., and DeBevoise, M.). Yale University Press, New Haven.

Morrison, R., Schleicher, S.M., Sun, Y., et al. (2011). Targeting the mechanisms of resistance to chemotherapy and radiotherapy with the cancer stem cell hypothesis. *Journal of Oncology*, 2011:941876.

Newton, I. (1960). *Mathematical Principles of Natural Philosophy*, 3rd ed. University of California Press, Berkley.

Nguyen, L.V., Vanner, R., Dirks, P., et al. (2012). Cancer stem cells: an evolving concept. *Nature Reviews Cancer*, **12**, 133–43.

Nowell, P.C. (1976). The clonal evolution of tumor cell populations. *Science*, **194**, 23–8.

Persidis, A. (1999). Cancer multidrug resistance. *Nature Biotechnology*, **17**, 94–5.

Pradeu, T. (2009). *Les limites du soi: immunologie et identité biologique*. Presses Universitaires de Montréal, Montréal—Vrin, Paris.

Pradeu, T. (2011). Philosophie de la biologie. In A. Barberousse, D. Bonnay & M. Cozic, eds., *Précis de philosophie des sciences*. Vuibert, Paris.

Rahman, M., Deleyrolle, L., Vedam-Mai, V., et al. (2011). The cancer stem cell hypothesis: failures and pitfalls. *Neurosurgery*, **68**, 531–45.

Recker, D. (1987). Causal efficacy: the structure of Darwin's argument strategy in the origin of species. *Philosophy of Science*, **54**, 147–75.

Reya, T., Morrison, S.J., Clarke, M.F., et al. (2001). Stem cells, cancer, and cancer stem cells. *Nature*, **414**, 105–11.

Rivera, S., Rivera, C., Loriot, Y., et a. (2011). Cellules souches cancéreuses: nouvelle cible thérapeutique dans le traitement des cancers bronchopulmonaires. *Cancer Radiothérapie*, **15**, 355–64.

Rosenberg, A., and McShea, D.W. (2008). *The Philosophy of Biology. A Contemporary Introduction*. Routledge, New York and London.

Ruse, M. (1975). Darwin's debt to philosophy: an examination of the influence of the philosophical ideas of John F.W. Herschel and William Whewell on the development of Charles Darwin's theory of evolution. *Studies in History and Philosophy of Science*, **6**, 159–81.

Ruse, M. (1973) *The Philosophy of Biology*. Hutchinson University Press, London.

Shipitsin, M., and Polyak, K. (2008). The cancer stem cell hypothesis: in search of definitions, markers, and relevance. *Laboratory Investigation*, **88**, 459–63.

Smart, J.J.C. (1963). *Philosophy and Scientific Realism*. Routledge & Kegan Paul, London.

Sober, E. (2006). Two outbreaks of lawlessness in recent philosophy of biology. In E. Sober, ed., *Conceptual Issues in Evolutionary Biology*. MIT Press, Cambridge and London, pp. 249–58.

Southam, C.M., and Brunschwig, A. (1961). Quantitative studies of autotransplantation of human cancer. Preliminary report. *Cancer*, **14**, 971–8.

Stevens, L.C. (1960). Embryonic potency of embryoid bodies derived from a transplantable testicular teratoma of the mouse. *Developmental Biology*, **2**, 285–97.

Suppe, F. (1977). The search for philosophical understanding of scientific theories. In F. Suppe, ed., *The Structure of Scientific Theories*, 2nd ed. University of Illinois Press, Champaign.

Suppes, P. (1960). A comparison of the meaning and use of models in mathematics and the empirical sciences. *Synthese*, **12**, 287–301.

Suppes, P. (1967). What is a scientific theory ? In S. Morgenbesser, ed., *Philosophy of Science Today*. Basic Books, New York, pp. 55–67.

Tai, M.H., Chang, C.C., Kiupel, M., et al. (2005). Oct4 expression in adult human stem cells: evidence in support of the stem cell theory of carcinogenesis. *Carcinogenesis*, **26**, 495–502.

Tan, B.T., Park, C.Y., Ailles, L.E., et al. (2006). The cancer stem cell hypothesis: a work in progress. *Laboratory Investigation*, **86**, 1203–7.

Thagard, P. (1978). The best explanation: criteria for theory choice. *Journal of Philosophy of Science*, **37**, 325–39.

Therasse, P., Arbuck, S.G., Eisenhauer, E.A., et al. (2000). New guidelines to evaluate the response to treatment in solid tumors. European Organization for Research and Treatment of Cancer, National Cancer Institute of the United States, National Cancer Institute of Canada. *Journal of the National Cancer Institute*, **92**, 205–16.

van Fraassen, B.C. (1980). *The Scientific Image*. Oxford University Press, Oxford.

van Fraassen, B.C. (1987). The semantic approach to scientific theories. In N.J. Nersessian, ed., *The Process of Science: Contemporary Philosophical Approaches to Understanding Scientific Practice*. Martinus Nihoff, Dordrecht, pp. 105–124.

Vermeulen, L., de Sousa e Melo, F., Richel, D.J., et al. (2012). The developing cancer stem-cell model: clinical challenges and opportunities. *Lancet Oncology*, **13**, e83–e89.

Vorms, M. (2011). *Qu'est-ce qu'une théorie scientifique?* Vuibert, Paris.

Waters, K. (1986). Taking analogical inference seriously: Darwin's argument from artificial selection. *PSA: Proceedings of the Biennial Meeting of the Philosophy of Science Association*, **1**, 502–13.

Williams, M.B. (1981) Similarities and differences between evolutionary theory and the theories of physics. *PSA: Proceedings of the Biennial Meeting of the Philosophy of Science Association*, **2**, 385–96.

Woodward, W.A., Chen, M.S., Behbod, F., et al. (2007). WNT/beta-catenin mediates radiation resistance of mouse mammary progenitor cells. *Proceedings of the National Academy of Sciences USA*, **104**, 618–23.

Yoo, M.H., and Hatfield, D.L. (2008). The cancer stem cell theory: is it correct? *Molecules and Cells*, **26**, 514–16.

CHAPTER 17

Animal development in a microbial world

Spencer V. Nyholm and Margaret J. McFall-Ngai

Introduction

Animal developmental biology has largely focussed on the autonomous processes, and their underlying mechanisms, whereby the zygote is transformed into the tissues, organs, and organ systems that make up the individual. The last few years have seen the emergence of a new area, 'eco-devo', in which the research community is beginning to address the question of how the environment, past and present, has been involved in inducing the patterns of development that we can observe. Whereas abiotic forces, such as temperature, salinity, and oxygen tension, have long been implicated in driving developmental patterns, only recently has the role of the biotic environment been considered with any depth. The application of newly developed genomic technologies and studies in a variety of systems have put into sharp focus the fact that the individual animal is not alone; rather, it develops within the context of a complex and highly interactive microbial world. The emerging data reveal a rich variety of animal–microbe associations, from relatively loose (i.e. transient or of low specificity) to the tightly coupled partnerships of obligate symbioses (McFall-Ngai et al., 2013). Each of these associations often has the challenge of assuring that, with fidelity, the association persists between generations although the animal partner, in sexual reproduction, reverts to a single-celled zygote at the onset of development of a new individual. In studying these arenas, developmental biologists face two challenges: (i) defining basic principles across animal systems in the context of diversity of interactions through the ontogeny of an individual, as well as diversity of types of associations across the animal kingdom; and (ii) integrating these ideas into the strong existing framework of current thinking in developmental biology as well as other fields in the greater discipline of biology. This chapter highlights our current understanding of the influences of the microbial world on the development of animals and discusses the strategies being formulated to address these challenges.

The historical influence of microbes on animal development—the tell-tale signs of microbial influence on the evolution of developmental patterns of the digestive and immune systems

The most conspicuous interactions between animals and microbes involve nutrition. How did this relationship evolve, and how did it influence developmental patterns in animals? In addressing these questions, it is useful to consider the environment in which animals arose. The eukaryotic ancestors of animals arose in a world that had been dominated by Archaea and Bacteria for well over a billion years. That new domain of life, the Eukarya, itself arose by an endosymbiotic event in which an archaean ancestor had incorporated the α-proteobacterium that would become the mitochondrion. Molecular evidence suggests that this endosymbiosis arose once, a milestone that lead not only to the radiation of the diverse eukaryotic microbes, animals, plants, and fungi, but also to a metabolic 'strategy' that would likely benefit from alliances. Specifically,

Towards a Theory of Development. Edited by Alessandro Minelli and Thomas Pradeu

the Eukarya might be thought of as metabolic specialists. Whereas Archaea and Bacteria use a wide variety of electron donors and acceptors in energy acquisition, the eukaryotes rely entirely on oxygen as the terminal electron acceptor for respiration. As such, forming alliances with the Archaea and Bacteria would greatly expand the metabolic scope of an animal. Certain extant symbiotic associations, such as mutualisms with luminous bacteria, appear not to have a nutritional base, but even those alliances are thought to have originated as nutritional symbioses.

The contribution of mitochondria to nutrition, as the site of oxidative phosphorylation in the eukaryotic cell, is conspicuous and has historically been the principal focus of study of this organelle's form and function. However, data continue to emerge demonstrating that the mitochondria have profound effects on a much wider variety of activities of the eukaryotic cell, including developmental processes. Over the past 20 years, several studies have documented the central role of mitochondria in apoptosis, and more recently their role in autophagy has been recognized (Hamasaki et al., 2013). In addition, new data reveal that the mitochondria are powerful controllers of cell cycle (Mitra et al., 2009) and cell differentiation (Mitra et al., 2012). Taken together, these findings are providing evidence that the mitochondria are likely to be pivotal players in many, if not most, critical aspects of development.

The relationship between nutrition and bacteria-induced development may extend as far back as the origin of animals. Evidence from phylogenetic relationships suggests that choanoflagellate protists and animals are sister groups and that these clades arose from a common choanoflagellate-like ancestor. Extant choanoflagellates are bacterivorous, and some representatives have orthologues of molecules that mediate signalling between bacteria and animals. Further, data are emerging that implicate bacteria in the induction of multicellularity in some choanoflagellate species. King and coworkers are characterizing the interactions between the choanoflagellate *Salpingoeca rosetta* and bacteria in their environment (Alegado et al., 2012; Dayel et al., 2011). They have demonstrated that sulfonolipids from the surfaces of certain species of bacteria in the phylum Bacteroidetes induce the choanoflagellate to form multicellular rosettes. The behaviour of colony development in this species, as well as its underlying molecular mechanisms, suggests a means by which bacteria could have contributed to the evolution of multicellular animal forms. This basal group also provides an example of a link between nutrition and induction of development.

The sponges, which are a basal metazoan lineage, also use choanoflagellate-like cells to feed on microbes in the water column. Sponges are at the cellular grade of animal evolution; that is, they do not have true tissues, such as epithelia. The assemblage of cells that make up the sponge body can be, by weight, up to 40% bacterial. These bacterial cells occur both intracellularly and extracellularly, and phylogenetic analysis provides evidence for host–symbiont co-evolution in some instances, as do sponge-specific phyla of bacteria, such as the Poribacteria. In some sponge species, the associations with their bacterial partners are established very early in development (Hentschel et al., 2012). *Xestospongia bocatorensis* incorporates the cells of *Oscillatoria spongeliae*, a cyanobacterium, within intracellular vacuoles in developing oocytes. In other sponge species, dense layers of bacterial filaments form over the developing embryos. While the function of these relationships remains to be explored, might symbiotic bacteria be required to maintain intercellular adhesion in these sponge embryos, as is the case in the multicellular colonies of their relatives, the choanoflagellates?

The relationship of the sponges to the other metazoan phyla is controversial. However, it is clear that the most important subsequent steps in animal evolution involved the appearance of two interrelated features: (i) differentiation of cells and tissues to produce true epithelia, that is, with occluding cell junctions; and (ii) the process of gastrulation. The environmental features that drove these events are not known, but their 'invention' did correlate with some interesting changes in the relationship of animals to environmental bacteria. Unlike sponges, the tissues underlying epithelial layers are generally low in microbes. Where they do occur within these tissues, they are most often intracellular, such as in the bacteriome of insects, the trophosome of hydrothermal vent tubeworms, and the lung macrophages of tuberculosis patients. Thus, the formation of epithelia may be the evolutionary step that excluded colonization by dense bacterial populations in internal tissues in all animals that are diplo- and triploblastic. The process of gastrulation produces the inner epithelium that faces the archenteron. Coincident with an inner epithelium in animals is the ability of the mature tissue to carry out digestive and absorptive functions, activities that would occur efficiently in such a confined space. Most theories concerning the evolutionary origin of gastrulation posit that the precursor state or 'gastraea' arose from ancestors that were bacterivores, such as the homoscleromorph sponges or the placozoans. Regardless of the origin, once mechanisms for embryonic gut formation have evolved, a space is created not only for extracellular digestion, but also for the housing of a consortium of bacterial partners that could participate in the breakdown of food.

Because of their ubiquity, diversity, and abundance, it is almost certain that once a gut cavity was formed, microbes were present there, as food or as transient or persistent partners in the process of food digestion. With the advent of the one-way gut, regionalization could occur, and the appearance of the coelom, which separates the gut from the body wall, allowed for elongation and folding of the gut within the body cavity. The extent to which bacterial partners participated in driving the evolution of developmental programs that result in these features will be difficult to determine. Systems such as the rumen of ungulate mammals and the termite hindgut, which have conspicuous adaptations for housing symbionts, have allowed for the evolution of a unique metabolism (i.e. the degradation of lignocellulose) and the expansion of these hosts into specialized niches that otherwise may have been inaccessible. Even if evolution did not play a hand in these major milestones, mechanisms to interface with the microbial associates (e.g. expression of receptors on epithelial surfaces, etc.) must have been built into developmental pathways over evolution of the animal gut. Studies of extant animals indicate that, during embryogenesis, the host develops features that allow for interactions with microbes upon hatching or birth, and, where systems have been analysed, microbes participate in postembryonic maturation of these tissues (see 'Influences on the immune system').

Microbes also associate with the other epithelia of both aquatic and terrestrial animals. Interestingly, unlike terrestrial taxa, many aquatic animals absorb nutrients not just from the gut, but also directly across the body wall. How the microbial communities on these epithelia affect the carbon balance (i.e. increase the efficiency of the uptake of dissolved organic carbon (DOC)) or compete with the animal tissue has not been addressed. In the background of these interactions, animals must have evolved mechanisms to protect their surfaces from microbial overgrowth. Epithelia are often rich in antimicrobial molecules, which are derived either from the animal cells themselves or from the bacterial partners residing on the apical surfaces. These challenges affect the newly hatched larval or juvenile animal, and thus the molecules that mediate these interactions must be provided during embryogenesis. To avoid desiccation and because air does not contain high levels of DOC, the ability to take up DOC across the body wall was lost in the water-to-land transition.

Whether environmental bacteria might have participated in driving selection that gave rise to the gastrulation process will be difficult to determine, but some intriguing molecular data suggest that molecules associated with animal–microbe interactions do participate in this process. Specifically, gastrulation in extant animals is coincident with regulation of several genes that encode proteins of immune pathways (STAT3, TGF-β) (Messler et al., 2011; Saina et al., 2009; Yamashita et al., 2002). Dedicated immune genes are also implicated. One study has shown that, in the sea urchin, the gene encoding the complement protein C3, *Sp064*, is expressed during embryogenesis and that its expression peaks during gastrulation; in addition, *Sp064* expression in the embryos is upregulated upon exposure to bacteria. In this study it was posited that production of immune genes by the embryo is a defense mechanism (Shah et al., 2003); but why then would the expression of an immune gene peak at gastrulation? Another interpretation might be that the first role of these immune genes was to mediate interactions with environmental bacteria and that they were later recruited into a developmental role, in much the same way as genes in the Toll-receptor/NF-κB pathway.

The influence of microbial partners on the development patterns of extant animals

Generally, there are two mechanisms by which animal hosts can acquire their symbionts (Bright & Bulgheresi, 2010; McFall-Ngai, 2002). The first is horizontal transmission, in which colonization from the environment occurs in each generation. This mode is by far the most common way in which hosts acquire their microbiota. For example, most gut associations are established shortly after an animal is born or emerges from its egg. In some associations, initiation of symbiosis can be difficult because the microbial partner(s) may be in low abundance and/or establishment may require a high degree of specificity; for example, whereas the symbiosis between the light organ in the Hawaiian bobtail squid (*Euprymna scolopes*) and the bioluminescent bacterium *Vibrio fischeri* is also established via horizontal transmission, in the squid–vibrio association, colonization is multifactorial, requiring both host and symbiont input to ensure initiation and specificity. Hatchling squid must specifically recognize *V. fischeri*, which make up less than 0.1% of the total bacterial population, and prevent non-symbiotic bacteria from colonizing the light organ (Nyholm & McFall-Ngai, 2004). For the initiation of symbiosis between a host and a microbial consortium, the need to establish specificity may not be as stringent as with a binary association, but specificity and a regulated microbial succession do occur. For example, in the mammalian gut, *Bacteroides thetaiotaomicron* is a dominant member of the microbiota and often the first colonizer (Bry et al., 1996).

The other type of symbiont transmission is vertical, wherein the microbial partner is passed down directly from parent to offspring, usually through the egg. Such a mode of transmission assures transfer to the offspring, but at some cost: the bacterial symbionts do not have an opportunity for genetic exchange, and this results in an accumulation of deleterious mutations and gene loss (Moran et al., 2008). Genome reduction in vertically transmitted symbionts has been most extensively studied in insects, where extreme genome reduction has been documented (McCutcheon & Moran, 2010). In some cases, these symbiont genomes are smaller than the genomes of some plant organelles (McCutcheon & Moran, 2012). In addition, a correlation exists between the geological age of the association and the extent of reduction of the symbiont genome (Heddi et al., 1998), which reflects the gradual nature of genome loss.

The influence of microorganisms on host developmental stages and organ systems

Protection of eggs

Many animals lay their eggs in the environment, where successful embryogenesis depends on minimizing fouling by microorganisms and/or predation. Aquatic/marine organisms are especially susceptible to fouling, since their eggs are under constant exposure to high densities of microorganisms. Given that embryogenesis can often take weeks and biofilms can generally form in a matter of hours or days, mechanisms must be present to prevent microbial overgrowth in the developing embryos. To overcome this threat, animals often incorporate endogenous antimicrobial compounds in their eggs or elicit the cooperation of microorganisms, for example, by directly incorporating bacteria and/or bacterial products in their eggs that prevent fouling by other microorganisms.

A number of both invertebrates and vertebrates are known to use antimicrobial compounds to protect their eggs and/or embryos from infection. The cuticle of a chicken eggshell contains lysozyme and other antimicrobial factors and has been shown to inhibit the growth of some bacteria (Rose-Martel et al., 2012; Wellman-Labadie et al., 2008, 2010). Zebrafish use the amidase and broad-spectrum bactericidal activities of peptidoglycan recognition proteins (PGRPs) to protect developing embryos from bacterial infections during the short embryonic period (Li et al., 2007). In *Hydra*, the use of antimicrobial peptides goes one step further, as these compounds are thought to select for the colonization of stage-specific bacteria during embryogenesis (Fraune et al., 2010). These bacteria may then provide antifungal protection for the developing embryos (T. Bosch, pers. comm.). The shrimp *Palaemon macrodactylus* also uses an antimicrobial compound (2,3-indolinedione) from a bacterial *Alteromonas* sp. to prevent fungal infection of developing embryos

(Gil-Turnes et al., 1989). The female members of many cephalopods, including squids and cuttlefish, have a reproductive organ, the accessory nidamental gland (ANG), which harbours a dense consortium of bacteria (Bloodgood, 1977). Like other symbiotic organs such as the light organ and gut, in the ANG extracellular bacteria are housed in epithelium-lined crypts. Culture-dependent and independent methods have identified the dominant members of these microbial communities, and all ANGs examined to date are dominated by α-proteobacteria, usually members of the Roseobacter clade within Rhodobacterales (Barbieri et al., 2001; Collins et al., 2012a). Bacteria from the ANG are deposited directly into the egg cases (Barbieri et al., 2001; Collins et al., 2012a), where they are hypothesized to provide a protective role for the developing embryos, possibly preventing the growth of fungal biofilms (Biggs & Epel, 1991; S.V. Nyholm, unpubl.). Efforts are currently underway to understand how the microbiota of the ANG may prevent fouling and provide other forms of protection for developing eggs.

Microorganisms may also provide egg protection from environmental stresses other than fouling and infection. For example, eggs of the spotted salamander *Ambystoma maculatum* harbour symbiotic photosynthetic algae (*Oophila amblystomatis*) that are thought to provide oxygen to developing embryos in gelatinous egg masses, where natural oxygen diffusion may be limiting (Pinder & Friet, 1994). A more recent study also demonstrated that these intracapsular algae provide fixed carbon to the developing embryos (Graham et al., 2013). This symbiosis appears to be critical for the salamander host; optimal development occurs in the presence of the algae but is negatively impacted if the symbionts are removed (Tattersall, 2008). Furthermore, the algal symbiont is capable of invading host tissues and appears to be transmitted vertically (Kerney et al., 2011).

The gut microbiome and development

It is perhaps not surprising that microbes have a profound influence on the development of the gut. The human microbiome contains approximately 10^{14} bacteria, ten times the number of cells that one would conventionally count as 'self', and the majority of these partners are in the gut. The combined genomes of these microbial members are 100 times greater than that of our own, meaning that both the cellular and genomic complexity of the entire human 'ecosystem' is vast. Much of our understanding comes from the NIH-sponsored Human Microbiome Project, which has dramatically increased our knowledge of the human microbiota in recent years. As the result of this project, it has been revealed that humans have a diverse microbiota made up of over 1000 species but that the vast majority of these reside in the digestive tract and are dominated by just two bacterial phyla, Bacteroidetes and Firmicutes, with other members mainly distributed among Proteobacteria, Actinobacteria, and Verrucomicrobia (Lozupone et al., 2012; Tremaroli & Backhed, 2012). Although microbial species diversity can vary greatly between individuals, there does appear to be a set of shared functional genes that makes up the core microbiome of humans (Lozupone et al., 2012). Most of these microbes reside in our gut, where they are known to aid in food digestion and provide essential vitamins. Given that the gut tissues are in direct contact with the microorganisms that reside there, these bacteria are poised to directly influence the development of this symbiotic organ. The microbiota have a great influence on host metabolism. Polysaccharides, cholesterol, and the essential nutrient choline are all metabolized by the gut microbiota and lead to the generation of bioactive compounds such as short-chain fatty acids, bile acids, and trimethylamine-N-oxide, all of which can have both positive and negative effects on host health. Metabolites derived from gut microbiota can be found in both blood and urine and can have systemic effects on their hosts (Nicholson et al., 2012; Wikoff et al., 2009). The mechanisms of these effects have yet to be revealed. However, it is becoming evident that perturbations to the normal gut microbiota and the bioactive compounds they produce leads to dysbiosis and is linked to a number of diseases including chronic malnutrition (Smith et al., 2013), gastric ulcers, obesity, diabetes, cardiovascular disease, and inflammatory bowel disease (reviewed in Nicholson et al., 2012; Tremaroli & Backhed, 2012). Furthermore, it is now known that the gut microbiota influence a number of other organ systems such as the circulatory, immune, and neuroendocrine systems; some of these findings are discussed below.

Much of our understanding of the influence of the animal microbiota on the development of organ systems has come from the study of germ-free animals (e.g. mice, rats, rabbits, and more recently, zebrafish). The techniques necessary to raise germ-free animals were established in the 1920s and perfected over decades. Russell Schaedler, Rene Dubos, and colleagues first used germ-free models to show that there was an indigenous gut microbiota and that this microbiota could influence development of the digestive system. Work over the years has shown that, compared to conventionally raised animals, germ-free animals exhibit altered phenotypes that influence a number of functions in the gut (reviewed in Smith et al., 2007). The ability to colonize germ-free (GF) animals with a defined (gnotobiotic) microbiota has allowed researchers to ask how specific microbes influence animal physiology and development. The development of GF and gnotobiotic zebrafish revealed that a core microbiota have a profound influence on host gene expression and cellular responses like epithelial proliferation (Rawls et al., 2004). A study with gnotobiotic fish showed that the induction of epithelial proliferation is meditated by interactions between bacterial signals and components of the innate immune system, primarily MyD88, an adaptor for a Toll-like receptor (Cheesman et al., 2011). Recent work has also demonstrated that the zebrafish microbiota are instrumental in fatty acid metabolism (Roeselers et al., 2011).

It is also becoming clear that the composition of microbiota can change significantly during the life of the host. In humans, remodelling of the gut microbiota takes place during pregnancy, with significant changes in the microbiota apparent by the third trimester; these changes may be mediated by both immune and metabolic factors (Koren et al., 2012). In human infants, colonization of the gut begins at birth as newborns exit the birth canal. Babies delivered by Caesarean section have a gut microbial community that more resembles their mothers' skin microbiota than the vaginal microbial community found in those delivered by natural birth (Dominguez-Bello et al., 2010). Over the first few weeks, the gut microbiota is dominated by an anaerobic microbial community (Nicholson et al., 2012). The complex nature of human milk oligosaccharides may help select for certain gut bacteria early on during development (Zivkovic et al., 2011). The gut community stabilizes throughout childhood and adulthood, although recent evidence suggests that the gut microbiome may undergo a change during adolescence and changes to bacteria-derived metabolites have also been detected (Agans et al., 2011; Mariat et al., 2009). As humans age, a detectable change in the microbiota composition also occurs and is associated with alterations in the immune system and overall physiology of the host. Analyses of bacteria-derived metabolite profiles also show changes associated with age (Nicholson et al., 2012).

Influences on the immune system

Classically, the immune system has been viewed as an interface between an organism and pathogens. There is a long-standing view that the immune system evolved to combat pathogens. Traditionally this view states that innate immunity, which uses cellular (e.g. phagocytosis) and acellular tools (e.g. antimicrobial peptides, humoral factors, etc.) to clear an infection evolved first, and with the jawed vertebrates emerged adaptive immunity and the ability to target specific pathogens with an antibody response. Recent research shows that the immune system of animals plays a critical role in shaping associations with microorganisms (Lee & Mazmanian, 2010; McFall-Ngai, 2007; Nyholm & Graf, 2012), putting into question both the 'defensive' view of the immune system and the traditional self–nonself dichotomy (Pradeu, 2012).

One of the hallmarks of a pathogenic infection is the inflammation response (Strowig et al., 2012). Long thought of as a defense mechanism against pathogenic microorganisms, the process of inflammation has also recently been characterized as a part of normal development (Ramos, 2012). For example, in his treatise Ramos argues that the cell and tissue regeneration are mediated through inflammatory processes that occur regularly throughout an organism's life. As we shall discuss below, in many host/microbe interactions, host responses that would normally be associated with an immune or pathogen response are often involved with shaping the development of a normal microbiota.

The influence of the normal microbiota on the development of the immune system occurs at multiple levels (e.g. systemic, tissue, cellular, and molecular). In mammals, postembryonic development of components of the circulatory system is dependent on bacterial colonization of the gut. In mice normal capillary development is impaired in GF animals and can be restored by adding gut bacteria from a conventional mouse or by colonization with *Bacteroides thetaiotamicron*, an early colonizer of the mammalian gut (Stappenbeck et al., 2002). Gut-associated lymphoid tissue (GALT) is the main interface between the immune system and the digestive system of mammals. In GF mice GALT is not fully developed, and these animals also have deficiencies in their Peyer's patches, mesenteric lymph nodes, and lamina propria (reviewed in Smith et al., 2007). GF mice have several immune cell types that are also compromised or not fully functional; for example they have reduced numbers of CD4 $^+$ T cells (Niess et al., 2008). The gut microbiota are also known to influence the differentiation of different T-cell classes in GF mice. In GF rabbits, the gut microbiota also influences the expansion of B cells (Rhee et al., 2004). The gut microbiota induces the development of isolated lymphoid follicles, and signalling between the host (in the form of innate lymphoid cells of the adaptive immune system) and the symbionts leads to gut homeostasis (Eberl, 2005; Sawa et al., 2011). Some specific bacterial factors have been implicated in some of these processes. For example, *Bacteroides fragilis* polysaccharide A is known to induce T-cell proliferation in mice, but *B. fragilis* by itself fails to induce maturation of GALT (Mazmanian et al., 2005). However, addition of another bacterium, *Bacillus subtilis*, induces complete GALT development, suggesting that multiple bacterial signals may be necessary for this induction (Rhee et al., 2004).

Exposure to a normal microbiota early on in development can also have effects later in life. For example, in humans, children raised on farms have a significantly lower number of cases of asthma and allergies (reviewed in von Mutius & Vercelli, 2010). This protective effect has been explained in part in terms of the hygiene hypothesis, which states that intense exposure to microorganisms is required for the normal development of the immune system and a robust immune response. By the time they reach school age, children raised on farms in Europe displayed significantly higher gene expression for innate immunity factors such as TLR2 and CD14 than children raised in a non-farming environment. Furthermore, the number of farm animals a woman is exposed to during pregnancy may contribute to higher innate immunity gene expression as well as altered chemokine levels, suggesting that exposure to an abundance and diversity of farm-associated microorganisms may have long-term effects on the development of the immune system. A recent study also showed that colonization of GF mice with a conventional microbiota had a protective effect against inflammatory bowel disease (IBD) by regulating invariant natural killer T cells and the chemokine ligand CXCL16 (Olszak et al., 2012). This protection was based on time-dependent exposure to a conventional microbiota, with hosts that were colonized as neonates enjoying resistance to IBD as adults whereas GF mice that were colonized as adults lacked resistance to IBD.

Communication between animal hosts and their microbiota is often mediated by signalling via the immune system involving interactions between microbe-associated molecular patterns (MAMPs) and pattern recognition receptors (PRRs). Animal PRRs are evolutionarily conserved and found throughout eumetazoa. Toll-like receptors and PGRPs were first identified in *Drosophila* (Lemaitre & Hoffmann, 2007; Leulier & Lemaitre, 2008) and in the case of Toll were first associated with developmental regulation. Recent work has focussed on characterizing how *Drosophila* PRRs and downstream signalling pathways respond to both the normal microbiotia and pathogenic microorganisms (Cherry & Silverman, 2006; Pham et al., 2007; Ryu et al., 2010). Inducing septic injury in the fly gut leads to rapid systemic activation of immune effectors via the Toll and immune deficiency (IMD) pathways, especially in response to whole pathogens or MAMPS such as peptidoglycan (PGN; Ryu et al., 2008, 2010). To counteract the constant exposure of MAMPs in the gut from the normal microbiota, *Drosophila* employs a number of negative regulatory mechanisms

(Ryu et al., 2008; Zaidman-Remy et al., 2006). *Drosophila* uses several PGN-degrading PGRPs to reduce and maintain low levels of PGN (Zaidman-Remy et al., 2006, 2011). These PGRPs also act as negative regulatory molecules by inhibiting the IMD pathway and the induction of antimicrobial peptides and reactive oxygen species.

The mutualism between the Hawaiian bobtail squid and the bioluminescent bacterium *V. fischeri* is used as a model for understanding how bacteria influence animal development (McFall-Ngai, 2002; Nyholm & McFall-Ngai, 2004). The symbionts are housed in a specialized light organ and must colonize the host from the environment. Research has shown that *V. fischeri* initiates a developmental program in the host whereby the juvenile light organ undergoes morphogenesis, resulting in the apoptosis and regression of ciliated epithelial fields that assist in the colonization of the host (McFall-Ngai & Ruby, 1991). The bacterial signals that trigger these events include the bacterial outer membrane and cell wall components lipopolysaccharide and a monomer of PGN called trachaeal cytotoxin (TCT) (Foster et al., 2000; Koropatnick et al., 2004). Although these MAMPs have often been associated with pathogenic interactions, in the squid-vibrio association these bacterial compounds are used for signalling to the host and do not enable virulence. Although the host receptors for these specific ligands have not been described, the bobtail squid does have a number of PRRs (several TLRs and five PGRPs) and downstream members of the NFκB signalling cascade that are normally associated with the response to MAMPs (Collins et al., 2012b; Goodson et al., 2005; Troll et al., 2009, 2010; M. McFall-Ngai, pers. comm). Furthermore, two of these PGRPs are differentially expressed during development (Troll et al., 2009, 2010). For example, EsPGRP2 is secreted into the light-organ crypt spaces after morphogenesis and has been shown to have amidase activity, suggesting it may mediate the host response to TCT (Troll et al., 2010).

One intriguing finding in the squid-vibrio symbiosis is that luminescence of the symbionts is required for the host's full response to symbiont MAMPs; that is, mutants defective in light production show delay or attenuation in all phenotypes induced by MAMPs (Chun et al., 2008; McFall-Ngai et al., 2012). These data suggest that light can behave as a morphogen in this system. Since the symbiont induces loss of the tissues that facilitate colonization, these findings suggest that the animal sanctions these mutants, slowing the morphogenesis and, perhaps, giving the animal more time to continue recruiting symbionts to ensure quality partners.

Hematopoiesis begins early during embryogenesis in *Drosophila*, and haemocytes migrate throughout the developing embryos and larvae. Besides having a role in phagocytosing apoptotic bodies during development, they are also poised to respond to microbes during these early stages. In the tsetse fly, the development of the immune system is influenced by the presence of the bacterial symbiont *Wigglesworthia glossinidia* (Weiss et al., 2011). Flies raised without these bacteria had fewer haemocytes and decreased expression in transcription factors involved with haematopoiesis, as well as genes involved with the antimicrobial melanization response. Furthermore, adult tsetse flies that developed in the absence of *W. glossinidia* are more susceptible to bacterial infection. Both mosquitoes and fruit flies are exposed to environmental microorganisms throughout development, but if raised in a GF state, they have compromised humoral immune systems as adults, suggesting that MAMPs are necessary for the proper development of the immune system. Tsetse flies are unusual in that they engage in viviparous reproduction, in which single larvae develop in utero. The developing larvae are bathed in and nourished by 'mother's milk' that contains both lipids and proteins. A recent study showed that the milk secretions also contain a PGN recognition protein, PGRP-LB, that appears to help regulate *Wigglesworthia* transmission and protect the embryos from protozoan parasites (Wang & Aksoy, 2012).

The amidase activity of PGRP-LB is thought to help prevent the larval immune system from attacking the symbionts. Flies raised with reduced PGRP-LB had lower fecundity, and larvae were susceptible to increased antimicrobial peptide activity. Mammals also incorporate components of the immune system, such as antibodies, in their milk. These antibodies are thought to protect neonates from infection and contribute to the maturation of the immune system. Given that neonatal humans also acquire gut bacteria, such as lactic acid bacteria, it would be interesting to see how components of the mother's immune system interact with the symbionts during the transfer and establishment of the gut microbiota. In zebrafish, the microbiota influence the abundance of neutrophils in the intestine, and GF fish have a significant reduction in the number of neutrophils (Bates et al., 2007). In the bobtail squid, exposure to the symbiont *V. fischeri* also leads to an infiltration of haemocytes into the superficial light organ within the first two hours of the association (Koropatnick et al., 2007). Although the role of host haemocytes in shaping early colonization of the light organ has not yet been assessed, in adults, the colonization state influences the ability of these cells to recognize *V. fischeri* (Nyholm et al., 2009) and also influences haemocyte gene expression of innate immunity factors, including a peptidoglycan recognition protein (EsPGRP 5), complement-related proteins, and a nitric oxide synthase (Collins et al., 2012b).

Influences on the brain and behaviour

Recent evidence suggests that an animal's microbiota can have a profound influence on development of the central nervous system, as well as long-term effects on behaviour (for a more extensive analysis see Forsythe & Kunze, 2013). For example, GF mice have been found to have a decrease in behaviours associated with anxiety and increased motor activity compared to conventional animals (Diaz Heijtz et al., 2011; Neufeld et al., 2011). A number of studies also indicate that gut microbiota can modulate animal behaviours by directly influencing neural pathways as well as regulating hormone and metabolite levels (Hsiao et al., 2013; Fetissov et al., 2008; Lesniewska et al., 2006). Bacteria have also been shown to produce a number of compounds normally associated with mammalian nervous system signalling, including melatonin, serotonin, acetylcholine, GABA, and nitric oxide (Forsythe & Kunze, 2013). Bacteria also have receptors that recognize mammalian-derived compounds. For example, *E. coli* O157:H7 has a receptor that recognizes mammalian epinephrine/norepinephrine and regulates expression of virulence factors (Clarke et al., 2006). Taken together, these data suggest that gut microbiota can directly respond to and influence components of the animal nervous system and influence behaviour directly.

Among insects, bacteria like *Wolbachia* have been shown to influence insect populations by manipulating developmental outcomes during embryogenesis (reviewed in Werren et al., 2008). In a phenomenon called cytoplasmic incompatibility, female flies infected with *Wolbachia* can mate with either infected or uninfected males, resulting in viable offspring with maternally transmitted bacteria. However, if uninfected females mate with infected males, the offspring do not develop. Therefore, infected females have a greater mate choice, and infections can rapidly sweep through insect populations as a result. *Wolbachia* can also rescue sterility in female *Drosophila melanogaster* that are mutant for the gene *Sex-lethal* and thus are unable to produce eggs. Recent evidence suggests that a developmental gene normally involved with axis determination may help regulate *Wolbachia* titer in *Drosophila* (Serbus et al., 2011). In the parasitic wasp *Asobora tabida*, *Wolbachia* is required for egg production and has therefore become an obligate symbiont for reproduction (Dedeine et al., 2001). Recent evidence also suggests that gut microbiota can also have an influence on behaviours involved with mate choice (Sharon et al., 2010). Fruit flies fed on a diet of either molasses or starch prefer to mate with flies that have been raised on a similar diet. These mating preferences are established within one generation and can be maintained for multiple generations. Mating preference is lost with antibiotic treatment but is reestablished with the exposure to the single bacterium *Lactobacillus plantarum*. Furthermore, this bacterium appears to influence specific pheromone levels in the fly, suggesting that members of the gut microbiota can modify fly behaviour by influencing the production of sex pheromones.

Conclusion

The long-standing view that the principal interactions between microbes and animals are antagonistic and that animal groups evolved mechanisms to prevent and exclude intimate interactions with (pathogenic) microbes is giving way to the view that a co-evolved microbiota has had a sweeping influence on the development and evolution of all animals.

A growing body of evidence suggests that these microbes impact the development of multiple organ systems at all life stages of their hosts (Figure 17.1). It has been proposed that selection may even act on hosts and their co-evolved microbiota as a single ecosystem or unit (Rosenberg et al., 2007, 2009). This 'holobiome theory of evolution' posits that forming associations with microbes has given animal hosts access to a huge repertoire of genetic and physiological diversity that has allowed this group to adapt to changing environments and stresses. Although aspects of this idea still need to be tested, evidence presented in this review suggests that microbes have played and continue to play a primary role in animal evolution. Gilbert and colleagues have recently proposed that the whole idea of the individual biological self needs to be re-examined in light of the long co-evolutionary history between animals and microorganisms (Gilbert et al., 2012). Proper development, once thought to be under the strict regulation of an individual's genome, is now known to also require interactions with the environment, including numerous symbionts and their collective genomes. Although much work still needs to be done to understand the specific mechanisms of signalling between partners that initiate and sustain developmental processes, a number of model associations are starting to decode this dialogue. MAMPs and other microbial compounds that under some contexts may be considered to result in virulence and pathogenesis are in many cases, as outlined in this chapter, instead involved with 'normal' development. On the host side, the immune system, in addition to its role in defending against pathogens, is also essential in the normal dialogue necessary for the establishment and maintenance of beneficial associations.

This is an exciting time for those studying host/microbe interactions. The huge technological advances of the 'omics' era have presented scientists with a wealth of information about animal microbiomes. However, with these massive datasets comes the challenge of deciphering how specific

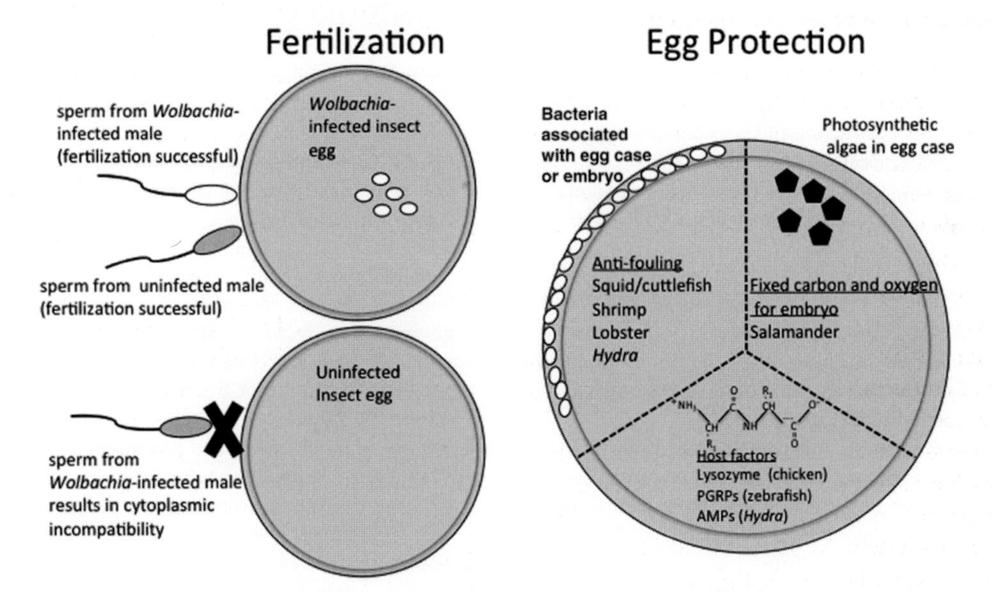

Figure 17.1 Influence of bacteria on different life stages. Bacterial influences on animal development can occur as early as fertilization. For example, in many insects infection with the bacterial *Wolbachia* sp. profoundly influences reproductive outcomes. Uninfected females are at a disadvantage, as a union with an infected male leads to a phenomenon called cytoplasmic incompatibility and death of the embryos. Because *Wolbachia* is maternally transmitted, infected females can freely mate with both infected and uninfected males. This often leads to a population bias favouring infected individuals. Organisms that lay their eggs externally must protect the developing embryos from environmental assaults such as biofilm formation and pathogens while ensuring sufficient oxygen delivery. Egg protection often involves a combination of host factors, such as components of the immune system, as well as the use of microbial symbionts that may prevent biofouling. In the case of the salamander, algae in the eggs are thought to provide both oxygen and photosynthetic products to the developing embryos.

microbes influence animal ontogeny. The continued development of model systems to analyse these interactions will be key. No association by itself can answer all of these questions. However, by incorporating animal systems representing varying levels of interactions with microbes (from binary to consortial) and from diverse taxa, we will build a more inclusive view of how these associations have evolved.

References

Agans, R., Rigsbee, L., Kenche, H., et al. (2011). Distal gut microbiota of adolescent children is different from that of adults. *FEMS Microbiology Ecology*, **77**, 404–12.

Alegado, R.A., Brown, L.W., Cao, S., et al. (2012). A bacterial sulfonolipid triggers multicellular development in the closest living relatives of animals. *elife*, **1**, e00013.

Barbieri, E., Paster, B.J., Hughes, D., et al. (2001). Phylogenetic characterization of epibiotic bacteria in the accessory nidamental gland and egg capsules of the squid *Loligo pealei* (Cephalopoda: Loliginidae). *Environmental Microbiology*, **3**, 151–67.

Bates, J.M., Akerlund, J., Mittge, E., et al. (2007). Intestinal alkaline phosphatase detoxifies lipopolysaccharide and prevents inflammation in zebrafish in response to the gut microbiota. *Cell Host and Microbe*, **2**, 371–82.

Biggs, J.E., and Epel D. (1991). Egg capsule sheath of *Loligo opalescens* Berry: structure and association with bacteria. *Journal of Experimental Zoology*, **259**, 263–7.

Bloodgood, R.A. (1977). The squid accessory nidamental gland: ultrastructure and association with bacteria. *Tissue and Cell*, **9**, 197–208.

Bright, M., and Bulgheresi, S. (2010). A complex journey: transmission of microbial symbionts. *Nature Reviews Microbiology*, **8**, 218–30.

Bry, L., Falk, P.G., Midtvedt, T., et al. (1996). A model of host-microbial interactions in an open mammalian ecosystem. *Science*, **273**, 1380–3.

Cheesman, S.E., Neal, J.T., Mittge, E., et al. (2011). Epithelial cell proliferation in the developing zebrafish intestine is regulated by the Wnt pathway and microbial signaling via Myd88. *Proceedings of the National Academy of Sciences USA*, **108** Suppl 1, 4570–7.

Cherry, S., and Silverman, N. (2006). Host-pathogen interactions in drosophila: new tricks from an old friend. *Nature Immunology*, **7**, 911–17.

Chun, C.K., Troll, J.V., Koroleva, I., et al. (2008). Effects of colonization, luminescence, and autoinducer on host transcription during development of the squid-vibrio association. *Proceedings of the National Academy of Sciences USA*, **105**, 11323–8.

Clarke, M.B., Hughes, D.T., Zhu, C., et al. (2006). The QseC sensor kinase: a bacterial adrenergic receptor. *Proceedings of the National Academy of Sciences USA*, **103**, 10420–5.

Collins, A.J., Labarre, B.A., Won, B.S., et al. (2012a). Diversity and partitioning of bacterial populations within the accessory nidamental gland of the squid *Euprymna scolopes*. *Applied and Environmental Microbiology*, **78**, 4200–8.

Collins, A.J., Schleicher, T.R., Rader, B.A., et al. (2012b). Understanding the role of host hemocytes in a squid/vibrio symbiosis using transcriptomics and proteomics. *Frontiers in Immunology*, **3**, 91.

Dayel, M.J., Alegado, R.A., Fairclough, S.R., et al. (2011). Cell differentiation and morphogenesis in the colony-forming choanoflagellate *Salpingoeca rosetta*. *Developmental Biology*, **357**, 73–82.

Dedeine, F., Vavre, F., Fleury, F., et al. (2001). Removing symbiotic *Wolbachia* bacteria specifically inhibits oogenesis in a parasitic wasp. *Proceedings of the National Academy of Sciences USA*, **98**, 6247–52.

Diaz Heijtz, R., Wang, S., Anufar, F., et al. (2011). Normal gut microbiota modulates brain development and behavior. *Proceedings of the National Academy of Sciences USA*, **108**, 3047–52.

Dominguez-Bello, M. G., Costello, E. K., Contreras, M., Magris, M., Hidalgo, G., Fierer, N., and Knight, R. (2010) Delivery mode shapes the acquisition and structure of the initial microbiota across multiple body habitats in newborns. *Proceedings of the National Academy of Sciences USA*, **107**, 11971–5.

Eberl, G. (2005). Inducible lymphoid tissues in the adult gut: recapitulation of a fetal developmental pathway? *Nature Reviews Immunology*, **5**, 413–20.

Fetissov, S. O., Hamze Sinno, M., Coeffier, M., et al. (2008). Autoantibodies against appetite-regulating peptide hormones and neuropeptides: putative modulation by gut microflora. *Nutrition*, **24**, 348–59.

Forsythe, P., and Kunze, W.A. (2013). Voices from within: gut microbes and the CNS. *Cellular and Molecular Life Sciences*, **70**, 55–69.

Foster, J.S., Apicella, M.A. and McFall-Ngai, M.J. (2000). *Vibrio fischeri* lipopolysaccharide induces developmental apoptosis, but not complete morphogenesis, of the *Euprymna scolopes* symbiotic light organ. *Developmental Biology*, **226**, 242–54.

Fraune, S., Augustin, R., Anton-Erxleben, F., et al. (2010). In an early branching metazoan, bacterial colonization of the embryo is controlled by maternal antimicrobial peptides. *Proceedings of the National Academy of Sciences USA*, **107**, 18067–72.

Gilbert, S.F., Sapp, J., and Tauber, A.I. (2012). A symbiotic view of life: we have never been individuals. *Quarterly Review of Biology*, **87**, 325–41.

Gil-Turnes, M.S., Hay, M.E., and Fenical, W. (1989). Symbiotic marine bacteria chemically defend crustacean embryos from a pathogenic fungus. *Science*, **246**, 116–18.

Goodson, M.S., Kojadinovic, M., Troll, J.V., et al. (2005). Identifying components of the NF-κB pathway in the beneficial *Euprymna scolopes-Vibrio fischeri* light organ symbiosis. *Applied and Environmental Microbiology*, **71**, 6934–46.

Graham, E.R., Fay, S., Davey, A., et al. (2013) Intracapsular algae provide fixed carbon to developing embryos of the salamander *Ambystoma maculatum*. *Journal of Experimental Biology*, **216**, 452–9.

Hamasaki, M., Furuta, N., Matsuda, A., et al. (2013). Autophagosomes form at ER-mitochondria contact sites. *Nature*, **495**, 389–93.

Heddi, A., Charles, H., Khatchadourian, C., et al. (1998). Molecular characterization of the principal symbiotic bacteria of the weevil *Sitophilus oryzae*: a peculiar G + C content of an endocytobiotic DNA. *Journal of Molecular Evolution*, **47**, 52–61.

Hentschel, U., Piel, J., Degnan, S.M., et al. (2012). Genomic insights into the marine sponge microbiome. *Nature Reviews Microbiology*, **10**, 641–54.

Hsiao, E. Y., McBride, S. W., Hsien, S., Sharon, G., Hyde, E. R., McCue, T., Codelli, J. A., Chow, J., Reisman, S. E., Petrosino, J. F., Patterson, P. H., Mazmanian, S. K. (2013). Microbiota modulate behavioral and physiological abnormalities associated with neurodevelopmental disorders. *Cell*, 155, 1452–63.

Kerney, R., Kim, E., Hangartner, R.P., et al. (2011). Intracellular invasion of green algae in a salamander host. *Proceedings of the National Academy of Sciences USA*, **108**, 6497–502.

Koren, O., Goodrich, J.K., Cullender, T.C., et al. (2012). Host remodeling of the gut microbiome and metabolic changes during pregnancy. *Cell*, **150**, 470–80.

Koropatnick, T.A., Engle, J.T., Apicella, M.A., et al. (2004). Microbial factor-mediated development in a host-bacterial mutualism. *Science*, **306**, 1186–8.

Koropatnick, T.A., Kimbell, J.R., and McFall-Ngai, M.J. (2007). Responses of host hemocytes during the initiation of the squid-*Vibrio* symbiosis. *Biological Bulletin*, **212**, 29–39.

Lee, Y.K., and Mazmanian, S.K. (2010). Has the microbiota played a critical role in the evolution of the adaptive immune system? *Science*, **330**, 1768–73.

Lemaitre, B., and Hoffmann, J. (2007). The host defense of *Drosophila melanogaster*. *Annual Reviews of Immunology*, **25**, 697–743.

Lesniewska, V., Rowland, I., Cani, P.D., et al. (2006). Effect on components of the intestinal microflora and plasma neuropeptide levels of feeding *Lactobacillus delbrueckii*, *Bifidobacterium lactis*, and inulin to adult and elderly rats. *Applied and Environmental Microbiology*, **72**, 6533–8.

Leulier, F., and Lemaitre, B. (2008). Toll-like receptorstaking an evolutionary approach. *Nature Reviews Genetics*, **9**, 165–78.

Li, X., Wang, S., Qi, J., et al. (2007). Zebrafish peptidoglycan recognition proteins are bactericidal amidases essential for defense against bacterial infections. *Immunity*, **27**, 518–29.

Lozupone, C.A., Stombaugh, J.I., Gordon, J.I., et al. (2012). Diversity, stability and resilience of the human gut microbiota. *Nature*, **489**, 220–30.

Mariat, D., Firmesse, O., Levenez, F., et al. (2009). The Firmicutes/Bacteroidetes ratio of the human microbiota changes with age. *BMC Microbiology*, **9**, 123.

Mazmanian, S.K., Liu, C.H., Tzianabos, A.O., et al. (2005). An immunomodulatory molecule of symbiotic bacteria directs maturation of the host immune system. *Cell*, **122**, 107–18.

McCutcheon, J.P., and Moran, N.A. (2010). Functional convergence in reduced genomes of bacterial symbionts spanning 200 My of evolution. *Genome Biology and Evolution*, **2**, 708–18.

McCutcheon, J.P., and Moran, N.A. (2012). Extreme genome reduction in symbiotic bacteria. *Nature Reviews Microbiology*, **10**, 13–26.

McFall-Ngai, M.J. (2002). Unseen forces: the influence of bacteria on animal development. *Developmental Biology*, **242**, 1–14.

McFall-Ngai, M. (2007). Adaptive immunity: care for the community. *Nature*, **445**, 153.

McFall-Ngai, M., Hadfield, M.G., Bosch, T.C., et al. (2013). Animals in a bacterial world, a new imperative for the life sciences. *Proceedings of the National Academy of Sciences USA*, **110**, 3229–36.

McFall-Ngai, M., Heath-Heckman, E.A., Gillette, A.A., et al. (2012). The secret languages of coevolved symbioses: insights from the *Euprymna scolopes-Vibrio fischeri* symbiosis. *Seminars in Immunology*, **24**, 3–8.

McFall-Ngai, M.J., and Ruby, E.G. (1991). Symbiont recognition and subsequent morphogenesis as early events in an animal-bacterial mutualism. *Science*, **254**, 1491–4.

Messler, S., Kropp, S., Episkopou, V., et al. (2011). The TGF-β signaling modulators TRAP1/TGFBRAP1 and VPS39/Vam6/TLP are essential for early embryonic development. *Immunobiology*, **216**, 343–50.

Mitra, K., Rikhy, R., Lilly, M., et al. (2012). DRP1-dependent mitochondrial fission initiates follicle cell differentiation during *Drosophila* oogenesis. *Journal of Cell Biology*, **197**, 487–97.

Mitra, K., Wunder, C., Roysam, B., et al. (2009). A hyperfused mitochondrial state achieved at G1-S regulates cyclin E buildup and entry into S phase. *Proceedings of the National Academy of Sciences USA*, **106**, 11960–5.

Moran, N.A., McCutcheon, J.P., and Nakabachi, A. (2008). Genomics and evolution of heritable bacterial symbionts. *Annual Reviews of Genetics*, **42**, 165–90.

Neufeld, K.M., Kang, N., Bienenstock, J., et al. (2011). Reduced anxiety-like behavior and central neurochemical change in germ-free mice. *Neurogastroenterology and Motility*, **23**, 255–64.

Nicholson, J.K., Holmes, E., Kinross, J., et al. (2012). Host-gut microbiota metabolic interactions. *Science*, **336**, 1262–7.

Niess, J.H., Leithauser, F., Adler, G., et al. (2008). Commensal gut flora drives the expansion of proinflammatory CD4 T cells in the colonic lamina propria under normal and inflammatory conditions. *Journal of Immunology*, **180**, 559–68.

Nyholm S.V., and Graf, J. (2012). Knowing your friends: invertebrate innate immunity fosters beneficial bacterial symbioses. *Nature Reviews Microbiology*, **10**, 815–27.

Nyholm, S.V., and McFall-Ngai, M.J. (2004). The winnowing: establishing the squid-vibrio symbiosis. *Nature Reviews Microbiology*, **2**, 632–42.

Nyholm, S.V., Stewart, J.J., Ruby, E.G., et al. (2009). Recognition between symbiotic *Vibrio fischeri* and the haemocytes of *Euprymna scolopes*. *Environmental Microbiology*, **11**, 483–93.

Olszak, T., An, D., Zeissig, S., et al. (2012). Microbial exposure during early life has persistent effects on natural killer T cell function. *Science*, **336**, 489–93.

Pham, L.N., Dionne, M.S., Shirazu-Hiza, M., et al. (2007). A specific primed immune response in *Drosophila* is dependent on phagocytes. *PLoS Pathogens*, **3**, e26.

Pinder, A., and Friet, S. (1994). Oxygen transport in egg masses of the amphibians *Rana sylvatica* and *Ambystoma maculatum*: convection, diffusion and oxygen production by algae. *Journal of Experimental Biology*, **197**, 17–30.

Pradeu, T. (2012). *The Limits of the Self: Immunology and Biological Identity*. Oxford University Press, Oxford.

Ramos, G.C. (2012). Inflammation as an animal development phenomenon. *Clinical and Developmental Immunology*, **2012**, 983203.

Rawls, J.F., Samuel, B.S., and Gordon, J.I. (2004). Gnotobiotic zebrafish reveal evolutionarily conserved responses to the gut microbiota. *Proceedings of the National Academy of Sciences USA*, **101**, 4596–601.

Rhee, K.J., Sethupathi, P., Driks, A., et al. (2004). Role of commensal bacteria in development of gut-associated lymphoid tissues and preimmune antibody repertoire. *Journal of Immunology*, **172**, 1118–24.

Roeselers, G., Mittge, E.K., Stephens, W.Z., et al. (2011). Evidence for a core gut microbiota in the zebrafish. *ISME Journal*, **5**, 1595–608.

Rose-Martel, M., Du, J., and Hincke, M.T. (2012). Proteomic analysis provides new insight into the chicken eggshell cuticle. *Journal of Proteomics*, **75**, 2697–706.

Rosenberg, E., Koren, O., Reshef, L., et al. (2007). The role of microorganisms in coral health, disease and evolution. *Nature Reviews Microbiology*, **5**, 355–62.

Rosenberg, E., Sharon, G., and Zilber-Rosenberg, I. (2009). The hologenome theory of evolution contains Lamarckian aspects within a Darwinian framework. *Environmental Microbiology*, **11**, 2959–62.

Ryu, J.H., Ha, E.M., and Lee, W.J. (2010). Innate immunity and gut-microbe mutualism in *Drosophila*. *Developmental and Comparative Immunology*, **34**, 369–76.

Ryu, J.H., Kim, S.H., Lee, H.Y., et al. (2008). Innate immune homeostasis by the homeobox gene *caudal* and commensal-gut mutualism in *Drosophila*. *Science*, **319**, 777–82.

Saina, M., Genikhovich, G., Renfer, E., et al. (2009). BMPs and chordin regulate patterning of the directive axis in a sea anemone. *Proceedings of the National Academy of Sciences USA*, **106**, 18592–7.

Sawa, S., Lochner, M., Satoh-Takayama, N., et al. (2011). RORgammat + innate lymphoid cells regulate intestinal homeostasis by integrating negative signals from the symbiotic microbiota. *Nature Immunology*, **12**, 320–6.

Serbus, L.R., Ferreccio, A., Zhukova, M., et al. (2011). A feedback loop between *Wolbachia* and the *Drosophila* gurken mRNP complex influences *Wolbachia* titer. *Journal of Cell Science*, **124**, 4299–308.

Shah M., Brown K.M., and Smith, L.C. (2003). The gene encoding the sea urchin complement protein, SpC3, is expressed in embryos and can be upregulated by bacteria. *Developmental and Comparative Immunology*, **27**, 529–38.

Sharon, G., Segal, D., Ringo, J.M., et al. (2010). Commensal bacteria play a role in mating preference of *Drosophila melanogaster*. *Proceedings of the National Academy of Sciences USA*, **107**, 20051–6.

Smith, K., McCoy, K.D., and Macpherson, A.J. (2007). Use of axenic animals in studying the adaptation of mammals to their commensal intestinal microbiota. *Seminars in Immunology*, **19**, 59–69.

Smith, M.I., Yatsunenko, T., Manary, M.J., et al. (2013) Gut microbiomes of Malawian twin pairs discordant for kwashiorkor. *Science*, **339**, 548–54.

Stappenbeck, T.S., Hooper, L.V., and Gordon, J.I. (2002). Developmental regulation of intestinal angiogenesis by indigenous microbes via Paneth cells. *Proceedings of the National Academy of Sciences USA*, **99**, 15451–5.

Strowig, T., Henao-Mejia, J., Elinav, E., et al. (2012). Inflammasomes in health and disease. *Nature*, **481**, 278–86.

Tattersall, G.J., and Spiegelaar N. (2008). Embryonic motility and hatching success of *Ambystoma maculatum* are influenced by a symbiotic alga. *Canadian Journal of Zoology*, **86**, 1289–98.

Tremaroli, V., and Backhed, F. (2012). Functional interactions between the gut microbiota and host metabolism. *Nature*, **489**, 242–9.

Troll, J.V., Adin, D.M., Wier, A.M., et al. (2009). Peptidoglycan induces loss of a nuclear peptidoglycan recognition protein during host tissue development in a beneficial animal-bacterial symbiosis. *Cellular Microbiology*, **11**, 1114–27.

Troll, J.V., Bent, E.H., Pacquette, N., et al. (2010). Taming the symbiont for coexistence: a host PGRP neutralizes a bacterial symbiont toxin. *Environmental Microbiology*, 12, 2190–203.

von Mutius, E., and Vercelli, D. (2010). Farm living: effects on childhood asthma and allergy. *Nature Reviews Immunology*, **10**, 861–8.

Wang, J., and Aksoy, S. (2012). PGRP-LB is a maternally transmitted immune milk protein that influences symbiosis and parasitism in tsetse's offspring. *Proceedings of the National Academy of Sciences USA*, **109**, 10552–7.

Weiss, B.L., Wang, J., and Aksoy, S. (2011). Tsetse immune system maturation requires the presence of obligate symbionts in larvae. *PLoS Biology*, **9**, e1000619.

Wellman-Labadie, O., Lemaire, S., Mann, K., et al. (2010). Antimicrobial activity of lipophilic avian eggshell surface extracts. *Journal of Agricultural and Food Chemistry*, **58**, 10156–61.

Wellman-Labadie, O., Picman, J., and Hincke, M. T. (2008). Antimicrobial activity of the anseriform outer eggshell and cuticle. *Comparative Biochemistry and Physiology B Biochemistry and Molecular Biology*, **149**, 640–9.

Werren, J.H., Baldo, L., and Clark, M.E. (2008). *Wolbachia*: master manipulators of invertebrate biology. *Nature Reviews Microbiology*, **6**, 741–51.

Wikoff, W.R., Anfora, A.T., Liu, J., et al. (2009). Metabolomics analysis reveals large effects of gut microflora on mammalian blood metabolites. *Proceedings of the National Academy of Sciences USA*, **106**, 3698–703.

Yamashita, S., Miyagi, C., Carmany-Rampey, A., et al. (2002). Stat3 controls cell movements during zebrafish gastrulation. *Developmental Cell*, **2**, 363–75.

Zaidman-Remy, A., Hervé, M., Poidevin, M., et al. (2006). The *Drosophila* amidase PGRP-LB modulates the immune response to bacterial infection. *Immunity*, **24**, 463–73.

Zaidman-Remy, A., Poidevin, M., Hervé, M., et al. (2011). *Drosophila* immunity: analysis of PGRP-SB1 expression, enzymatic activity and function. *PLoS One*, **6**, e17231.

Zivkovic, A.M., German, J.B., Lebrilla, C.B., et al. (2011). Human milk glycobiome and its impact on the infant gastrointestinal microbiota. *Proceedings of the National Academy of Sciences USA*, **108** Suppl 1, 4653–8.

Index